Lecture Notes in Computer Science 15252

The series Lecture Notes in Computer Science (LNCS), including its subseries Lecture Notes in Artificial Intelligence (LNAI) and Lecture Notes in Bioinformatics (LNBI), has established itself as a medium for the publication of new developments in computer science and information technology research, teaching, and education.

LNCS enjoys close cooperation with the computer science R & D community, the series counts many renowned academics among its volume editors and paper authors, and collaborates with prestigious societies. Its mission is to serve this international community by providing an invaluable service, mainly focused on the publication of conference and workshop proceedings and postproceedings. LNCS commenced publication in 1973.

Tianqing Zhu · Jin Li · Aniello Castiglione
Editors

Algorithms and Architectures for Parallel Processing

24th International Conference, ICA3PP 2024
Macau, China, October 29–31, 2024
Proceedings, Part II

Editors
Tianqing Zhu
City University of Macau
Macau, China

Jin Li
Guangzhou University
Guangzhou, China

Aniello Castiglione
University of Salerno
Fisciano, Italy

ISSN 0302-9743 ISSN 1611-3349 (electronic)
Lecture Notes in Computer Science
ISBN 978-981-96-1527-8 ISBN 978-981-96-1528-5 (eBook)
https://doi.org/10.1007/978-981-96-1528-5

This Springer imprint is published by the registered company Springer Nature Singapore Pte Ltd.
The registered company address is: 152 Beach Road, #21-01/04 Gateway East, Singapore 189721, Singapore

Preface

On behalf of the Conference Committee, we welcome you to the proceedings of the 2024 International Conference on Algorithms and Architectures for Parallel Processing (ICA3PP 2024), which was held in Macau Special Administrative Region, China from October 29–31, 2024. ICA3PP 2024 was the 24th in this series of conferences (started in 1995) that are devoted to algorithms and architectures for parallel processing. ICA3PP is now recognized as the main regular international event that covers the many dimensions of parallel algorithms and architectures, encompassing fundamental theoretical approaches, practical experimental projects, and commercial components and systems. This conference provides a forum for academics and practitioners from countries around the world to exchange ideas for improving the efficiency, performance, reliability, security, and interoperability of computing systems and applications.

A successful conference would not be possible without the high-quality contributions made by the authors. This year, ICA3PP received a total of 265 submissions from authors in various countries and regions. Based on rigorous peer reviews by the Program Committee members and reviewers, 131 high-quality papers were accepted to be included in the conference proceedings and submitted for EI indexing. In addition to the contributed papers, distinguished scholars were invited to give keynote lectures, providing us with the recent developments in diversified areas in algorithms and architectures for parallel processing and applications.

Among the accepted papers, the papers with the highest weighted review mark in each round received the Best Paper Award. The Best Papers were *Updates Leakage Attack against Private Graph Split Learning,* Hao Yang, Zhuo Ma, Yang Liu, Xinjing Liu, Beiwei Yang, and Jianfeng Ma, *Data-Free Encoder Stealing Attack in Self-supervised Learning*, Chuan Zhang, Xuhao Ren, Haotian Liang, Qing Fan, Xiangyun Tang, Chunhai Li, Liehuang Zhu, and Yajie Wang, and *Federated Meta Continual Learning for Efficient and Autonomous Edge Inference* Bingze Li, Stella Ho, Youyang Qu, Chenhao Xu, Tom H. Luan, and Longxiang Gao. Best Student paper was *DIsFU: Protecting Innocent Clients in Federated Unlearning* Fanyu Kong, Xiangyun Tang, Yu Weng, Tao Zhang, Hongyang Du, Jiawen Kang, and Chi Liu.

We would like to take this opportunity to express our sincere gratitude to the 262 Program Committee members and reviewers for their dedicated and professional service. We highly appreciate the track chairs for their hard work in promoting this conference and organizing the reviews for the papers submitted to their tracks. We are so grateful to the publication chairs, Zuobin Ying and Sheng Wen, for their meticulous work in editing the conference proceedings. We must also say "thank you" to all the volunteers who helped us at various stages of this conference.

Moreover, we would like to extend our appreciation to the following chairs for their invaluable contributions:

Local Chairs
Wenjian Liu, Chris Chu, Max Kuok

Workshop Chairs
Jia Gu, Chaofeng Zhang, Fengshi Jin, Chi Liu

Publicity Chairs
Gengshen Wu, Lefeng Zhang

Registration Chairs
Kaiyao Jiang, Congcong Zhu

We were so honored to have many renowned scholars be part of this conference. Finally, we would like to thank all speakers, authors, and participants for their great contribution to and support for the success of ICA3PP 2024.

October 2024

Wanlei Zhou
Paulo Quaresma
Albert Zomaya
Willy Susilo
Tianqing Zhu
Jin Li
Aniello Castiglione

Organization

General Chairs

Wanlei Zhou	City University of Macau, China
Paulo Quaresma	University of Évora, Portugal
Albert Zomaya	University of Sydney, Australia
Willy Susilo	University of Wollongong, Australia

Program Chairs

Tianqing Zhu	City University of Macau, China
Jin Li	Guangzhou University, China
Aniello Castiglione	University of Salerno, Italy

Local Chairs

Wenjian Liu	City University of Macau, China
Chris Chu	City University of Macau, China
Max Kuok	City University of Macau, China

Publication Chairs

Zuobin Ying	City University of Macau, China
Sheng Wen	Swinburne University of Technology, Australia

Workshop Chairs

Jia Gu	City University of Macau, China
Chaofeng Zhang	Advanced Institute of Industrial Technology, Japan
Fengshi Jin	City University of Macau, China
Chi Liu	City University of Macau, China

Publicity Chairs

Gengshen Wu	City University of Macau, China
Lefeng Zhang	City University of Macau, China

Registration Chairs

Kaiyao Jiang	City University of Macau, China
Congcong Zhu	City University of Macau, China

Program Committee

Aniello Castiglione	Department of Management and Innovation Systems
Bangbang Ren	National University of Defense Technology
Bin Wu	Chinese Academy of Sciences
Bo Li	Swinbourn University of Technology
Bo Liu	University of Technology Sydney
Bowen Liu	Nanjing University
Chao Li	Beijing Jiaotong University
Chao Wang	University of Science and Technology of China
Chaokun Zhang	Tianjin University
Chen Zhang	City University of Hong Kong
Chentao Wu	Shanghai Jiao Tong University
Chi Liu	City University of Macau
Chris Chu	City University of Macau
Chuan Zhang	Beijing Institute of Technology
Chuang Hu	Wuhan University
Congcong Zhu	City University of Macau
Daniel Andresen	Kansas State University
Dayong Ye	University of Wollongong
Deze Zeng	China University of Geosciences
Dezun Dong	National University of Defense Technology
En Shao	Institute of Computing Technology
Faqian Guan	China University of Geosciences
Fei Lei	NUDT
Fuliang Li	Northeastern University
Fuyuan Song	Nanjing University of Information Science and Technology
Geng Sun	Jilin University

Shuai Zhou	City University of Macau
Shuang Chen	Huawei Cloud
Shujie Han	Peking University
Shuxin Zhong	Rutgers University
Simin Chen	UTD
Songwen Pei	Dept. of Computer Science and Engineering
Su Yao	Tsinghua University
Susumu Matsumae	Saga University
Tao Wu	National University of Defense Technology
Tianqing Zhu	University of Technology
Tianyi Liu	HUAWEI
Tie Qiu	Tianjin University
Tingwen Liu	Institute of Information Engineering
Vladimir Voevodin	RCC MSU
Wei Bao	The University of Sydney
Wei Wang	Central South University
Weibei Fan	Nanjng University of Posts and Telecommunications
Weihua Zhang	Fudan University
Weitian Tong	Georgia Southern University
Weixing Ji	Beijing Normal University
Weizhi Meng	Lancaster University
Wenjuan Li	The Education University of Hong Kong
Wenxin Li	Tianjin University
Wenzheng Xu	Sichuan University
Xiang Zhang	Nanjing University of Information Science and Technology
Xiangyu Kong	Dalian University of Technology
Xiangyun Tang	BIT
Xiangyun Tang	Minzu University of China
Xiaojie Zhang	Hunan First Normal University
Xiaoli Gong	Nankai University
Xiaolu Li	Huazhong University of Science and Technology
Xiaoyang Xie	Rutgers University
Xiaoyi Tao	Dalian University of Technology
Xiaoyong Tang	School of Computer and Communication Engineering
Xiaoyu Wang	Soochow University
Xin He	Nanjing University of Posts and Telecommunications
Xin Xie	The Hong Kong Polytechnic University
Xuan Liu	Hunan University

Xueqin Liang	Xidian University
Yajie Wang	BIT
Yajie Wang	Beijing Institute of Technology
Yang Du	University of Science and Technology of China
Yanyan Wang	Nanjing University
Yi Ding	University of Texas at Dallas
Yi Zhao	Beijing Institute of Technology
Yifei Zhu	Shanghai Jiao Tong University
Yitao Hu	Tianjin University
Yizhi Zhou	Dalian University of Technology
Yongkun Li	University of Science and Technology of China
Yongqian Sun	Nankai University
Youyang Qu	Qilu University of Technology
Youyou Lu	Tsinghua University
Yu Zhang	Huazhong University of Science and Technology
Yuan Cao	Ocean University of China
Yuben Qu	Shanghai Jiao Tong University
Yuchao Zhang	Beijing University of Posts and Telecommunications
Yueming Wu	Nanyang Technological University
Yukun Yuan	University of Tennessee at Chattanooga
Yunxia Lin	Yangzhou University
Yutong Gao	Minzu University of China
Ze Zhang	University of Michigan/Cruise
Zhaoyan Shen	Shandong University
Zhen Ling	Southeast University
Zhengkai Wu	Citadel Securities
Zhengxiong Li	The University of Colorado Denver
Zhenlin An	The Hong Kong Polytechnic University
Zhiqiang Li	University of Nebraska
Zhiquan Liu	Jinan University
Zhou Qin	Amazon
Zhuoxuan Du	Ant Group
Zichen Xu	Nanchang University
Zongheng Wei	School of Computer Science
Zonghua Gu	UMU
Zuobin Ying	City University of Macau

Contents – Part II

Updates Leakage Attack Against Private Graph Split Learning

Hao Yang, Zhuo Ma(✉), Yang Liu, Xinjing Liu, Beiwei Yang, and Jianfeng Ma

Xidian University, Xi'an, China
mazhuo@mail.xidian.edu.cn

Abstract. Recently, to promote private graph data sharing, a collaborative graph learning paradigm known as Graph Split Learning (GSL) is proposed. However, current security research about GSL focuses more on one-shot learning but ignores the fact that training models is usually an ongoing process in practice. Fresh data need to be added periodically to ensure the time-effectiveness of the trained model. In this paper, we propose the first attack against GSL, called Graph Update Leakage Attack (GULA), to show the vulnerability of GSL to privacy leakage attacks when running with updated training sets. Specifically, we systematically analyze the adversary's knowledge of GSL from three dimensions, leading to 8 different implementations of GULA. All 8 attacks demonstrate that a malicious server in GSL can leverage the posteriors received during the forward computation stage to reconstruct the update graph data of clients. Extensive experiments on 6 real-world datasets and 8 different GNN models show that for GSL, our attacks can effectively reveal the private links and node features in the update set.

Keywords: Updates leakage attack · Graph split learning · Privacy attacks and defense

1 Introduction

Graph Neural Networks (GNNs) [29] are designed to effectively process graph-structured data, such as social networks [12] and transportation systems [18]. However, their high computational requirements hinder deployment on resource-constrained devices. To mitigate this, the Graph Split Learning (GSL) framework [31,35] is proposed, distributing the computational load across multiple clients and a cloud server. In GSL, clients manage early-layer computations while the server executes subsequent layers to generate the final outputs. This configuration not only reduces the running overhead on clients but also fortifies data privacy, as only the activation values are transmitted to the server. Major cloud platforms, including AWS [1] and Azure [2], have adopted this framework to enhance model training efficiency and safeguard user data privacy.

However, in real-world applications, model training is usually not a one-shot process. To ensure the quality of service, the underlying models for business

T. Zhu et al. (Eds.): ICA3PP 2024, LNCS 15252, pp. 1–21, 2025.
https://doi.org/10.1007/978-981-96-1528-5_1

always need to be evolved periodically with fresh [4,13,21]. For instance, as reported in [13], recommender systems frequently update their models daily. Consequently, GSL must accommodate ongoing data updates across various applications. While, this requirement inevitably motivates us to think a question: *GSL can ensure graph data privacy as the training set is fixed, but what about the updated training data?*

Prior works [19,33] have ever discussed similar questions in other learning frameworks and proposed the Update Leakage Attack (ULA). Salem *et al.* [19] utilize an encoder-decoder attack to reconstruct data for updates in machine-learning-as-a-service platforms, while Zanella *et al.* [33] focus on natural language processing models to extract specific updated sentences or discourse fragments. However, current efforts on ULA primarily center on "ordered grid" data like images and text, but never step into the realm of graph data. Moreover, these approaches cannot be seamlessly applied to GSL due to the fundamental distinction stemming from the dependencies between nodes connected by links.

Our Contributions. In this paper, we take the first step to unveil the ULA problem of GSL. We propose the first attack designed to steal the update set against GSL, termed the Graph Updates Leakage Attack (GULA). Specifically, we refer to the newly added set for model training in GSL as *update set*. The goal of GULA is to reconstruct this update set as two versions of a GNN are obtained with GSL. This form of information leakage compromises the intellectual property and data privacy of GSL participants.

Attack Classification. In our work, we systematically investigate the adversary's background knowledge and leverage it to classify the GULA adversaries into 8 types. Within the standard GSL setting, where the server acts as the adversary to steal client's training data, we analyze two primary graph update scenarios: link-only and node-and-link update. We exclude the node-only update scenario from our analysis as adding isolated nodes without links is meaningless in most cases. Next, like prior works [10,20], we assume the adversary can access parts of the original training data as background knowledge, including the adjacency matrix and node features. Therefore, ranging from the background knowledge being all accessible to neither, we identify 4 knowledge combinations for each scenario. Above all, we define 8 types of GULA adversaries.

Link-Only Attack. In this scenario, GULA works similarly to conventional link inference attacks [5,10,28], which involve either modifying the graph structure [14,28] or inferring links based on posterior similarity between nodes [3,5,10]. However, the modification approach discords with the setting of GSL, while the latter basically follows a brute-force workflow, i.e., roughly traversing each pair of nodes and predicting links as long as the similarity exceeds a certain threshold. This method tends to significantly overestimate the presence of links, as seen in our evaluation with the Cora dataset, where the Link Stealing Attack (LSA) [10] identifies 236230 links, vastly exceeding the actual count of 1374 links.

In GULA, we design a link deletion strategy to address this problem. Through detailed analysis, we discover that links predominantly cluster around high-degree nodes, known as "hub nodes" [15]. These nodes often incorrectly link to numerous identical nodes when they share similar posteriors, a phenomenon we term Common-Neighbor-Pattern (CNP) errors. Our solution is to mark all CNP nodes by locating their second-order neighbors and propagate the link assessment from two-node to three-node. We correct superfluous links using a "summing and differencing" principle of link probabilities and analyzing posterior variations. Section 4.1 details the specific link correction rule and the extension of GULA to 4 different adversaries in this scenario.

Node-And-Link Attack. In this scenario, reconstructing the update set entails optimizing two different objects (links and node features) simultaneously, which dual focus makes the encoder-decoder style optimizer hard to converge. To address this, we downgrade the problem to reveal only one of them at one time through an alternate optimization strategy, executed in two iterative steps: The first is to leverage the above-mentioned method to reveal graph links. The second is to use a self-designed graph generator to reconstruct node features with links (the revealed ones) as known info. Here, the objective of the generator is to invert the mapping from node posteriors back to their features. Section 4.2 details the extension of GULA to 4 different adversaries in this scenario.

Evaluation. We comprehensively evaluate the effectiveness of GULA across 8 attacks using 6 widely used graph datasets and 8 GNN models (including GCN [11], GraphSAGE [9], GAT [23] and etc.). Experimental results show that Type I attacks improve the link accuracy and precision by an average of 36.95% and 6.50%, respectively. For the Cora dataset, Type II attacks achieve an average Mean Squared Error (MSE) of 0.012 for reconstructed node features. Moreover, our attacks ensure that 99.31% of nodes maintain identical predictions to their true states. We also assess possible defense mechanisms against our attacks.

Our contributions are concluded as follows.

- We propose GULA, the first update leakage attack against GSL.
- We systematically analyze the threat model of GULA and categorize it into 8 types according to the attack scenario and the adversary's knowledge. Also, we propose 8 methods to achieve these 8 types of attacks.
- We propose and evaluate the possible defense against GULA.
- We extensively evaluate our 8 attacks on 6 real-world datasets. Experimental results show that GULA effectively reconstructs the update graph data.

2 Preliminary

2.1 Graph Neural Network

GNNs are typically tailored for semi-supervised node classification tasks with graph data [9,30]. Given the adjacency matrix $\mathcal{A}$, node features $\mathcal{X}$ and labels $\mathcal{Y}$

from the training set, GNNs are trained to predict classes of unlabeled nodes. Specifically, GNNs utilize a propagation rule in each layer, formalized as follows:

$$H^{(l+1)} \leftarrow \sigma(\tilde{Q}^{-\frac{1}{2}}\tilde{\mathcal{A}}\tilde{Q}^{-\frac{1}{2}}H^{(l)}Z^{(l)}), \tag{1}$$

where $H^{(l+1)}$ is the activation matrix for the $(l+1)$-th layer, $H^{(0)} = \mathcal{X}$. $\sigma(\cdot)$ is the activation function such as ReLU. $\tilde{Q}^{-\frac{1}{2}}\tilde{\mathcal{A}}\tilde{Q}^{-\frac{1}{2}}$ is the symmetric normalized adjacency matrix, where $\tilde{Q}_{uu} = \sum_v \tilde{\mathcal{A}}_{uv}$ is a diagonal matrix and $\tilde{\mathcal{A}} = \mathcal{A} + I_{\mathcal{N}}$ is the adjacency matrix of $\mathcal{G}$ with added self-connections, $I_{\mathcal{N}}$ is the identity matrix. $Z^{(l)}$ is the trainable weight matrix for the l-th layer. As the most representative GNN, Graph Convolution Network (GCN) is further defined as follows:

$$\mathcal{M} \leftarrow \text{softmax}(\tilde{Q}^{-\frac{1}{2}}\tilde{\mathcal{A}}\tilde{Q}^{-\frac{1}{2}}(\text{ReLU}(\tilde{Q}^{-\frac{1}{2}}\tilde{\mathcal{A}}\tilde{Q}^{-\frac{1}{2}}\mathcal{X}Z^{(0)}))Z^{(1)}), \tag{2}$$

where $\mathcal{M}$ is the target model, $Z^{(0)}$ and $Z^{(1)}$ are the trainable weight matrix for 1-th and 2-th layers. The GCN typically comprises two layers [10,26], while recent studies have demonstrated the effectiveness of deeper GCNs [7]. Both aspects are discussed in this work. Note that $\mathcal{M}$ is partitioned into a client model $\mathcal{M}_c$ and a server model $\mathcal{M}_s$ in GSL. We use $\mathcal{M}(\cdot)$ to represent the collaborative inference $\mathcal{M}_s(\mathcal{M}_c(\cdot))$ in our paper.

2.2 Updates Leakage Attack

The ULA [19,33] focuses on the privacy leakage during service updates and aims to recover the update set. Adversaries often employ encoder-decoder models for data reconstruction, formulated as Eq. 3:

$$\mathcal{F}_{ULA} : \mathcal{M}_{ULA}(\mathcal{P}, \mathcal{P}') \rightarrow \mathcal{D}_{update}, \tag{3}$$

where $\mathcal{M}_{ULA}$ represents the attack model of ULA and $\mathcal{D}_{update}$ denotes the update set. The model update process is defined as $\mathcal{F}_{update} : \mathcal{M} \xrightarrow{\mathcal{D}_{update}} \mathcal{M}'$. $\mathcal{M}$ and $\mathcal{M}'$ are the target models before and after the update, respectively. $\mathcal{P}$ and $\mathcal{P}'$ denote the posteriors predicted by $\mathcal{M}$ and $\mathcal{M}'$ on the same input.

2.3 Link Stealing Attack

The LSA [10] finds that posteriors contain link information. An adversary can reveal private graph links with a binary classifier based on the fact that connected nodes tend to have similar posteriors, which can be expressed as Eq. 4:

$$\mathcal{F}_{LSA} : \mathcal{M}_{LSA}(p_u, p_v) \rightarrow \mathbb{P}_{uv}, \tag{4}$$

where $\mathcal{M}_{LSA}$ represents the attack model for LSA, which outputs the probability $\mathbb{P}_{uv}$ that a link e_{uv} exists between nodes u and v based on their posteriors p_u and p_v. If $\mathbb{P}_{uv} \geq \varepsilon$, link e_{uv} exists. Conversely, if $\mathbb{P}_{uv} < \varepsilon$, the link does not exist. Here, ε is the threshold for link existence determination.

3 Problem Description

3.1 Attack Scenario

As elucidated in Sect. 1, GSL usually requires periodic interactions between the client and server for updates, which can be executed in two ways. The first is adding a new graph to the original graph that involves updates on the dataset only, known as data update; The other is retraining the model on the updated graph dataset, which involves updates on both the dataset and the model, known as model update. We first delve into our attack in the first scenario, and then we show how to extend GULA to the second scenario in Sect. 5.4.

3.2 Adversary's Goal and Capability

Adversary's Goal. The adversary's goal is to find a functionality $\mathcal{F}_{GULA}$ to infer the update set in GSL, formulated as Eq. 5:

$$\mathcal{F}_{\text{GULA}} : \mathcal{M}_{\text{GULA}}(\mathcal{P}, \mathcal{P}') \rightarrow \mathcal{D}_{update}, \tag{5}$$

where $\mathcal{M}_{\text{GULA}}$ represents the attack model of GULA and $\mathcal{D}_{update}$ denotes the update set. The data update process is defined as $\mathcal{F}_{update} : \mathcal{D} + \mathcal{D}_{update} \rightarrow \mathcal{D}'$. $\mathcal{D}$ and $\mathcal{D}'$ are the training data before and after the update, respectively. $\mathcal{P}$ and $\mathcal{P}'$ denote the posteriors corresponding to $\mathcal{D}$ and $\mathcal{D}'$.

Adversary's Capability. Referring to existing graph split learning works [8, 32], we assume the adversary is a semi-honest server who honestly participates in model training and prediction but is curious about the training data privacy. More specifically, the adversary is prohibited from manipulating the training process, poisoning the client's model, or interacting with the client's data. According to GSL workflow [31], the adversary can obtain the posteriors of training data and access the model $\mathcal{M}$ with their data. Similar to prior works [10,20], the adversary has access to the original graph data as background knowledge, which is categorized into two cases.

- *Adjacency matrix of the original graph.* We use $\mathcal{A}$ to represent the linked and unlinked node pairs of the original dataset $\mathcal{D}$.
- *Node features of the original graph.* We use $\mathcal{X}$ to denote the node features of the original dataset $\mathcal{D}$.

When $\mathcal{A}$ is unavailable, it is a common assumption that the adversary can access public graphs from external sources, known as a shadow dataset $\mathcal{D}_{shadow}$ [10,26].

3.3 Attack Taxonomy

Graph data updates primarily occur in two scenarios: link-only update and node-and-link update. We propose corresponding attacks for each scenario:

- *Type I Attack.* When $\mathcal{D}_{update}$ involves only links, Type I attack is launched. The update process is defined as $\mathcal{F}_{update} : \mathcal{D}(\mathcal{A}, \mathcal{X}) + \mathcal{D}_{update}(\mathcal{A}_{update}) \rightarrow \mathcal{D}'(\mathcal{A}', \mathcal{X})$, where $\mathcal{D}_{update}(\mathcal{A}_{update})$ represents the update set for $\mathcal{A}$.
- *Type II Attack.* When $\mathcal{D}_{update}$ involves both nodes and links, Type II attack is launched. The update process is defined as $\mathcal{F}_{update} : \mathcal{D}(\mathcal{A}, \mathcal{X}) + \mathcal{D}_{update}(\mathcal{A}_{update}, \mathcal{X}_{update}) \rightarrow \mathcal{D}'(\mathcal{A}', \mathcal{X}')$, where $\mathcal{D}_{update}(\mathcal{A}_{update}, \mathcal{X}_{update})$ represents the update set of both $\mathcal{A}$ and $\mathcal{X}$.

Taking the background knowledge settings of $\mathcal{A}$ and $\mathcal{X}$ into account, we divide our attack into 8 categories, summarized in Table 1.

Table 1. Attack taxonomy. ✓(×) means the adversary has (does not have) the background knowledge.

Type I Attack	$\mathcal{A}$	$\mathcal{X}$	Type II Attack
Attack-I-0	✓	✓	Attack-II-0
Attack-I-1	✓	×	Attack-II-1
Attack-I-2	×	✓	Attack-II-2
Attack-I-3	×	×	Attack-II-3

4 Attack Framework

4.1 Type I Attack: Link-Only Update

To reveal the updated graph structure, Type I attacks of GULA proceed as the following three steps (also outlined in Algorithm 1). Note that we introduce our attacks using the link set $\mathcal{E}$ corresponding to $\mathcal{A}$.

• **Step-1:** Our first step is to identify all possible links with LSA, which is implemented by developing a MLP-based link filter model $\mathcal{M}_{LF}$ to score node pairs based on their posteriors. Here, the score reflects the existence probability of a link between two nodes. We select top-$\hat{k}$ scoring pairs as predicted links.

Specifically, training labels s_{link} and s_{unlink} ($s_{link} > s_{unlink}$) are assigned to linked and unlinked node pairs in the training set, respectively. Then, we calculate posterior distances similar to LSA as training features. Trained with the MSE loss, $\mathcal{M}_{LF}$ evaluates each pair in $\mathcal{E}_0$, the set of all node pairs in $\mathcal{V}$ expect links in $\mathcal{E}$. It assigns a score s_{uv} to each node pair (u, v), encapsulated in the score set $\mathcal{S}$. Node pairs are then ranked by their scores, and the top-$\hat{k}$ are designated as linked, forming the set $\mathcal{E}_1$. $\hat{k}$ is defined as the filter value.

• **Step-2:** In the second step, we propose the following two methods to reduce the number of superfluous links.

Algorithm 1. Type I attack

Input: Target GNN model $\mathcal{M}$; Link Filter attack model $\mathcal{M}_{LF}(\cdot)$; set of nodes $\mathcal{V}$, the adjacency matrix $\mathcal{A}$, node features $\mathcal{X}$ and posteriors $\mathcal{P}$ of the original graph; posteriors of the updated graph $\mathcal{P}'$; filter value $\hat{k}$; Maximum number of common neighbors $\mathcal{N}_{neighbor}^{max}$; set of hub nodes $\mathcal{V}_{hub}$

Output: the update link set $\mathcal{E}_3$

1: $\mathcal{E}_0 \leftarrow \{e_{uv} \mid u, v \in \mathcal{V}, u \neq v, \mathcal{A}_{uv} \neq 1\}$
  • **Step-1:**
2: Score predicted links: $\mathcal{S} = \mathcal{M}_{LF}(\mathcal{P}', \mathcal{E}_0)$
3: Select links in $\mathcal{E}_0$ with the top $\hat{k}$ of $\mathcal{S}$ as $\mathcal{E}_1$
  • **Step-2:**
4: $\mathcal{E}_t \leftarrow CND(\mathcal{V}, \mathcal{E}_1, \mathcal{S}, \mathcal{N}_{neighbor}^{max})$
5: Posterior accuracy: $\mathcal{L}_t = \mathcal{L}_{ce}(\mathcal{M}(\mathcal{A}_t, \mathcal{X}), \mathcal{P}')$
6: Select nodes in $\mathcal{V}$ with the top $p\%$ of $\mathcal{L}_t$ as $\mathcal{V}_{bad}$.
7: $\mathcal{E}_2, \mathcal{E}_{t'} \leftarrow \{e_{uv} \mid e_{uv} \in \mathcal{E}_t, u, v \notin \mathcal{V}_{bad}\}$
8: **for** *each link* $e_i \in \mathcal{E}_{t'}$ **do**
9:   $\mathcal{E}_i = \mathcal{E}_2 - e_i$
10:   **if** $\mathcal{L}_{ce}(\mathcal{M}(\mathcal{A}_2, \mathcal{X}), \mathcal{P}') - \mathcal{L}_{ce}(\mathcal{M}(\mathcal{A}_i, \mathcal{X}), \mathcal{P}') > 0$ **then**
11:     $\mathcal{E}_2 \leftarrow \mathcal{E}_i$
  • **Step-3:**
12: $\mathcal{E}_{hub} \leftarrow \{e_{vv_{hub}} \mid v \in \mathcal{V}_{update}, v_{hub} \in \mathcal{V}_{hub}, s_{vv_{hub}} > s_{link}\}$
13: **for** *each link* $e_i \in \mathcal{E}_{hub}$ **do**
14:   $\mathcal{E}_i = \mathcal{E}_2 + e_i$
15:   **if** $\mathcal{L}_{ce}(\mathcal{M}(\mathcal{A}_2, \mathcal{X}), \mathcal{P}') - \mathcal{L}_{ce}(\mathcal{M}(\mathcal{A}_i, \mathcal{X}), \mathcal{P}') > 0$ **then**
16:     $\mathcal{E}_2 \leftarrow \mathcal{E}_i$
17: **return** $\mathcal{E}_3 = \mathcal{E}_2$

Common Neighbors Based Deletion. To mitigate superfluous links caused by CNP errors, we leverage a three-node link deletion strategy based on the "summing and differencing" principle. This strategy degrades the task of deleting erroneous links across the entire link set $\mathcal{E}_1$ to addressing specific three-node groups. If w_0 is a common neighbor of nodes u and v, the prediction method in Step-1 might erroneously connect nodes u and v with all nodes $\{w_1, ..., w_i, ..., w_n\}$ with similar posterior to w_0. The deletion strategy focuses on each group formed by u, v, and a common neighbor w_i, with links related to w_i segmented into three distinct scenarios for precise removal.

- Case-1: w_i connects both nodes u and v.
- Case-2: w_i connects either node u or v.
- Case-3: w_i connects neither node u nor v.

For nodes u and v, each common neighbor w_i falls into one of the three cases based on their link scores to delete superfluous links in Case-2 and Case-3. In Case-1, both scores s_{uw_i} and s_{vw_i} are high; in Case-2, one score is higher than the other; and in Case-3, both scores are lower. Intuitively, we can calculate the summing values $s_{uw_i} + s_{vw_i}$ for all common neighbors and set thresholds for categorization, but this is labor-intensive for large datasets and inadequate as they fail to recognize Case-2. Here, we introduce the differencing values $|s_{uw_i} - s_{vw_i}|$ to aid classification, where Case-2 shows a larger difference than Case-1 and Case-3. Specifically, we first calculate the summing set $\mathbb{S}$ and the differencing set $\mathbb{D}$ for all w_i in $\mathbb{W}_{uv} = \{w_0, ...w_i, ..., w_n\}$, following the relationships in Eq. 6.

$$\begin{cases} s_{uw_i^1} + s_{vw_i^1} > s_{uw_i^2} + s_{vw_i^2} > s_{uw_i^3} + s_{vw_i^3} \\ |s_{uw_i^2} - s_{vw_i^2}| > |s_{uw_i^1} - s_{vw_i^1}| \approx |s_{uw_i^3} - s_{vw_i^3}|, \end{cases} \tag{6}$$

where w_i^1, w_i^2, and w_i^3 correspond to common neighbors of three cases, respectively. We apply uniform partition ratios $\{p_1, p_2, p_3\}$ across all groups, sort $\mathbb{S}$ and $\mathbb{D}$ in ascending order, and divide them into $\mathbb{S} = [\mathbb{S}_{p_3}, \mathbb{S}_{p_2}, \mathbb{S}_{p_1}]$ and $\mathbb{D} = [\mathbb{D}_{p_1+p_3}, \mathbb{D}_{p_2}]$. Nodes at the intersection of $\mathbb{S}_{p_1}$ and $\mathbb{D}_{p_1+p_3}$ are classified as Case-1, preserving their links to u and v. Case-3 includes nodes at the intersection of $\mathbb{S}_{p_3}$ and $\mathbb{D}_{p_1+p_3}$, with the rest categorized as Case-2. The result is $\mathcal{E}_t$ and the process is outlined in Algorithm 2.

Algorithm 2. Common Neighbors based Deletion (CND)

Input: Set of nodes $\mathcal{V}$; link set $\mathcal{E}_1$; score set $\mathcal{S}$; maximum number of common neighbors $\mathcal{N}_{neighbor}^{max}$
Output: the link set $\mathcal{E}_t$
1: $\mathcal{W}_{uv} \leftarrow \{w_i \,|\, w_i \in \mathcal{V}, e_{uw_i}, e_{vw_i} \in \mathcal{E}_1\}, u, v \in \mathcal{V}$
2: $\mathbb{S} \leftarrow \{\mathbb{S}_i | \mathbb{S}_i = s_{uw_i} + s_{vw_i}, w_i \in \mathcal{W}_{uv}, s_{uw_i}, s_{vw_i} \in \mathcal{S}\}$
3: $\mathbb{D} \leftarrow \{\mathbb{D}_i | \mathbb{D}_i = |s_{uw_i} - s_{vw_i}|, w_i \in \mathcal{W}_{uv}, s_{uw_i}, s_{vw_i} \in \mathcal{S}\}$
4: Sort $\mathbb{S}$ and $\mathbb{D}$ in ascending order.
5: Divide $\mathbb{S} = [\mathbb{S}_{p3}, \mathbb{S}_{p2}, \mathbb{S}_{p1}]$
6: Divide $\mathbb{D} = [\mathbb{D}_{p_1+p_3}, \mathbb{D}_{p2}]$
7: $\mathcal{V}_{Case-1}^{uv} \leftarrow \{v_i \,|\, v_i \in \mathcal{W}_{uv}, \mathbb{S}_i \in \mathbb{S}_{p1}, \mathbb{D}_i \in \mathbb{D}_{p1+p3}\}$
8: $\mathcal{V}_{Case-3}^{uv} \leftarrow \{v_i \,|\, v_i \in \mathcal{W}_{uv}, \mathbb{S}_i \in \mathbb{S}_{p3}, \mathbb{D}_i \in \mathbb{D}_{p1+p3}\}$
9: $\mathcal{V}_{Case-2}^{uv} \leftarrow \{v_i \,|\, v_i \in \mathcal{W}_{uv}, v_i \notin \mathcal{V}_{Case-1}^{uv}, v_i \notin \mathcal{V}_{Case-3}^{uv}\}$
10: **if** $|\mathcal{W}_{uv}| > \mathcal{N}_{neighbor}^{max}$ **then**
11: **for** *each node* $w_i \in \mathcal{V}_{Case-1}^{uv}$ **do**
12: Preserve both links e_{uw_i} and e_{vw_i}
13: **for** *each node* $w_i \in \mathcal{V}_{Case-2}^{uv}$ **do**
14: Preserve e_{tw_i} in $\mathcal{E}$, $s_{tw_i} = \max(s_{uw_i}, s_{vw_i})$
15: **for** *each node* $w_i \in \mathcal{V}_{Case-3}^{uv}$ **do**
16: Delete both links e_{uw_i} and e_{vw_i}
17: **return** $\mathcal{E}_t = \mathcal{E}_1$

Posterior Accuracy Based Deletion. To further remove superfluous links, we use a method based on *posterior accuracy*, measured by the cross-entropy loss: $\mathcal{L}_i = \mathcal{L}_{ce}(\mathcal{P}_i, \mathcal{P}')$, where $\mathcal{P}'$ is true posteriors. Here, $\mathcal{P}_i = \mathcal{M}(\mathcal{A}_i, \mathcal{X})$ denotes the predicted posteriors, and $\mathcal{A}_i$ is the adjacency matrix with the update link set $\mathcal{E}_i$. The underlying principle is that higher posterior accuracy correlates with more precise link predictions.

First, we query $\mathcal{M}$ to obtain $\mathcal{P}_t = \mathcal{M}(\mathcal{A}_t, \mathcal{X})$ and calculate $\mathcal{L}_t = \mathcal{L}_{ce}(\mathcal{P}_t, \mathcal{P}')$. Nodes with large losses indicate potential error links. We select $p = 10\%$ nodes with the highest loss as $\mathcal{V}_{bad}$ and their links are removed from $\mathcal{E}_t$ to form $\mathcal{E}_{t'}$. Then we iteratively set $\mathcal{E}_i = \mathcal{E}_{t'} - e_i$ for each link $e_i \in \mathcal{E}_{t'}$ and recalculate $\mathcal{P}_i = \mathcal{M}(\mathcal{A}_i, \mathcal{X})$. Links leading to reduced loss are deemed erroneous and permanently deleted; otherwise, the deletions are reverted. The objective function is formulated as Eq. 7:

$$\mathcal{L}_{ce}(\mathcal{P}_{i-1}, \mathcal{P}') - \mathcal{L}_{ce}(\mathcal{P}_i, \mathcal{P}') > 0, \tag{7}$$

where $\mathcal{P}_{i-1} = \mathcal{M}(\mathcal{A}_{i-1}, \mathcal{X})$ denotes the posteriors before deleting e_i, and we can obtain $\mathcal{E}_2$ after eliminating superfluous links.

• **Step-3:** In this step, we aim to identify the missing links by applying the concept of *Preferential Attachment* [17,22], where nodes or links tend to connect with nodes having higher degrees (i.e., hub nodes). Therefore, we identify hub nodes within the original graph and regard them as potential connection points. Leveraging posterior accuracy as the guiding criterion, we discern links between hub nodes and update nodes for addition.

Specifically, we first designate $p_{hub}\%$ highest-degrees nodes as $\mathcal{V}_{hub}$. Then we utilize node pairs between $\mathcal{V}_{update}$ and $\mathcal{V}_{hub}$ to build $\mathcal{E}_{hub}$, where $\mathcal{V}_{update} = \mathcal{V}$ in the link-only update scenario. For each link $e_i \in \mathcal{E}_{hub}$, we calculate $\mathcal{P}_i$ after its addition. If the condition in Eq. 7 is satisfied, the link is retained; otherwise, we evaluate the next one. Note that we only focus on the node pairs in $\mathcal{E}_{hub}$ with scores surpassing s_{link} as possible links. With all steps completed, the server can obtain the predicted link set $\mathcal{E}_3$ and the corresponding adjacency matrix $\bar{\mathcal{A}}'$.

Attack Extension. The following shows how to extend the above attack workflow to the four Type I attacks listed in Table 1.

Attack-I-0. This scenario grants the adversary has access to both the adjacency matrix $\mathcal{A}$ and node features $\mathcal{X}$. We use links from $\mathcal{A}$ as ground truth labels and calculate their feature and posterior distances for training $\mathcal{M}_{LF}$. Subsequently, $\mathcal{M}_{LF}$ scores each pair of nodes in the target graph.

Attack-I-1. In this attack, the adversary only possesses the adjacency matrix $\mathcal{A}$. Consequently, $\mathcal{M}_{LF}$ is trained only based on the posterior distances. The absence of node features $\mathcal{X}$ necessitates skipping the Posterior Accuracy based Deletion in Step-2 and modifying Step-3 to randomly add links in $\mathcal{E}_{hub}$.

Attack-I-2. In this case, the adversary has the knowledge of node features $\mathcal{X}$. Without $\mathcal{A}$, we need to rely on a shadow dataset $\mathcal{D}_{shadow}$ to build a shadow target model $\mathcal{M}_{shadow}$. Then we derive the distances of node features and posteriors of $\mathcal{D}_{shadow}$ to train $\mathcal{M}_{LF}$, and predict both original and update links.

Attack-I-3. In the last scenario, neither $\mathcal{X}$ nor $\mathcal{A}$ is known to the adversary. Similar to Attack-I-2, we rely on the graph $\mathcal{D}_{shadow}$ and build the training set with the distances of the posteriors. Additionally, as in Attack-I-1, adjustments are made in Step-2 and Step-3, which involve manipulating features.

4.2 Type II Attack: Node-And-Link Update

In this scenario, the adversary's objective is to infer both node features $\mathcal{X}_{update}$ and adjacent matrix $\mathcal{A}_{update}$ used for update. We follow the link-only scenario employing Type I attacks to reveal links and design an alternate process of link revelation and feature reconstruction. During this process, the adversary reconstructs node features after each Type I attack step to support subsequent steps. In particular, we propose a two-step feature inference method named Graph Feature Inversion (GFI). The first step generates node features, followed by a refinement process. Detailed description of this method ensues.

• **Step-1:** We design a feature inversion model to reconstruct the features of newly added nodes by reverse engineering the mapping from node posteriors to their respective features. However, unlike structured data like images and text where each sample is predicted independently, each graph node is interconnected, necessitating the incorporation of neighbor node information in feature reconstruction. Our inversion model adopts a GNN architecture, which considers both node posteriors and adjacency matrix as inputs and produces node features.

Specifically, our objective is to implement the feature fitting strategy by minimizing the distance between the predicted features $\hat{\mathcal{X}}$ and true features $\mathcal{X}$, which can be quantified by the MSE loss function, as shown in Eq. 8.

$$\begin{aligned} \min_{\hat{\mathcal{X}}} \mathcal{L}_{inv} &= \mathcal{L}_{mse}(\hat{\mathcal{X}}, \mathcal{X}) + \mathcal{L}_f + \alpha \|\hat{\mathcal{X}}\|_F \\ s.t. \quad \hat{\mathcal{X}} &= \mathcal{M}_{inv}(\mathcal{A}, \mathcal{P}) \\ \mathcal{L}_f &= \sum_{u,v=1}^{\mathcal{N}} \mathcal{A}_{uv}(\hat{x}_u - \hat{x}_v)^2, \end{aligned} \tag{8}$$

where $\mathcal{L}_{mse}(\cdot)$ is the MSE function and $\mathcal{M}_{inv}$ denotes the feature inversion model. Reflecting patterns observed in real-world graphs like social networks and knowledge graphs, where connected nodes often exhibit similar features, we incorporate the restriction $\mathcal{L}_f$ [34]. Here, $\mathcal{N}$ is the number of nodes, $\mathcal{A}_{uv}$ is the connection between nodes u and v, and $(\hat{x}_u - \hat{x}_v)^2$ measures their difference in predicted features. Besides, we introduce an F-norm term $\alpha\|\hat{\mathcal{X}}\|_F$ to stabilize the convergence process, where α is the hyper-parameter controlling its weight.

In Attack-II-1 and Attack-II-3, where the adversary lacks access to $\mathcal{X}$ required for the above attack, a posterior simulation strategy is employed. Features of each node are reconstructed by minimizing the difference in their posteriors compared to real nodes, quantified using cross-entropy loss as detailed in Eq. 9.

$$\begin{aligned} \min_{\hat{\mathcal{X}}} \mathcal{L}_{inv} &= \gamma \mathcal{L}_{ce}(\hat{\mathcal{P}}, \mathcal{P}) + \mathcal{L}_f + \alpha \|\hat{\mathcal{X}}\|_F \\ s.t. \quad \hat{\mathcal{P}} &= \mathcal{M}(\mathcal{A}, \hat{\mathcal{X}}) \\ \hat{\mathcal{X}} &= \mathcal{M}_{inv}(\mathcal{A}, \mathcal{P}), \end{aligned} \tag{9}$$

where $\hat{\mathcal{P}}$ represents the predicted posteriors and γ is the hyper-parameter controlling the weight of posterior simulation. We also use $\mathcal{L}_f$ and the F-norm term to enhance feature generation.

Once the feature inversion model is trained by one of the above two methods, the adversary feeds the predicted adjacency matrix $\bar{\mathcal{A}}'$ and posteriors $\mathcal{P}'$ to it, and obtains the predicted node features $\hat{\mathcal{X}}_{update}$.

• **Step-2:** We design a feature optimization model to refine the predicted features using node posteriors. Specifically, we adopt an MLP model designed to minimize the cross-entropy loss between the posteriors $\bar{\mathcal{P}}'$ and the true posteriors $\mathcal{P}'$, as shown in Eq. 10.

$$\begin{aligned} \min_{\bar{\mathcal{X}}_{update}} \ & \mathcal{L}_{opt} = \beta \mathcal{L}_{ce}(\bar{\mathcal{P}}', \mathcal{P}') + \mathcal{L}_{mse}(\bar{\mathcal{X}}_{update}, \hat{\mathcal{X}}_{update}) \\ s.t. \ & \bar{\mathcal{X}}_{update} = \mathcal{M}_{opt}(\hat{\mathcal{X}}_{update}) \\ & \qquad \bar{\mathcal{P}}' = \mathcal{M}(\bar{\mathcal{A}}', \mathcal{X} + \bar{\mathcal{X}}_{update}), \end{aligned} \tag{10}$$

where $\mathcal{M}_{opt}$ represents the feature optimization model and $\bar{\mathcal{X}}_{update}$ denotes the node features generated by $\mathcal{M}_{opt}$. $\mathcal{X} + \bar{\mathcal{X}}_{update}$ is the union of $\mathcal{X}$ and $\bar{\mathcal{X}}_{update}$, i.e., the predicted features of all nodes. β is the hyper-parameter controlling the weight of posteriors.

Through alternate revealing links and node features, the server can reconstruct the updated graph. The GFI method integrates the feature inversion model and feature optimization model, adapting its training strategy according to the adversary's background knowledge. Specifically, we refer to the GFI method utilizing the feature fitting strategy as GFI-FF and the one employing the posterior simulation strategy as GFI-PS.

Attack Extension. The following shows how to combine the two GFI methods with Type I attacks and extend the workflow to the 4 scenarios listed in Table 1. Given the necessity to reconstruct the features of newly added nodes, we omit the pruning of $\mathcal{V}_{bad}$ during the Type I attack process.

Attack-I-0. In this attack, the adversary has access to both the adjacency matrix $\mathcal{A}$ and node features $\mathcal{X}$. The framework from Attack-I-0 is used to reveal links, but only posterior distances derived from $\mathcal{A}$ are used to train $\mathcal{M}_{LF}$. Node features $\mathcal{X}$ serve as ground truth labels for training the GFI-FF attack model.

Attack-II-1. In this case, the adversary has the knowledge of the adjacency matrix $\mathcal{A}$. Since $\mathcal{X}$ is unavailable, the GFI-PS method is employed for this attack and the generated features can be used for Type I attacks.

Attack-II-2. In this scenario, the adversary only possesses node features $\mathcal{X}$. Utilizing the shadow dataset $\mathcal{D}_{shadow}$, the adversary predicts both original and new links, employing only node posterior distances. The GFI-FF method is used to restore features of newly added nodes.

Attack-II-3. In this scenario, the adversary lacks any prior knowledge of the original graph. With the shadow dataset $\mathcal{D}_{shadow}$, posterior distances are built to reveal links. The GFI-PS method is employed in this attack.

5 Evaluation

In this section, we provide a comprehensive evaluation for GULA. We first introduce our experimental setup and then present the detailed results for each attack. All experiments are repeated five times, with averages calculated accordingly.

5.1 Experimental Setup

Attack Setting. In our experiments, we first utilize the GCN as the default model. Then, we extend our attack to 10-layer GCN, GAT, and GraphSAGE in Sect. 5.3. Experiments are conducted on 6 public datasets, including Cora [11], Citeseer [20], ACM [25], UAI [24], Amazon [16] and Pubmed [26].

Table 2. Selection of shadow datasets for all 6 datasets.

target dataset	shadow dataset	dataset	shadow dataset
Cora	Citeseer	UAI	Cora
Citeseer	Cora	Amazon	ACM
ACM	Cora	Pubmed	UAI

To simulate attack scenarios, we follow the split learning setting involving clients and a malicious server [8,32]. For Type I attack, we randomly sample 5%–25% of links from each dataset to create the update link set $\mathcal{E}_{update}$, while the remaining forms the original graph link set $\mathcal{E}$. For Type II attack, the update dataset $\mathcal{D}_{update}$ consists of 5%–25% randomly sampled nodes, along with their associated links, and the remaining forms the original dataset $\mathcal{D}$. Our experiments default to a 15% update rate and a filter value of k. We evaluate various filter values $\hat{k} = \{k/4, k/2, k, 2k, 4k\}$ for link inference [28], where k is the actual number. The scores for linked and unlinked nodes are set at $s_{link} = 500$ and $s_{unlink} = 0$ respectively. The ratio p_1 of Case-1 is twice $\mathcal{N}^{max}_{neighbor}$, with the remaining nodes split equally between Case-2 and Case-3. We perform a comprehensive grid search for other hyper-parameters and select the shadow dataset based on LSA, detailed in Table 2.

Evaluation Metric. To comprehensively evaluate GULA, we use several standard metrics: Adjacency Matrix Accuracy (AMA) [6] evaluates accuracy across all node pairs, Average Precision (AP) [26] quantifies the proportion of true links among revealed links, Mean Squared Error (MSE) [19] measures the distance between predicted and true features, Classification Accuracy (CA) and Impact of Classification Accuracy (ICA) [9,30] assesses the attack performance by querying the target model with the reconstructed graph. The formulas for CA and ICA are detailed in Eq. 11.

$$CA = \frac{\mathcal{N}^{c}_{update}}{\mathcal{N}_{update}}, \quad ICA = 1 - \frac{\mathcal{N}^{c}}{\mathcal{N}}, \tag{11}$$

where $\mathcal{N}$ and $\mathcal{N}_{update}$ represent the number of nodes in the original and update graphs, respectively, while $\mathcal{N}^c$ and $\mathcal{N}^c_{update}$ denote the number of nodes classified correctly in each graph. Note that $\mathcal{N}_{update} = \mathcal{N}$ in the link-only update scenario. In general, higher scores for AMA, AP, and CA scores and lower scores for MSE and ICA scores reflect better attack performance.

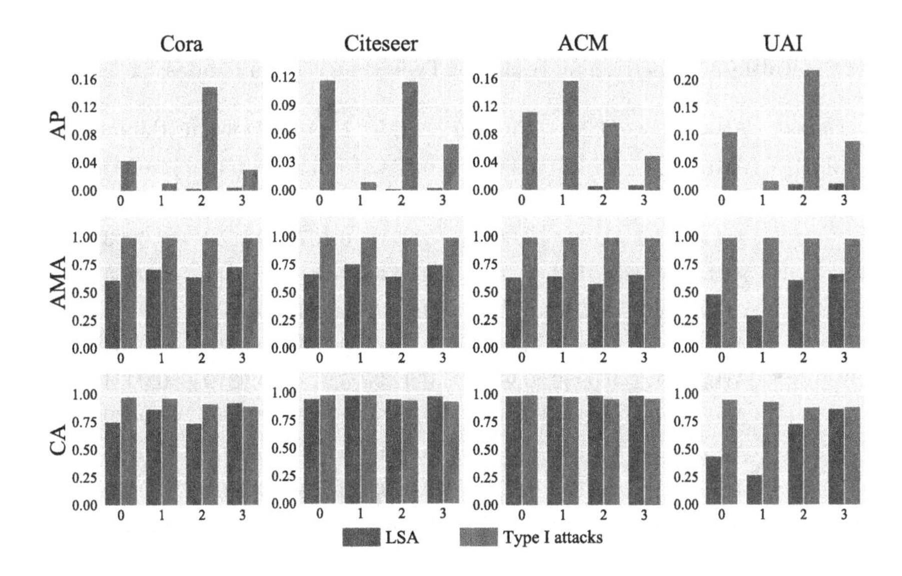

Fig. 1. Experimental results of Type I attacks.

5.2 Benchmark Performance

Type I Attack. We use AMA, AP, and CA to evaluate the effectiveness of the predicted adjacency matrix $\hat{\mathcal{A}}'$.

Accuracy. Due to space limitations, Fig. 1 presents the comparison results between Type I attacks and LSA across 4 dataset. Our attacks demonstrate notable improvements with an average improvement of 36.95% in AMA scores, 6.50% in AP scores, and 8.20% in A scores across six datasets. Notably, LSA's AP scores are nearly zero, exemplified by its failure on the Cora dataset where it inferred 236230 links in a graph with only 1374 links, resulting in an AP score of 0.5%. Moreover, we find that Attack-I-0 and Attack-I-2 with $\mathcal{X}$ outperform Attack-I-1 and Attack-I-3, especially in AP scores. This is anticipated as links often connect nodes with similar features (e.g., common interests lead to friendships). Overall, GULA effectively enhances link revelation across diverse situations.

Robustness. We further investigate Type I attacks across 5 update rates (5%–25%) and 5 filter values ($k/4, k/2, k, 2k, 4k$). Due to space constraints, only AMA and CA scores for the Citeseer dataset are shown in Fig. 2. The results demonstrate that Type I attacks maintain high AMA and CA scores even with increasing update rates and imprecise filter values. For instance, Attack-I-0 achieves a CA score of 99.39% at a 5% update rate and 97.64% at 25%. These findings indicate that the attacks are not overly dependent on exact filter values, highlighting GULA's robustness under various conditions.

Table 3. Experimental results of Type II attacks on 6 datasets.

Dataset	Attack	Cora	Citeseer	ACM	UAI	Amazon	Pubmed
MSE↓	Attack-II-0	0.0117	0.0081	0.0342	0.1135	0.2113	0.0003
	Attack-II-1	0.0123	0.0084	0.0411	0.1267	0.3155	0.0004
	Attack-II-2	0.0116	0.0081	0.0341	0.1077	0.2252	0.0003
	Attack-II-3	0.0122	0.0084	0.0430	0.1315	0.3211	0.0005
CA↑	Attack-II-0	0.9946	0.9960	0.9700	0.4848	0.9932	0.9912
	Attack-II-1	0.9882	0.9976	0.9863	0.7752	0.9916	0.9801
	Attack-II-2	0.9956	0.9980	0.9965	0.9783	0.9819	0.9914
	Attack-II-3	0.9941	0.9980	0.9978	0.9965	0.9846	0.9889
ICA↓	Attack-II-0	0.0144	0.0152	0.0035	0.0418	0.0105	0.0225
	Attack-II-1	0.0160	0.0156	0.0033	0.0453	0.0102	0.0229
	Attack-II-2	0.0142	0.0147	0.0066	0.0466	0.0135	0.0253
	Attack-II-3	0.0151	0.0149	0.0067	0.0481	0.0133	0.0297

Type II Attack. We use MSE, CA, and ICA to assess our attacks, where ↑ (↓) signifies a positive (negative) correlation with attack performance.

Accuracy. Table 3 summarizes results of four Type II attacks on 6 datasets, demonstrating excellent performances. For instance, on the Cora dataset, four attacks achieve an average CA score of 99.31%, with ICA and MSE scores at 1.42% and 0.0120, respectively. This is not only attributed to Type I attacks, our GFI methods also perform very well in feature reconstruction. Furthermore, Attack-II-0 and Attack-II-2, which leverage $\mathcal{X}$ to implement the GFI-FF method, achieve better performance in node feature inference. For instance, on the ACM dataset, they achieve an average MSE score of 0.0342, compared to 0.0421 in Attack-II-1 and Attack-II-3.

Comparison. We compare our GFI methods with two other node feature inference methods: *Random Generation Attack (RGA)* and *Neighbor Feature Synthesis Attack (NFSA)* [27]. RGA randomly generates node features within the target dataset's value range. NFSA synthesizes unknown nodes' features by combining features of their 1-hop and 2-hop neighbor nodes, leveraging the intuition that neighbor nodes tend to share similar features. We assume the adversary

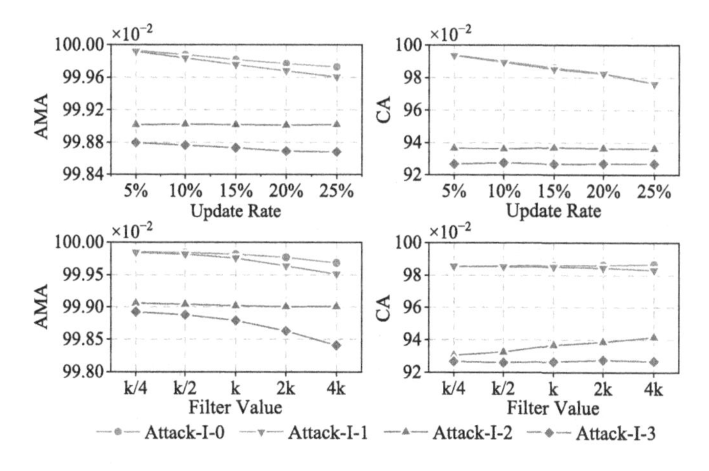

Fig. 2. The AMA and CA scores of Type I attacks at different update rates and filter values on the Citeseer dataset.

knows both the updated adjacency matrix $\mathcal{A}'$ and the original node features $\mathcal{X}$. For each dataset, we randomly select 50% of the nodes as the target data and set the adjustment factor for NFSA attack at 0.8 [27]. Results in Table 4 demonstrate that both GFI-FF and GFI-PS outperform NFSA and RGA across all datasets. For example, on the ACM dataset, two GFI methods register an average CA score improvement of 15.25% over NFSA and 42.96% over RGA.

Table 4. Experimental results of four feature inference methods on 6 datasets.

Dataset	Method	Cora	Citeseer	ACM	UAI	Amazon	Pubmed
MSE↓	RGA	0.5000	0.5001	0.4998	0.5000	0.4999	0.5273
	NFSA	0.0127	0.0085	0.0453	0.1379	0.3503	**0.0003**
	GFI-FF	**0.0118**	**0.0081**	**0.0344**	**0.1046**	**0.2065**	0.0004
	GFI-PS	0.0124	0.0084	0.0415	0.1178	0.3162	0.0004
CA↑	RGA	0.3049	0.3559	0.5602	0.6227	0.7333	0.4774
	NFSA	0.8811	0.8197	0.8373	0.8944	0.9618	0.8130
	GFI-FF	**0.9823**	**0.9879**	**0.9901**	0.9116	0.9778	0.9473
	GFI-PS	0.9694	0.9841	0.9894	**0.9317**	**0.9797**	**0.9478**
ICA↓	RGA	0.6245	0.3073	0.1685	0.2557	0.1682	0.4130
	NFSA	0.0391	0.0349	0.0145	0.0685	0.0193	0.0622
	GFI-FF	**0.0160**	**0.0083**	**0.0042**	0.0517	**0.0145**	0.0389
	GFI-PS	0.0225	0.0130	0.0059	**0.0403**	0.0157	**0.0336**

Ablation Studies. To understand the effectiveness of the two steps employed by GFI, we show CA scores achieved after each step in Fig. 3, comparing them

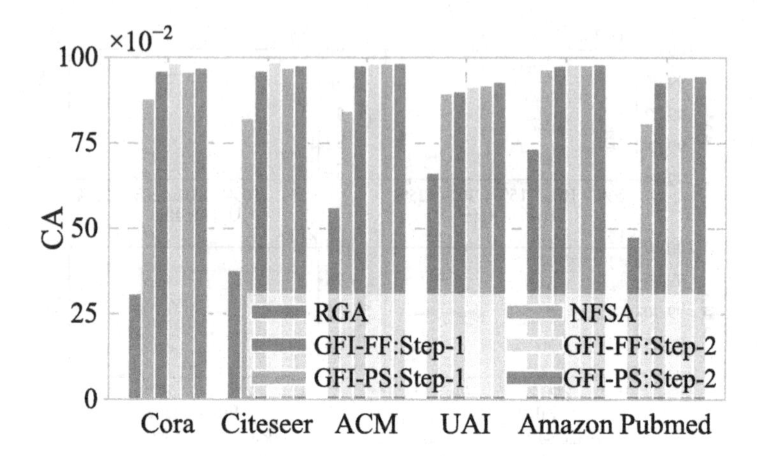

Fig. 3. The CA scores of different feature inference methods on 6 datasets.

against baseline methods. It is straightforward to see that both GFI-FF and GFI-PS outperform baselines significantly after Step-1, with further improvements observed in Step-2.

Table 5. Experimental results of Attack-II-0 on the Cora dataset when the layer number increases.

Model	Model Accuracy	AMA↑	AP↑	MSE↓	CA↑	ICA↓
2-layer	0.9438	0.9995	0.0707	0.0117	0.9946	0.0144
4-layer	0.9184	0.9995	0.0128	0.0118	0.9680	0.0225
6-layer	0.8527	0.9996	0.0286	0.0118	0.9852	0.0194
8-layer	0.7611	0.9996	0.0191	0.0119	0.9650	0.0486
10-layer	0.6403	0.9996	0.0109	0.0119	0.9433	0.0599

5.3 Attacks on Other GNNs

Multi-layer GNNs. To gauge the generality of GULA, we first assess its performance on multi-layer GNNs. Table 5 presents the performance of Attack-II-0 on the Cora dataset when the target model's layer number increases from 2 to 10. We observe better performance with lower layer numbers, CA scores drop from 99.46% to 94.33% as layers increase from 2 to 10. This decline can be attributed to the misclassified nodes due to reduced model accuracy, leading to varied posterior distributions among linked nodes. Consequently, GULA's link revelation ability worsens, evident in changes in AP and ICA scores.

Table 6. Experimental results of Attack-II-0 when using GraphSAGE or GAT as the target model on 6 datasets.

Dataset	GNN Model	AMA↑	AP↑	MSE↓	CA↑	ICA↓
Cora	GraphSAGE	0.9995	0.0949	0.0119	0.9813	0.0149
	GAT	0.9995	0.0987	0.0117	0.9773	0.0164
Citeseer	GraphSAGE	0.9997	0.0498	0.0082	0.9848	0.0205
	GAT	0.9997	0.0171	0.0081	0.9868	0.0207
ACM	GraphSAGE	0.9983	0.0158	0.0343	1.0000	0.0027
	GAT	0.9988	0.0494	0.0342	0.9379	0.0055
UAI	GraphSAGE	0.9974	0.0366	0.1119	0.6839	0.0241
	GAT	0.9974	0.0254	0.1130	0.8426	0.0389
Amazon	GraphSAGE	0.9985	0.0074	0.2249	0.4139	0.0223
	GAT	0.9981	0.0975	0.2136	0.9885	0.0090
Pubmed	GraphSAGE	0.9999	0.0093	0.0004	0.9934	0.0235
	GAT	0.9999	0.0134	0.0003	0.9934	0.0240

Different GNNs. We further extend our investigation of GULA to GNN models other than GCN, focusing on GraphSAGE [9] and GAT [23] models. Both models employ the same two-layer network architecture as the target GCN model, including hidden layers, activation functions, and training processes, with the GAT model featuring 4 heads. Table 6 shows the results of Attack-II-0 on 6 datasets with GraphSAGE or GAT as the target model. From the results, GULA achieves excellent performance across these models, with Attack-II-0 achieving CA scores of 98.13% and 97.73% on the Cora dataset for GraphSAGE and GAT, respectively (compared to 99.46% for GCN). This further demonstrates GULA's broad effectiveness across different GNN models.

Table 7. Experimental results of our 8 attacks of the Cora dataset in the model update scenario.

Scenario	Attack	AMA↑	AP↑	MSE↓	CA↑	ICA↓
Type I	Attack-I-0	0.9997	0.0432	-	0.9815	-
	Attack-I-1	0.9996	0.0160	-	0.9798	-
	Attack-I-2	0.9979	0.1423	-	0.9264	-
	Attack-I-3	0.9977	0.0308	-	0.9360	-
Type II	Attack-II-0	0.9996	0.0235	0.0117	0.9798	0.0166
	Attack-II-1	0.9996	0.0537	0.0122	0.9645	0.0220
	Attack-II-2	0.9979	0.1269	0.0117	1.0000	0.0172
	Attack-II-3	0.9981	0.0553	0.0122	0.9995	0.0183

Table 8. Experimental results of Attack-II-0 and Attack-II-1 when the server can only access to the prediction label.

Dataset	Attack	AMA↑	AP↑	MSE↓	CA↑	ICA↓
Cora	Attack-II-0	0.9993	0.0136	0.0118	1.0000	0.0123
	Attack-II-1	0.9993	0.0138	0.0126	0.3842	0.0126
Citeseer	Attack-II-0	0.9996	0.0032	0.0081	0.9984	0.0133
	Attack-II-1	0.9996	0.0024	0.0085	0.1623	0.0127
ACM	Attack-II-0	0.9983	0.0013	0.0342	1.0000	0.0016
	Attack-II-1	0.9983	0.0000	0.0449	0.6454	0.0016
UAI	Attack-II-0	0.9968	0.0572	0.1119	0.9996	0.0390
	Attack-II-1	0.9969	0.0611	0.1436	0.3739	0.0417
Amazon	Attack-II-0	0.9979	0.0454	0.2163	1.0000	0.0115
	Attack-II-1	0.9979	0.0454	0.3457	0.4024	0.0114
Pubmed	Attack-II-0	0.9999	0.0005	0.0003	1.0000	0.0170
	Attack-II-1	0.9999	0.0002	0.0003	0.7096	0.0189

5.4 Attacks on Model Update Scenario

We assess the effectiveness of GULA in the model update scenario where clients update both graph data and model. Initially, the client and server collaboratively train model $\mathcal{M}$ with the original graph $\mathcal{D}$. Upon updating $\mathcal{D}$ with $\mathcal{D}_{update}$ to form $\mathcal{D}'$, they retrain a new model $\mathcal{M}'$ with $\mathcal{D}'$. The server then performs GULA to reconstruct the update set $\mathcal{D}_{update}$ by analyzing the posteriors $\mathcal{P}$ and $\mathcal{P}'$ from $\mathcal{M}$ and $\mathcal{M}'$, respectively. Experimental results on the Cora dataset, as shown in Table 7, reveal that GULA performs exceptionally well in this setup, achieving an average CA score of 97.09%. This indicates significant potential for information leakage from GSL updates.

5.5 Possible Defenses

To counter GULA, we restrict the server to obtain only prediction labels instead of complete posterior distributions. Here, we assume that the adversary knows the total number of classes for each dataset. Results from this defense are illustrated in Table 8, where it shows that although the defense reduces GULA's effectiveness, Attack-II-0 continues to successfully reconstruct the updated graph across all datasets. For instance, on the Cora dataset, Attack-II-0 maintains a CA score of 100%, while the CA score for Attack-II-1 drops from 98.82% to 38.42%. Besides, the average AP score for both attacks drops from 7.95% to 1.37%. This reduction is attributed to the difficulty of extracting node information from prediction results when multiple nodes share the same label. Note that this method limits the model's utility as each node uniquely corresponds to its posteriors, which reflects the prediction confidence across multiple categories.

6 Conclusion and Future Work

In this paper, we undertake the inaugural security risk assessment against GSL by examining it through the prism of update leakage attacks. We systematically define a threat model and categorize it into 8 types based on two data update scenarios and the background knowledge known by the adversary. Additionally, we propose 8 methods to achieve each type of attack. Extensive experiments on 6 widely used datasets and 3 popular GNN model architectures demonstrate that our attacks can successfully steal the adjacency matrix and node features of the update set. Moreover, the attacks are still effective even the adversary has no background knowledge. Our work takes the first insight into the security problem of collaborative learning frameworks with updated training data. Recall that periodically updating business models with fresh data is a common operation for many AI companies, and privacy-preserving collaborative learning is becoming a mainstream model training method. Thus, interesting future work includes generalizing our attacks to more collaborative learning frameworks and defending against these attacks.

Acknowledgement. This work was supported by the National Key Research and Development Program of China (Program No. 2023YFE0111100), the National Natural Science Foundation of China (Program No. U21A20464, Program No. 62261160651, Program No. U23A20307, Program No. U23A20306), the Fundamental Research Funds for the Central Universities (Program No. QTZX24081).

References

1. Amazon AWS. https://aws.amazon.com
2. Microsoft Azure. https://azure.microsoft.com/en-us
3. Backes, M., Humbert, M., Pang, J., Zhang, Y.: walk2friends: inferring social links from mobility profiles. In: Proceedings of the 2017 ACM SIGSAC Conference on Computer and Communications Security, pp. 1943–1957 (2017)
4. Bhardwaj, R., et al.: Ekya: continuous learning of video analytics models on edge compute servers. In: 19th USENIX Symposium on Networked Systems Design and Implementation (NSDI 2022), pp. 119–135. USENIX Association, Renton, WA (2022)
5. Ding, R., Duan, S., Xu, X., Fei, Y.: Vertexserum: poisoning graph neural networks for link inference. In: Proceedings of the IEEE/CVF International Conference on Computer Vision (ICCV), pp. 4532–4541. IEEE, Piscataway, NJ (2023)
6. Duddu, V., Boutet, A., Shejwalkar, V.: Quantifying privacy leakage in graph embedding. In: MobiQuitous 2020-17th EAI International Conference on Mobile and Ubiquitous Systems: Computing, Networking and Services, pp. 76–85 (2020)
7. Gallicchio, C., Micheli, A.: Fast and deep graph neural networks. In: Proceedings of the AAAI Conference on Artificial Intelligence, vol. 34, no. 04, pp. 3898–3905. AAAI Press, Palo Alto, California USA (2020)
8. Gao, X., Zhang, L.: PCAT: functionality and data stealing from split learning by Pseudo-Client attack. In: 32nd USENIX Security Symposium (USENIX Security 2023), pp. 5271–5288. USENIX Association, Anaheim, CA (2023)

9. Hamilton, W., Ying, Z., Leskovec, J.: Inductive representation learning on large graphs. In: Advances in Neural Information Processing Systems, vol. 30 (2017)
10. He, X., Jia, J., Backes, M., Gong, N.Z., Zhang, Y.: Stealing links from graph neural networks. In: 30th USENIX Security Symposium (USENIX Security 2021), pp. 2669–2686. USENIX Association, Vancouver, B.C., Canada (2021)
11. Kipf, T.N., Welling, M.: Semi-supervised classification with graph convolutional networks. In: International Conference on Learning Representations (2016)
12. Knoke, D., Yang, S.: Social Network Analysis. SAGE Publications (2019)
13. Lee, H., Yoo, S., Lee, D., Kim, J.: How important is periodic model update in recommender system? In: Proceedings of the 46th International ACM SIGIR Conference on Research and Development in Information Retrieval, pp. 2661–2668. Association for Computing Machinery, New York, NY, USA (2023)
14. Li, K., et al.: Towards practical edge inference attacks against graph neural networks. In: ICASSP 2023-2023 IEEE International Conference on Acoustics, Speech and Signal Processing (ICASSP), pp. 1–5. IEEE (2023)
15. Liu, Z., Fang, Y., Liu, C., Hoi, S.C.: Relative and absolute location embedding for few-shot node classification on graph. In: Proceedings of the AAAI Conference on Artificial Intelligence, vol. 35,5, pp. 4267–4275. AAAI, Menlo Park (2021)
16. McAuley, J., Targett, C., Shi, Q., Van Den Hengel, A.: Image-based recommendations on styles and substitutes. In: Proceedings of the 38th International ACM SIGIR Conference on Research and Development in Information Retrieval, pp. 43–52. Association for Computing Machinery, New York, NY, USA (2015)
17. Mislove, A., Marcon, M., Gummadi, K.P., Druschel, P., Bhattacharjee, B.: Measurement and analysis of online social networks. In: Proceedings of the 7th ACM SIGCOMM Conference on Internet Measurement, pp. 29–42. Association for Computing Machinery, New York, NY, USA (2007)
18. Nelson, Q., Steffensmeier, D., Pawaskar, S.: A simple approach for sustainable transportation systems in smart cities: a graph theory model. In: 2018 IEEE Conference on Technologies for Sustainability (SusTech), pp. 1–5. IEEE, Piscataway, NJ (2018)
19. Salem, A., Bhattacharya, A., Backes, M., Fritz, M., Zhang, Y.: {Updates-Leak}: data set inference and reconstruction attacks in online learning. In: 29th USENIX security symposium (USENIX Security 2020), pp. 1291–1308. Association for Computing Machinery, New York, NY, USA (2020)
20. Shen, Y., He, X., Han, Y., Zhang, Y.: Model stealing attacks against inductive graph neural networks. In: 2022 IEEE Symposium on Security and Privacy (SP), pp. 1175–1192. IEEE, Piscataway, NJ (2022)
21. Sima, C., et al.: Ekko: a {Large-Scale} deep learning recommender system with {Low-Latency} model update. In: 16th USENIX Symposium on Operating Systems Design and Implementation (OSDI 2022), pp. 821–839. USENIX Association, Carlsbad, CA (2022)
22. Vázquez, A.: Growing network with local rules: preferential attachment, clustering hierarchy, and degree correlations. Phys. Rev. E **67**(5), 056104 (2003)
23. Veličković, P., Cucurull, G., Casanova, A., Romero, A., Liò, P., Bengio, Y.: Graph attention networks. In: International Conference on Learning Representations (2018)
24. Wang, W., Liu, X., Jiao, P., Chen, X., Jin, D.: A unified weakly supervised framework for community detection and semantic matching. In: Phung, D., Tseng, V.S., Webb, G.I., Ho, B., Ganji, M., Rashidi, L. (eds.) PAKDD 2018. LNCS (LNAI), vol. 10939, pp. 218–230. Springer, Cham (2018). https://doi.org/10.1007/978-3-319-93040-4_18

25. Wang, X., et al.: Heterogeneous graph attention network. In: The World Wide Web Conference, pp. 2022–2032. Association for Computing Machinery, New York, NY, USA (2019)
26. Wang, X., Wang, W.H.: Group property inference attacks against graph neural networks. In: Proceedings of the 2022 ACM SIGSAC Conference on Computer and Communications Security, pp. 2871–2884. Association for Computing Machinery, New York, NY, USA (2022)
27. Wu, B., Yang, X., Pan, S., Yuan, X.: Model extraction attacks on graph neural networks: taxonomy and realisation. In: Proceedings of the 2022 ACM on Asia Conference on Computer and Communications Security, pp. 337–350. Association for Computing Machinery, New York, NY, USA (2022)
28. Wu, F., Long, Y., Zhang, C., Li, B.: Linkteller: recovering private edges from graph neural networks via influence analysis. In: 2022 IEEE Symposium on Security and Privacy (SP), pp. 2005–2024. IEEE (2022)
29. Wu, Z., Pan, S., Chen, F., Long, G., Zhang, C., Philip, S.Y.: A comprehensive survey on graph neural networks. IEEE Trans. Neural Networks Learn. Syst. **32**(1), 4–24 (2020)
30. Xu, K., Hu, W., Leskovec, J., Jegelka, S.: How powerful are graph neural networks? In: International Conference on Learning Representations (2018)
31. Xu, X., Lyu, L., Dong, Y., Lu, Y., Wang, W., Jin, H.: Splitgnn: splitting GNN for node classification with heterogeneous attention. arXiv preprint arXiv:2301.12885 (2023)
32. Yin, Y., et al.: Ginver: generative model inversion attacks against collaborative inference. In: Proceedings of the ACM Web Conference 2023, pp. 2122–2131. Association for Computing Machinery, New York, NY, USA (2023)
33. Zanella-Béguelin, S., et al.: Analyzing information leakage of updates to natural language models. In: Proceedings of the 2020 ACM SIGSAC Conference on Computer and Communications Security, pp. 363–375. Association for Computing Machinery, New York, NY, USA (2020)
34. Zhang, Z., Liu, Q., Huang, Z., Wang, H., Lee, C.K., Chen, E.: Model inversion attacks against graph neural networks. IEEE Trans. Knowl. Data Eng. **35**, 8729–8741 (2022)
35. Zheng, L., Zhou, J., Chen, C., Wu, B., Wang, L., Zhang, B.: Asfgnn: automated separated-federated graph neural network. Peer-to-Peer Network. Appl. **14**(3), 1692–1704 (2021)

Incremental Board Learning System Based on Feature Selection

Yi Xu[(✉)], HuaiDing Qu, Yin Liu, and ZhengYue Pan

Anhui University, Hefei 230039, Anhui, China
xuyi1023@126.com

Abstract. Board learning has similar learning ability and higher learning efficiency as deep learning. The Incremental Board Learning System (IBLS) as an important model of board learning consists of two components: Basic board learning and incremental board learning. However, both components are learned based on all features of the input data, and there are two possible limitations of this method: (1) since features may be redundant, the data set with a large sample size and a large number of features will affect the learning efficiency. (2) Since both components of IBLS are based on all features, the lack of complementarity between features will affect the learning accuracy. For this reason, we propose a new approach: incremental board learning system based on feature selection. First, the feature selection method is used to divide all features into two levels, and the selected features are used as the significant feature layer and the remaining features are used as the normal feature layer. Second, the significant feature layer is used as input for basic board learning to reduce feature redundancy and improve learning efficiency. Third, the normal feature layer is used as the input for incremental board learning to reduce the information loss caused by feature selection and to make full use of the complementarity between the two components of IBLS. Experiments using the proposed method on ten commonly used classification datasets show that it has better learning accuracy and higher classification efficiency than the classic IBLS.

Keywords: Board learning system · Feature selection · Incremental learning · Multilevel

1 Introduction

Nowadays, deep neural networks have made breakthroughs in various fields [10,12], especially in large-scale data processing [2,3]. However, deep learning is a time-consuming training process for most deep learning models due to the many iterations of hyperparameter tuning involved. Therefore Chen et al. [1] proposed the board Learning System (BLS). The BLS draws on the Random Vector Functional Link Neural Network [RVFLNN], therefore, the model parameters can be determined without iterations, which makes the training process of IBLS fast and efficient. And the incremental board learning system (IBLS) is an

T. Zhu et al. (Eds.): ICA3PP 2024, LNCS 15252, pp. 22–35, 2025.
https://doi.org/10.1007/978-981-96-1528-5_2

important model of BLS, which is composed of two main parts: basic board learning and incremental board learning. The structure of IBLS is very flexible, and when there are updates to the input data, IBLS can quickly adjust the network through incremental algorithms without retraining the entire network. Thanks to the incremental board learning, the classification accuracy can be improved by continuously adding new intermediate layer mappings. Since its introduction, IBLS has received increasing attention from academia and industry due to its good learning efficiency and powerful learning capabilities, and it has been successfully applied to many practical problems, such as time series testing [8,9,22], industrial fault detection [17], mask wearing detection [20], visual soccer analysis [18], text classification [4], object classification [5], etc.

The current related work on IBLS can be briefly summarized in two aspects. First, the model structure is optimized to obtain higher classification accuracy and training speed. For this purpose, researchers have provided different strategies, such as an algorithm for faster construction of basic board learning in [13], and in [15,16] is the combination of a board learning system with deep learning to combine the speed and efficiency of the model. In [14], a stacked BLS adaptive structure is proposed to improve the performance. The IBLS model capable of learning and fusing two modalities is proposed in [6,11], which uses a two-board learning system that first coarsely extracts data from both modalities via feature and enhancement nodes, maps the extracted features to the typical relevance analysis layer for relevance learning, matching and dimensionality reduction, and then performs nonlinear fusion via fusion nodes, and finally connects to the output layer for classification output. Second, feature extraction methods is optimized to obtain better feature representation. The paper [7] replaces the feature mapping approach in IBLS-based board learning with a set of depth and sparse learning approaches that combine depth and board learning to obtain a more advanced feature representation. The sparse radial basis network is also used in the paper [21] to feature the input data for the base board learning to improve the generalization ability of the model. A board learning system based on local receptive fields for feature extraction is proposed in the paper [19], which helps to extract features such as edges, textures, light and dark with the help of the idea that local receptive fields can maintain the translation invariance of image data in convolutional neural networks and keep the local features of images during feature extraction. However, existing feature optimization methods have a learning process based on all features of the input data in both parts of the IBLS, and there are two possible limitations to this approach: (1) since features may be redundant, their learning efficiency will be affected on datasets with large sample size and number of features. (2) Since both basic board learning and incremental board learning are based on all features, the lack of complementarity between features will affect their learning accuracy.

Therefore, in this paper, the features of input data are divided into two layers by feature selection method in IBLS, where the selected features are used as the significant feature layer and the remaining features are used as the normal feature layer. The significant feature layer is used in basic board learning and the

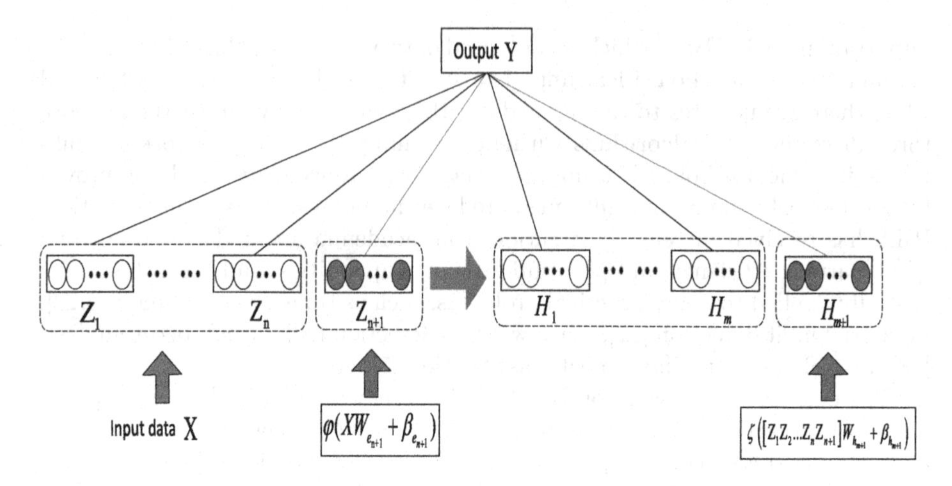

Fig. 1. The classic structure of IBLS.

normal feature layer is used in incremental board learning, which can reduce the impact of feature redundancy on model training and improve the efficiency of basic board learning, and also use different features to learn in two parts of IBLS, which increases the complementarity of features and improves the accuracy of the whole learning system.

The rest of the paper is organized as follows: Sect. 2 reviews the theory of IBLS. Section 3 gives the incremental board learning system based on feature selection. In Sect. 4, the effectiveness of the improved IBLS is experimentally verified with ten commonly used datasets. Finally, we conclude our work in Sect. 5.

2 Related Work

IBLS is a new neural network learning model that does not require a deep architecture, and it focuses on improving the model learning capability by extending the board structure. The typical structure of IBLS is shown in Fig. 1, which consists of two components: basic board learning and incremental board learning.

2.1 Basic Board Learning

The basic board learning system is a modified model based on random vector function linked neural network (RVFLNN). The input X of the basic board learning model is a given training dataset with labels, and the output Y is the classification result of the dataset. The purpose of training the model is to obtain the weight matrix W^m from the intermediate layer to the output layer. It is important to note that the intermediate layer of the model has only one layer, which is specifically composed of two other mapping layer links, the exact expression of which will be given below. Firstly, given a training dataset, we represent the it as

$$(X, Y)|X \in R^{(N \times d)}, Y \in R^{(N \times c)}, \tag{1}$$

where N is the number of training samples, d and c are the dimensions of the input and output, respectively. Then the input data X is mapped to the n group random feature space, the random initialized weight matrix $W_{e_i} \in R^{(d \times L)}$ and the bias matrix $\beta_{e_i} \in R^{(N \times L)}$ are needed, and L is the number of feature nodes in each group. And the feature mapping function can be expressed as φ, such that the feature mapping of the ith group can be expressed as:

$$Z_i = \varphi(XW_{e_i} + \beta_{e_i}), i = 1, ..., n \tag{2}$$

After the feature mapping of n groups, the cascade of n groups of mapped nodes can be expressed as $Z^n = [Z_1, ..., Z_n]$. Then the obtained Z^n is further randomly mapped by the nonlinear activation function ξ to generate the augmented nodes of group m. The weight matrix W_{h_m} and the bias matrix β_{h_m} are still randomly generated. where the augmented nodes in the jth group can be expressed as

$$H_j = \xi(Z^n W_{h_j} + \beta_{h_j}) \tag{3}$$

Similarly, the cascade of m groups of augmented nodes is represented as $H^m \equiv [H_1, ..., H_m]$. Finally, all groups of mapped feature nodes and augmented feature nodes are linked together and expressed as

$$Z_1, ..., Z_n|\xi(Z^n W_{h_1} + \beta_{h_1}), ..., \xi(Z^n W_{h_m} + \beta_{h_m}), \tag{4}$$

so that the output Y of basic board learning can be expressed as:

$$\begin{aligned} Y &= [Z_1, ..., Z_n|\xi(Z^n W_{h_1} + \beta_{h_1}), ..., \xi(Z^n W_{h_m} + \beta_{h_m})]W^m \\ &= [Z_1, ..., Z_n|H_1, ..., H_m]W^m \\ &= [Z^n|H^m]W^m \end{aligned} \tag{5}$$

where W^m is the final output layer weight to be calculated.

For the convenience of the later formula representation, we denote the link $[Z^n|H^m]$ of feature mapping nodes and enhancement nodes as A_n^m, then the formula (5) can be written as $Y = A_n^m W^m$, from which W^m can be expressed as:

$$W^m = (A_n^m)^+ Y \tag{6}$$

where A^+ is the pseudo-inverse of the matrix A. To avoid overfitting, the ridge regression algorithm is usually used to calculate W^m, so that Eq. (6) can again be expressed as:

$$W^m = ((A_n^m)^T A_n^m + \lambda I)^{-1} (A_n^m)^T Y \tag{7}$$

where λ is the ridge parameter and I is the unit matrix.

2.2 Incremental Board Learning

Based on the base board learning, the model can be quickly updated by incremental learning methods to adapt it to new learning tasks. There are three major incremental learning methods: (1) adding new input data features, when there is an update to the input data, the model can be updated quickly by the incremental learning method of the input data; (2) adding new feature mapping nodes, the addition of feature mapping nodes is achieved by adding a new feature mapping while the input data remains unchanged; (3) adding new enhancement nodes, the same as adding new feature mapping nodes, the addition of new enhancement nodes is achieved by adding a new set of nonlinear mappings while the feature mapping nodes remain unchanged. Because the input data used by the new feature mapping node and the input data of the basic board learning are both the original input data, it is convenient for the hierarchical input of features to better achieve the feature complementarity of the two components of IBLS. For the above reasons, we consider that the improved incremental learning method is implemented based on the addition of new feature mapping nodes. And each time a new feature mapping node is added, the whole model needs to be updated, which also means that an incremental learning is performed. Therefore, this paper focuses on the basic board learning training based on how to quickly update the output weights W_{n+1}^m of the model after adding a feature mapping node Z_{n+1}.

Based on the base board learning, we insert a new feature mapping node Z_{n+1} into the base framework, and this new node can be represented as:

$$Z_{n+1} = \varphi(XW_{e_{n+1}} + \beta_{e_{n+1}}) \tag{8}$$

where $\varphi(.)$ is the same activation function as the basic board learning, and $W_{e_{n+1}}$ and $\beta_{e_{n+1}}$ are the randomly generated weight matrix and bias matrix, respectively.

The new set of enhanced nodes H_{ex_m} is obtained by updating according to the newly generated feature mapping nodes Z_{n+1}, which can be expressed as follows:

$$H_{ex_m} = [\xi(Z_{n+1}W_{ex_1} + \beta_{ex_1}), ..., \xi(Z_{n+1}W_{ex_m} + \beta_{ex_m})] \tag{9}$$

where $\xi(.)$ is also the same as the activation function used in basic board learning to obtain the augmented nodes, and W_{ex_i} and β_{ex_i} are the randomly generated weight matrix and bias matrix, respectively.

In basic board learning we denote $A_n^m = [Z^n|H^m]$, since we now add a new feature mapping node Z_{n+1}, we have to update A_n^m to A_{n+1}^m, which can be expressed as follows:

$$A_{n+1}^m = [A_n^m|Z_{n+1}|H_{ex_m}] \tag{10}$$

By updating A_{n+1}^m, the update of the corresponding pseudo-inverse matrix can be expressed as:

$$(A_{n+1}^m)^+ = [\begin{matrix}(A_n^m)^+ - DB^T \\ B^T\end{matrix}] \tag{11}$$

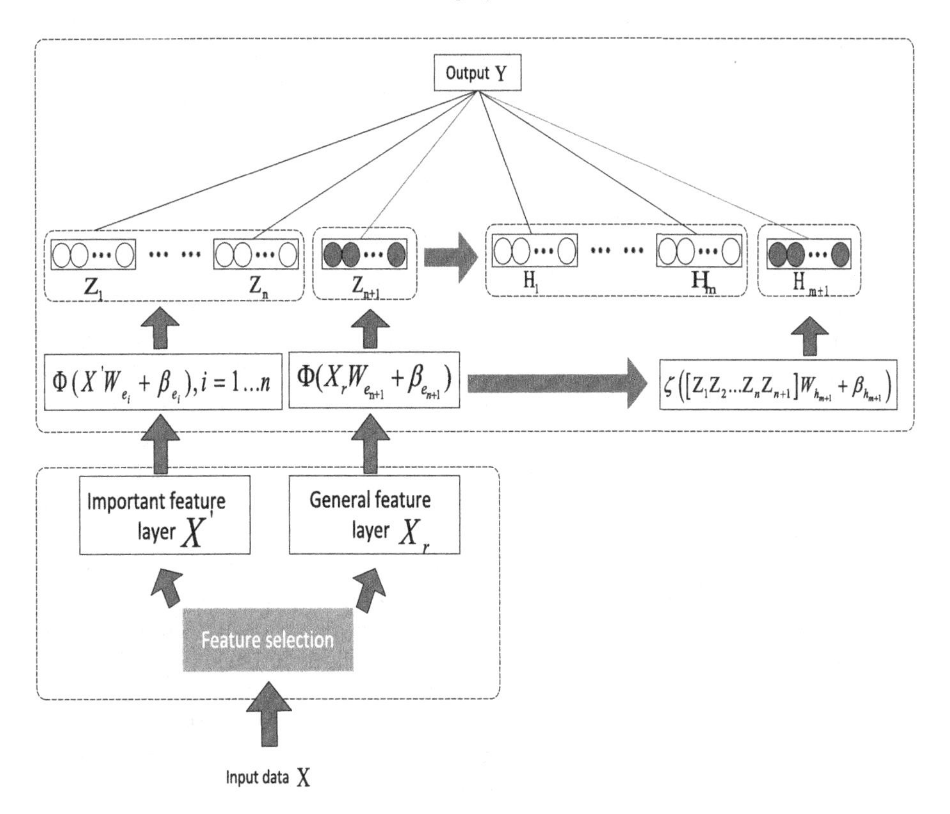

Fig. 2. An overall framework for improving incremental board learning.

where D $= (A_n^m)^+[Z_{n+1}|H_{ex_m}]$, and B^T can be expressed as:

$$B^T = \begin{cases} (C)^+ \\ (1 + D^T D)^{(-1)} D^T (A_n^m)^+ \end{cases} \tag{12}$$

And where $C = [Z_{(n+1)}|H_{ex_m}] - A_n^m D$, by calculation, we can finally get the updated new weights by Eq. (11). Thus by adding a new feature mapping node Z_{n+1}, an incremental board learning is completed.

$$W_{n+1}^m = \begin{bmatrix} W_n^m - DB^T Y \\ B^T Y \end{bmatrix} \tag{13}$$

3 Incremental Board Learning System Based on Feature Selection

Since in the classic IBLS, both basic board learning and incremental board learning for the input data feature mapping layer is based on all features, this will make some data sets with large data volume and large number of features have feature redundancy, thus reducing the accuracy of the model. In addition, although the feature selection method can select significant features and thus reduce feature redundancy, it ignores the importance of feature complementarity, resulting in the loss of some information. Therefore, in this section, incremental board learning system based on feature selection is proposed. Then the overall framework of this learning system is presented.

Figure 2 illustrates the overall framework of incremental board learning based on feature selection. As shown in the figure, we make the following two improvements to IBLS: (1) On the basis of the classic IBLS model, stratifying the features of the input data into a significant feature layer and a normal feature layer, this allows features of high importance to be selected by feature selection methods, while retaining the remaining normal features. (2) The significant feature layer is used as the input feature for basic board learning in IBLS, and the normal feature layer is used as the input feature for incremental board learning in IBLS. The significant feature layer contains features with higher importance, which is more favorable for basic board learning and can improve the efficiency of basic board learning. And the normal feature layer is used as complementary features in incremental board learning, which can avoid the loss of original information. The specific implementation of feature selection methods in basic board learning and incremental board learning is described next.

The feature selection method assumed to be used is $f(.)$, the input data set $X \in R^{(N \times d)}$, where N is the number of training samples and d denotes the number of features of the input data. We denote the features by feature selection as M', its corresponding dataset can be expressed as $X' \in R^{N \times M'}$, and the remaining features are expressed as M_r, its corresponding dataset can be expressed as $X_r \in R^{N \times M_r}$. The specific formula is shown in (12).

$$X' = f(X), X_r = X - X' \tag{14}$$

After the feature selection process, the features of the input data can be divided into two levels,X' and X_r: the selected features are used as the significant feature layer X' and the remaining features are used as the normal feature layer X_r. This allows the input features to be layered and learned in two separate parts of the IBLS, reducing feature redundancy and improving feature complementarity without increasing the time complexity of the model. The specific algorithm of the incremental board learning system based on feature selection is presented next.

First, given the training data set $\{(X, Y)|X \in R^{N \times d}, Y \in R^{N \times c}\}$, Y is the data set classification result. In the incremental board learning system based on feature selection, the input data used for basic board learning is $X' \in R^{N \times M'}$ containing M' features, and incremental board learning is $X_r \in R^{N \times M_r}$ containing M_r features. The purpose of training the model is still to determine the weight matrix W^m from the intermediate layer to the output layer.

In basic board learning, when mapping the input data projection into an n-group random feature space, we use the significant feature layer $X' \in R^{N \times M'}$,

Algorithm 1: Incremental of n+1 Residual Mapped Features Based on IBLS

Input: training samples X
Output: W
1 X' is selected by X, the remaining feature X_r;
2 **while** *the training error threshold is not satisfied* **do**
3 Random $W_{e_{n+1}}$, $\beta_{e_{n+1}}$;
4 Calculate $Z_n = [\varphi(X_r W_{e_{n+1}} + \beta_{e_{n+1}})]$;
5 Random $W_{ex_i}, \beta_{ex_i}, i = 1, ...m$;
6 Calculate H_{ex_m}=
7 $[\xi(Z_{n+1} W_{ex_1} + \beta_{ex_1}), ..., \xi(Z_{n+1} + W_{ex_m} + \beta_{ex_m})]$;
8 Update A^m_{n+1};
9 Update $(A^m_{n+1})^+$ and W^m_{n+1} by Eq;
10 n = n + 1;
11 **end**
12 Set $W = W^m_{n+1}$;

thus the feature mapping Z_i' corresponding to the ith group is expressed as:

$$Z_i' = \varphi(X'\mathrm{W}_{\mathrm{e_i}} + \beta_{e_i}), i = 1, ..., n \tag{15}$$

where W_{e_i} and β_{e_i} are still the randomly generated weight and bias matrices. The resulting cascade of n sets of feature mappings is represented as $Z^{n'} = [Z_1', ..., Z_n']$.

The subsequent steps are the same as the classic IBLS, based on the obtained $Z^{n'}$, further non-linear mapping is done to generate m sets of augmented nodes, and finally the output weights $W^{m'}$ we need are calculated by the ridge regression algorithm. Thus the first part of incremental board learning system based on feature selection for basic board learning is trained.

After training the basic board learning model, the accuracy of the results is generally not sufficient, so we need to continue with incremental board learning to train the model. Similar to the process in Sect. 2, we need to add a new feature mapping node Z_{n+1}'. Here we differ from the classic IBLS in that we use the normal feature laye X_r as the input for incremental board learning, so the new feature mapping node Z_{n+1}' can be expressed as:

$$Z_{n+1}' = \varphi(X_r W_{e_{n+1}} + \beta_{e_{n+1}}) \tag{16}$$

where $\varphi(.)$ is the determined activation function, then the enhancement node H_{ex_m}' is recomputed by Z_{n+1}', the connection node is updated to A^m_{n+1} by Z_{n+1}' and H_{ex_m}', and finally the output matrix $W^m_{n+1}{}'$ is updated, so that incremental board learning system based on feature selection is finished training, combined with the above analysis, the full process of improving IBLS is given in Algorithm 1.

4 Experiment

In this section, we use 10 commonly used data sets obtained from the UCI machine learning knowledge base to verify the effectiveness of incremental board learning system based on feature selection. On these commonly used data sets, we compare the accuracy and efficiency of the proposed method with the classic IBLS method. The specific description of the dataset is shown in Table 1. These are 10 commonly used data sets, covering different domains and with different quantities and characteristics. The feature selection methods used in this paper $f(.)$ include chi-square, variance selection and correlation coefficient, and three experiments are conducted for each data set through three common feature selections as a way to verify the validity of the experiments more effectively.

We implement our experiment on a PC with Microsoft Windows 10, Inter (R) Core (TM) i7-6700H CPU @ 3.70 GHz, 4.0 GB memory, and the language is Python.

Table 1. Specification of the used dataset

Dataset	instance	attribute
Iris	150	4
Spambase	4597	57
Absenteeism	740	21
Avila	20867	10
Hepatitis	155	19
Breast Cancer	286	9
Glass Identification	214	10
Log Data	65532	12
Red Wine	1599	12
Mnist	62235	784
White Wine	4898	12

Our experiments are divided into two parts. In the first part, we compare the incremental board learning system based on feature selection with the classic IBLS in accuracy. For each dataset, we use 70% of the samples of the input dataset as training samples to train the model, and then 30% of the samples of the input dataset as test samples to validate the accuracy of the model. In this experiment, in order to verify the effect of incremental learning times on model training under three different feature selection methods, our experiments are based on the accuracy of performing incremental learning 5, 20 and 50 times for each dataset recorded under each of the three feature selection methods. The specific experimental data are given in Tables 2, 3 and 4. In the second part, we compare the incremental board learning system based on feature selection

Table 2. Comparison results of variance selection, chi square and correlation coefficient on 5 times

Data set	IBLS	Variance selection	Chi square	Correlation coefficient
Spambase	0.831	**0.838**	**0.849**	**0.835**
Avila	0.631	**0.642**	**0.634**	**0.636**
White Wine	0.502	**0.509**	**0.51**	**0.521**
Red Wine	0.621	**0.63**	**0.635**	**0.623**
Log2Data	0.918	**0.926**	**0.921**	**0.925**
Absenteeism	0.684	**0.697**	**0.689**	**0.688**
Breast	0.876	**0.891**	**0.88**	**0.89**
Glass	0.824	**0.843**	**0.833**	**0.833**
Hepatitis	0.907	**0.94**	**0.913**	**0.933**
Sonar Data	0.957	**0.961**	**0.971**	**0.971**

Table 3. Comparison results of variance selection, chi square and correlation coefficient on 20 times

Data set	IBLS	Variance selection	Chi square	Correlation coefficient
Iris	0.88	**0.933**	**0.893**	**0.9**
Avila	0.609	**0.624**	**0.631**	**0.614**
White Wine	0.68	**0.698**	**0.689**	**0.692**
Absenteeism	0.673	**0.688**	**0.678**	**0.686**
Red Wine	0.686	**0.687**	**0.687**	**0.691**
Spambase	0.673	**0.681**	**0.691**	**0.688**
Hepatitis	0.72	**0.723**	**0.733**	**0.733**
Breast	0.558	**0.565**	**0.567**	**0.561**
Glass	0.719	**0.729**	**0.724**	**0.733**
Sonar Data	0.715	**0.739**	**0.725**	**0.72**

with the classic IBLS in efficiency. The specific comparison is the time taken to train the model based on 70% of the samples in the input dataset. Similarly, for the completeness of the experiments, we record the time to perform incremental learning as 5, 20, and 50 times under three different feature selection methods, respectively. In order to reduce the error of the experiment, our experimental data are taken as the result of the average of 50 times.

The specific parameters of the experiments will be given below. Firstly, the base experimental settings for this algorithm are: number of feature mapping node groups $N_2 = 10$, number of nodes in each feature mapping group $N_1 = 10$, number of augmented nodes $N_3 = 600$, shrinkage factor s=0.8 and regularization factor $c = 2^{-30}$. In the learning system proposed in this paper, the parameters of data feature selection are set as follows respectively: (1) In the variance

Table 4. Comparison results of variance selection, chi square and correlation coefficient on 50 times

Data set	IBLS	Variance selection	Chi square	Correlation coefficient
Iris	0.867	**0.9**	**0.88**	**0.893**
Avila	0.532	**0.547**	**0.554**	**0.537**
White Wine	0.512	**0.53**	**0.521**	**0.521**
Absenteeism	0.596	**0.603**	**0.608**	**0.601**
Red Wine	0.609	**0.61**	**0.614**	**0.61**
Spambase	0.766	**0.771**	**0.784**	**0.781**
Log2Data	0.862	**0.862**	**0.863**	**0.863**
Breast	0.652	**0.655**	**0.658**	**0.655**
Glass	0.814	**0.82**	**0.824**	**0.824**
Hepatitis	0.9	**0.9133**	**0.907**	**0.907**

Table 5. The training time (in seconds) of each algorithm on 5 times incremental learning

Data set	IBLS	Variance selection	Chi square	Correlation coefficient
Mnist	18.34	**16.71**	**16.90**	**17.41**
Spambase	1.25	**1.20**	**1.16**	**1.20**
Avila	2.80	**2.68**	**2.14**	**2.30**
White Wine	1.30	**1.23**	**1.19**	**1.22**
Red Wine	0.44	**0.41**	**0.39**	**0.40**
Log2Data	14.36	**13.78**	**13.56**	**14.12**
Absenteeism	0.31	**0.30**	**0.29**	**0.30**
Breast	0.23	**0.21**	**0.21**	**0.23**
Glass	0.13	**0.13**	**0.12**	**0.13**
Hepatitis	0.14	**0.12**	**0.12**	**0.12**

selection, the variance value of each feature of the input data set is first calculated by $s^2 = \frac{\sum_{i=1}^{n}(X_i - \bar{X})^2}{n}$, after which the threshold value of variance is set by us to determine that a threshold value of 2.8 has a better performance. (2) The chi-square validation is a method of filtering features by calculating the chi-square value, and we have determined through several experiments that the chi-square has a better performance when the threshold k = 2.0. (3) In the correlation coefficient method, we set the parameter of the regression model to 0.5 based on the results of multiple experiments, and the results of each experiment are the average of 50 experiments.

Tables 2, 3 and 4 show the comparison results of accuracy after 5, 20, and 50 times of incremental learning under the three feature selection methods,

Table 6. The training time (in seconds) of each algorithm on 20 times incremental learning

Data set	IBLS	Variance selection	Chi square	Correlation coefficient
Mnist	88.35	**80.06**	**80.65**	**82.41**
Spambase	3.94	**3.59**	**3.61**	**3.63**
Avila	6.26	**6.15**	**5.41**	**5.93**
White Wine	6.53	**6.42**	**6.40**	**5.61**
Red Wine	2.56	**2.16**	**2.27**	**2.20**
Log2Data	84.38	**80.92**	**82.76**	**76.92**
Absenteeism	1.23	**1.13**	**1.14**	**1.20**
Breast	1.27	**1.13**	**1.20**	**1.23**
Glass	0.59	**0.47**	**0.53**	**0.50**
Hepatitis	0.43	**0.42**	**0.42**	**0.43**

Table 7. The training time (in seconds) of each algorithm on 50 times incremental learning

Data set	IBLS	Variance selection	Chi square	Correlation coefficient
Mnist	150.98	**125.53**	**123.65**	**130.41**
Spambase	6.17	**6.14**	**6.15**	**6.15**
Avila	14.51	**14.48**	**14.41**	**14.39**
White Wine	27.50	**23.42**	**25.36**	**23.68**
Red Wine	9.70	**8.53**	**7.97**	**8.35**
Log2Data	112.36	**106.78**	**105.56**	**106.12**
Absenteeism	5.06	**4.68**	**5.01**	**4.86**
Breast	4.60	**3.96**	**4.05**	**4.11**
Glass	1.35	**1.03**	**1.12**	**1.08**
Hepatitis	4.55	**4.25**	**4.23**	**4.23**

respectively. As can be seen from the following table, feature selection can improve the accuracy of IBLS. On ten common datasets, the features selected by the three feature selection methods are not always the same, but the incremental board learning system based on feature selection is more accurate than the classical IBLS under the three different feature selection methods. The main reason is that the feature selection method selects most of the features that are beneficial for model training, allowing basic board learning and incremental board learning to make better use of feature complementarity to train the model, and therefore the model will be trained better.

Tables 5, 6 and 7 show, respectively, under the three feature selection methods, comparative results of time after 5, 20 and 50 times of incremental learning.

The following three tables compare the time consumed by the model's incremental board learning system based on feature selection and the classic IBLS training on each of the ten commonly used datasets. As can be seen from the table, although we added the feature selection method, the training time of the whole model is still slightly lower than that of the classic IBLS. The main reason is that the feature selection method selects most of the features that are beneficial for model training, which allows the basic board learning to be trained more quickly, and the incremental board learning based on the basic board learning can also be trained with fewer features, reducing the time by a certain amount, so in the final result, our method is more efficient than the classic IBLS.

The above experimental results conclude that the incremental board learning system based on feature selection proposed in this paper is better than the classic IBLS in terms of both accuracy and efficiency.

5 Conclusion

In this paper, we propose an incremental board learning system based on feature selection, aiming to improve the accuracy and efficiency of IBLS by combining the feature selection method with the incremental board learning system. Compared with the classic IBLS, we add a feature selection method to divide the features into two layers: an significant feature layer and a normal feature layer. On the one hand, to improve the efficiency of basic board learning, we use the significant feature layer as input for basic board learning, and on the other hand, to improve the accuracy of basic board learning and to utilize the complementarity of features, we use the normal feature layer as input for incremental board learning. By learning features in a layered manner, it helps to improve the accuracy and efficiency of IBLS. Our experiments on ten commonly used datasets validate the effectiveness of incremental board learning systems based on feature selection. In the process of feature selection in this paper, we use only simple and common methods, and in future research work, we will study more effective feature extraction algorithms for incremental board learning systems. In addition, for the case of new input data features and new feature mapping nodes, how to design a more effective incremental board learning system based on feature selection is also an important research direction in the next step.

References

1. Chen, C.L.P., Liu, Z.: Broad learning system: an effective and efficient incremental learning system without the need for deep architecture. IEEE Trans. Neural Networks Learn. Syst. **29**, 10–24 (2018)
2. Cun, Y.L., Boser, B., Denker, J.S., Henderson, D., Jackel, L.D.: Handwritten digit recognition with a back-propagation network. Adv. Neural. Inf. Process. Syst. **2**(2), 396–404 (1990)
3. Denker, J., et al.: Neural network recognizer for hand-written zip code digits, pp. 323–331 (1988)

4. Du, J., Vong, C.M., Chen, C.P.: Novel efficient RNN and LSTM-like architectures: recurrent and gated broad learning systems and their applications for text classification. IEEE Trans. Cybern. **51**(3), 1586–1597 (2020)
5. Gao, S., Guo, G., Philip Chen, C.: Event-based incremental broad learning system for object classification (2019)
6. Gong, X., Zhang, T., Chen, C.L.P., Liu, Z.: Research review for broad learning system: algorithms, theory, and applications. IEEE Trans. Cybern. **52**, 8922–8950 (2021)
7. Guo, W., Chen, S., Yuan, X.: H-BLS: a hierarchical broad learning system with deep and sparse feature learning. Appl. Intell. **53**, 153–168 (2022)
8. Han, M., Feng, S., Chen, C.L.P., Xu, M., Qiu, T.: Structured manifold broad learning system: a manifold perspective for large-scale chaotic time series analysis and prediction. IEEE Trans. Knowl. Data Eng. **31**, 1809–1821 (2019)
9. Han, M., Li, W., Feng, S., Qiu, T., Chen, C.P.: Maximum information exploitation using broad learning system for large-scale chaotic time-series prediction. IEEE Trans. Neural Networks Learn. Syst. **32**(6), 2320–2329 (2020)
10. Igelnik, B., Pao, Y.: Stochastic choice of basis functions in adaptive function approximation and the functional-link net. IEEE Trans. Neural Networks **6**(6), 1320–9 (1995)
11. Jia, C., Liu, H., Xu, X., Sun, F.: Multi-modal information fusion based on broad learning method. CAAI Trans. Intell. Syst. **14**(1), 154–161 (2019)
12. Leshno, M., Lin, V., Pinkus, A., Schocken, S.: Multilayer feedforward networks with a non-polynomial activation function can approximate any function. New York University Stern School of Business Research Paper Series (1991)
13. Leshno, M., Lin, V.Y., Pinkus, A., Schocken, S.: Multilayer feedforward networks with a nonpolynomial activation function can approximate any function. Neural Netw. **6**(6), 861–867 (1993)
14. Liu, Z., Chen, C.L.P., Feng, S., Feng, Q., Zhang, T.: Stacked broad learning system: from incremental flatted structure to deep model. IEEE Trans. Syst. Man Cybern. Syst. **51**, 209–222 (2021)
15. Pao, Y.H., Takefuji, Y.: Functional-link net computing: theory, system architecture, and functionalities. Computer **25**(5), 76–79 (1992)
16. Pao, Y.H., Park, G.H., Sobajic, D.J.: Learning and generalization characteristics of the random vector functional-link net. Neurocomputing **6**(2), 163–180 (1994)
17. Pu, X., Li, C.: Online semisupervised broad learning system for industrial fault diagnosis. IEEE Trans. Industr. Inf. **17**, 6644–6654 (2021)
18. Sheng, B., Li, P., Zhang, Y., Mao, L., Chen, C.P.: Greensea: visual soccer analysis using broad learning system. IEEE Trans. Cybern. **51**(3), 1463–1477 (2020)
19. Simonyan, K., Zisserman, A.: Very deep convolutional networks for large-scale image recognition. arXiv preprint arXiv:1409.1556 (2014)
20. Wang, B., Zhao, Y., Chen, C.L.P.: Hybrid transfer learning and broad learning system for wearing mask detection in the covid-19 era. IEEE Trans. Instrum. Meas. **70**, 1–12 (2021)
21. Wang, J., Lyu, S., Chen, C.P., Zhao, H., Lin, Z., Quan, P.: SPRBF-ABLS: a novel attention-based broad learning systems with sparse polynomial-based radial basis function neural networks. J. Intell. Manuf. **34**(4), 1779–1794 (2023)
22. Xu, M., Han, M., Chen, C.L.P., Qiu, T.: Recurrent broad learning systems for time series prediction. IEEE Trans. Cybern. **50**, 1405–1417 (2020)

High-Capacity Image Hiding via Compressible Invertible Neural Network

Changguang Wang[1,2], Haoyi Shi[1], Qingru Li[1,2], Dongmei Zhao[1,2], and Fangwei Wang[1,2(✉)]

[1] College of Computer and Cyber Security, Hebei Normal University, Shijiazhuang 050024, China
{qingruli,fw_wang}@hebtu.edu.cn

[2] Hebei Key Laboratory of Network and Information Security, Shijiazhuang 050024, China

Abstract. Image steganography involves the concealment of confidential information within images, rendering it undetectable to unauthorized observers, and subsequently retrieving the hidden information following secure transmission. Numerous investigations into image steganography employ invertible neural networks. Nevertheless, the intricate architecture of these models poses significant challenges to both their training and inference processes. Consequently, this paper proposes a high-capacity image steganography scheme utilizing compressible modules. We propose a multi-scale attention module that compresses the model structure through structural reparameterization post-training, thereby enhancing inference speed. Additionally, we employ a pre-trained image autoencoder to extract deep features from large-scale images, facilitating high-capacity steganography within invertible neural networks by concealing key features. Experimental results indicate that our approach offers superior image quality and model inference speed, while significantly enhancing security relative to existing methodologies.

Keywords: Image Steganography · Structural Reparameterization · Attention Mechanism · Invertible Neural Networks

1 Introduction

With the rapid growth of network technology, data transmission surges exponentially. Digital images, the key medium for online communication, gain significant value. However, these valuable images remain vulnerable to attacks like monitoring, intercepting, and tampering during transmission. So image security becomes a major public concern.

Image steganography is a mainstream technique for image security [1]. It involves embedding secret data into the cover image without detection by third parties, thereby avoiding the possibility of attacks [2]. Receiver can use the corresponding extraction algorithm to recover the hidden information, completing the

T. Zhu et al. (Eds.): ICA3PP 2024, LNCS 15252, pp. 36–53, 2025.
https://doi.org/10.1007/978-981-96-1528-5_3

secure and effective transmission of information [3]. Traditional image steganography schemes are mainly designed by humans. They use pixel transformations in images to hide secret information [4]. The Least Significant Bit (LSB) technique [5] is a representative scheme. Some schemes focus on the image frequency domain, hidding secret information by modifying frequency domain coefficients [6,7]. However, such schemes typically only apply to hiding a small amount of information, and there are also problems of image artifacts and distortion, which are not conducive to the security of images.

Compared to artificially designed schemes, neural networks offer more flexibility in their choices and strategies, allowing the cover image to achieve a higher steganographic capacity. Since the introduction of the first steganography model [8], researchers gradually integrate mainstream network structures such as U-Net and ResNet into image steganography tasks. The HiNet model [9] introduces the concept of invertible neural networks into image steganography tasks. By utilizing multiple sets of invertible submodules, the HiNet model completes both image steganography and extraction tasks in a single set of network parameters. However, most existing deep-learning steganography schemes struggle to balance the relationship between image security, model convenience, and steganographic capacity. They often improve steganography performance by adding depth and dimension to the model. This approach compromises both the model's inference speed and training complexity.

We propose a large-capacity compressible image steganography scheme. This scheme increases steganography capacity while compressing the structure of the steganography network. We design a multi-scale attention module to help model select reasonable embedding positions by multi-scale convolution and spatial channel attention mechanisms. After training, we use structural reparameterization to compress the invertible submodules, boosting reasoning speed without sacrificing performance. In addition, this scheme uses a pre-trained encoder to encode large-scale secret images, enabling large-capacity steganography of RGB images. The main contributions of this paper are as follows:

1) We design a multi-scale attention module and integrate it into an invertible steganography network. This design helps the model select embedding positions for secret information more effectively in the frequency domain of the cover image.
2) The trained model uses structural reparameterization to compress the multi-scale attention module, which can improve the inference speed of the model without affecting the performance of the model.
3) We design an autoencoder specifically for large-scale images and pre-trained it with vector quantization techniques. This approach converts image features into vector representations while also adjusting the feature dimensions to align with the input requirements of steganography network.

The rest of this study is listed as follows. Section 2 introduces the image steganography, invertible neural networks and structural reparameterization. Section 3 introduces the specific design of the proposed scheme. Section 4 conducts experimental verification of the proposed scheme to evaluate image similarity and anti-attack capability. Section 5 concludes this paper.

2 Related Work

2.1 Image Steganography

The traditional scheme is first applied to the spatial domain of images and embed secret information by modifying the pixel values of the image. The LSB [5] represents the most classic spatial domain steganography scheme. Traditional schemes often lead to systematic changes in the statistical features of the image. These changes are easily identifiable by steganalysis. In contrast, some schemes transfer the embedding of secret information to the transform domain of the image. The various transformation forms between spatial domain and transform domain provide more options for steganography. However, these schemes do not improve the low capacity and poor security of traditional steganography [10].

The widespread adoption of deep neural networks (DNN) brings new development opportunities to the field of image steganography. Baluja et al. [8] pioneered the application of DNN in image steganography tasks. They used three network models to deal with the pre-processing, steganography, and extraction stages. Duan et al. [11] adopted the U-Net as the backbone network of the steganography and extraction model. They completed the steganography task by hiding and extracting the wavelet components of secret images. In subsequent research, Duan et al. [12] used a double-layer U-Net network, combined with an attention mechanism, to achieve a comprehensive fusion of image features. To improve the data capacity, they combined the U-Net model with a flow encoder [13], successfully embedding a large amount of binary secret data in color images. Kumar et al. [14] incorporated skip connections into the U-Net to meet the requirement of preserving details in the cover image. Yao et al. [15] integrated generative adversarial networks (GANs) with discrete wavelet transform (DWT) to embed binary secret information in the frequency domain. They introduced a discriminator to improve the authenticity of stego images. Bui et al. [16] encoded binary secret information into the latent embeddings of images and injected the secret information directly into the latent encoding of the secure autoencoder. Zhang et al. [17] proposed a joint adjustment model that batched different resolutions of the same image into the encoder model. Li et al. [18] fed the jointly encoded image features into the Swin transformer module, effectively integrating high-resolution spatial features and global self-attention features, but with lower steganography capacity than general image steganography schemes. Wang et al. [19] adopted the transformer module as the primary structure for the autoencoder, combined with image encryption technology, to effectively enhance security and resilience against attacks. Although these methods achieve good results, few of them can simultaneously meet the requirements for image security, capacity, and model simplicity.

2.2 Invertible Neural Network

Dinh et al. [20] proposed an invertible neural network that can achieve bidirectional computations for specific tasks in a set of network model parameters. In the forward process, variable y is input to the invertible neural network, and the output result x is obtained through forward inference. In the backward inference stage, the result x is input from the tail of the model, and the variable y is reduced by backward inference. This model can obtain a bidirectional mapping between the data distribution x and the latent distribution y at the same time, that is, $f_\theta(y) = x$ and $f_\theta^{-1}(x) = y$, where f_θ and f_θ^{-1} share the same set of model parameters. The invertible neural networks gain widespread adoption in various computer vision tasks due to its exceptional performance. Liu et al. [21] employed a lightweight invertible neural network to achieve super-resolution operations on unpaired images. Ardizzone et al. [22] incorporated conditional mechanisms into an invertible neural network to guide image generation and coloring processes.

Invertible neural networks can achieve perfect bidirectional mapping between data, which is just suitable for the needs of image steganography tasks. Jing et al. [9] first introduced an invertible neural network into the field of image steganography. The HiNet model takes DenseBlock as an invertible submodule to realize image steganography and extraction. Liu et al. [23] mapped secret information into a binary sequence, achieving lossless steganography of information in an invertible neural network. Feng et al. [24] replaced the invertible submodules in HiNet with a Swin Transformer block, but the image quality and security decreased. To ensure the robustness of the invertible steganography network, Shang et al. [25] added an attack module and discriminator network to HiNet to improve the anti-attack capability. A similar approach is the PRIS model by Yang et al. [26], which added an image enhancement module into HiNet.

Although invertible neural networks achieve good results in the field of image steganography, most studies do not upgrade the main structure of HiNet. These schemes still suffer from issues such as unstable training, slow inference speed, and fixed capacity. Our scheme abandons the invertible submodule based on DenseBlock in HiNet and introduces a multi-scale attention module. This module efficiently selects embedding positions in the image frequency domain and performs structural reparameterization compression of the trained model, thereby improving inference speed. The introduction of a pre-trained autoencoder allows for hiding high-resolution images within low-resolution images, increasing capacity without altering the training difficulty, and ensuring the robustness.

2.3 Structural Reparameterization

When neural networks perform deep learning tasks, they require training model parameters to complete inference. In some research, network models often comes at the cost of sacrificing model simplicity. Such models have high training difficulty, slow inference speed, and require large memory during inference. The emergence of structural reparameterization improves this situation. Ding et al. [27] for the first time used structural reparameterization to compress three asymmetric convolution nuclei into a simple one 3×3 convolution operation. Subsequently, Ding et al. proposed a RepVGG module [28], which simplifies multi-branch convolutional structures into single-branch convolutions, continuously optimizing structural reparameterization technology in subsequent research.

The application of structural reparameterization designs the model as a multi-branch structure during the training phase, thereby enhancing the model's representation capability. A single-branch structure effectively improves the parallelism of model computation and enhances GPU utilization during the inference process. Compared to multi-branch structures, a single-branch structure requires only one access to input features, and the number and types of operators in a single-branch structure are fewer, which reduces the inference difficulty of the model. Using compressible modules as the main structure of invertible sub-modules not only enhances the hiding capability but also effectively reduces the time required for steganography and extraction.

3 Proposed Method

3.1 Overall Model

The overall model structure we designed is illustrated in Fig. 1. Here, x_{cov}, x_{sec}, x_{ste} and x_{ex} respectively represent the cover, secret, stego, and extracted images. The image is preprocessed before it is fed into the invertible neural network. The cover image undergoes a Haar wavelet transform [29], resulting in four sub-bands of x_{cov}. These sub-bands are then combined along the channel dimension and the scale of x_{cov} from $R^{c \times h \times w}$ to $R^{4c \times \frac{h}{2} \times \frac{w}{2}}$, where c denotes channels in the image, h, and w denotes the width and height of the image. Wavelet components can avoid artifacts and color distortion during steganography, while better embedding information into high-frequency subbands. For the x_{sec}, we design two processing methods. When the size of the x_{sec} is the same as the size of x_{cov}, we also convert it to the image frequency domain. For high-capacity steganography, where the size of the image x_{sec} is larger than that of the image x_{cov}, we use pre-trained encoder E and decoder D to map the image x_{sec} and the potential encoding each other, maintaining consistency in size with the wavelet-transformed images.

In Fig. 1, we present an overview of the invertible module, where ϕ, ρ, and η can be any network modules with the same structure. We use the compressible multi-scale attention module. The processed image is input into the invertible neural network, passing through multiple sub-modules, we obtain two types of output data: the wavelet components of the x_{ste} and the corresponding output r of the x_{sec}. In the process of image steganography, the secret image cannot be completely hidden. The output data r represents the lost information that cannot be hidden. For the i_{th} invertible module of the steganography process, the input is x_{cov}^{i} and x_{sec}^{i}, the output is x_{cov}^{i+1} and x_{sec}^{i+1} of the same size. After passing through the last invertible block, we obtain the wavelet components of

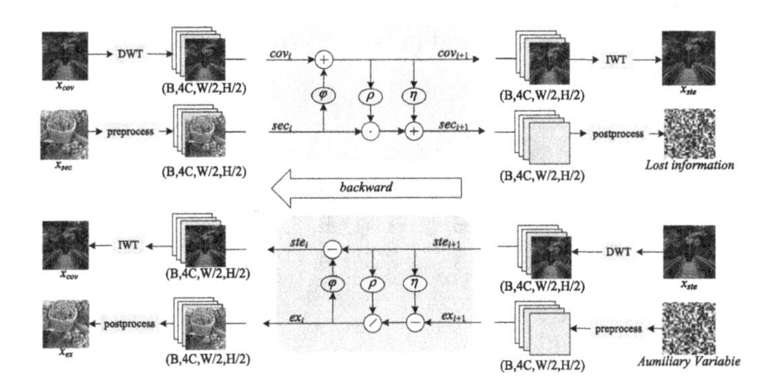

Fig. 1. The overall network structure includes the HAAR wavelet transform (DWT) and its inverse transform (IWT). ϕ, ρ, and η represent multi-scale attention modules.

x_{ste} and the lost information r. The reasoning formula for the steganography process is as follows:

$$\begin{aligned} x_{cov}^{i+1} &= x_{cov}^{i} + \phi(x_{sec}^{i}), \\ x_{sec}^{i+1} &= x_{sec}^{i} \odot \exp(\alpha(\rho(x_{cov}^{i+1}))) + \eta(x_{cov}^{i+1}). \end{aligned} \tag{1}$$

The output is transformed by inverse wavelet to obtain stego image. This kind of wavelet transform has strict symmetry, which reduces the training cost of the model without affecting end-to-end training.

The inverse process is similar to the forward process, and the information is passed from module $i+1$ to module i through the inverse function. To maintain input symmetry, we generate a random noise z with Gaussian distribution as the auxiliary variable of the inverse process to simulate the lost information. The inference formula for the inverse process is as follows:

$$\begin{aligned} z^{i} &= (z^{i+1} - \eta(x_{ste}^{i+1})) \odot exp(-\alpha(\rho(x_{ste}^{i+1}))), \\ x_{ste}^{i} &= x_{ste}^{i+1} - \phi(z^{i}). \end{aligned} \tag{2}$$

3.2 Multi-scale Attention Module

The introduction of residual connections effectively enhances the representational capacity but increases the inference cost. Multi-scale convolution in the steganography process expands the perception field of the image and helps the model better choose embedding positions in the cover image. Taking these considerations into account, we design a multi-scale attention module. This module leverages various scales of convolution and identity mapping to learn different ranges of image features. We also adopt the convolutional attention module (CBAM) [30], which concatenates channel attention mechanisms and spatial attention mechanisms to enhance the incoming feature layers. The structure of the multi-scale attention module we design is depicted in Fig. 2. In the convolutional module, we use two different sizes of convolutional kernels and a residual connection. This design reduces redundant features and lowers the computational cost of the model.

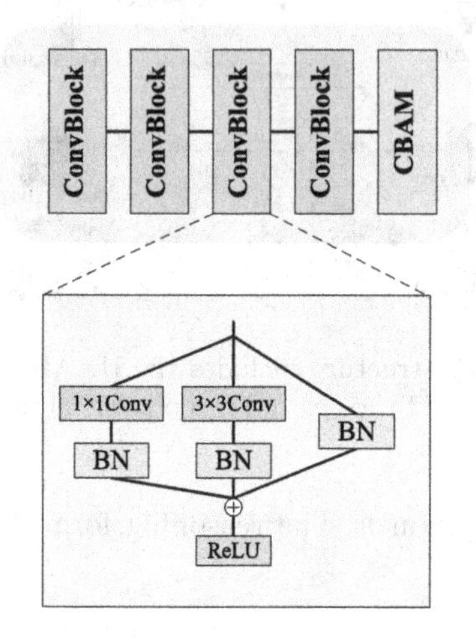

Fig. 2. The structure of the multi-scale attention module consists of multiple ConvBlocks. In these blocks, different-scale convolutions are used to learn image features from different perspectives.

The channel attention mechanism helps the model distinguish the importance of different channels in features by assigning different weights to channels. The spatial attention mechanism complements the channel attention mechanism by focusing on the importance of each feature point within the overall features. These two attention mechanisms enhance the invisibility of steganography scheme. Figure 3 shows the internal structure of the attention module. The calculation formula for the channel attention mechanism is as follows:

$$M_c(F) = \sigma(MLP(AvgPool(F)) + MLP(MaxPool(F))). \tag{3}$$

The spatial attention mechanism is as follows:

$$M_s(F) = \sigma(f^{7\times7}([AvgPool(F); MaxPool(F)])). \tag{4}$$

The computation result of the enhanced features output by the attention module is as follows:

$$F = M_s(M_c(F)). \tag{5}$$

Most image steganography schemes based on deep-learning employ image-adaptive techniques, hiding the feature information of the secret image within intricate texture regions of the cover image throughout the model training process. The introduction of attention mechanisms can enhance the rationality and accuracy of embedding positions.

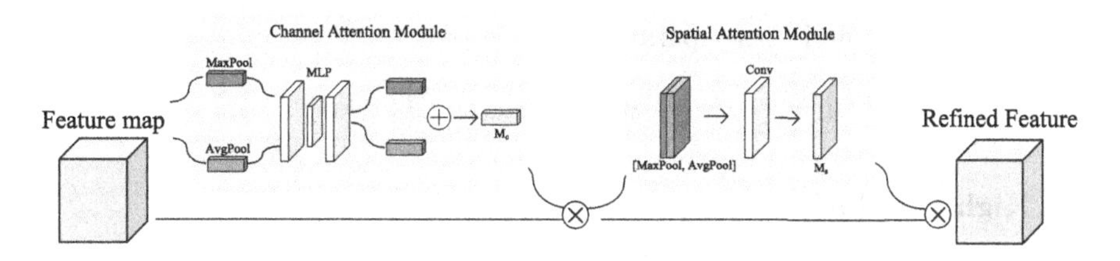

Fig. 3. The structure diagram of the convolutional attention module consists of two concatenated attention mechanisms. The refined feature size remains consistent with the input feature size.

3.3 Structural Reparameterization

In this paper, the invertible submodules adopt multi-scale convolution design scheme. In the training process, we use 1×1 convolutions, 3×3 convolutions, and residual branches. This design ensures high computational density of the model, and the residual branches can deepen the depth of model. However, such a design slows down the inference speed of the model. Therefore, we adopt structural reparameterization techniques to merge the parameters of the model. This compression does not affect model performance while speeding up the processes of image steganography and extraction.

In the invertible submodule, the input and output features share identical dimensions in terms of height, width, and number of channels. We first merge the 3×3 convolution layer and the BN layer, and the processing of BN layer is as follows:

$$y_i = \frac{x_i - \mu_i}{\sqrt{\sigma_i^2 + \epsilon}} \cdot \gamma_i + \beta_i, \tag{6}$$

where x_i is the input of the BN layer, y_i is the output, μ, σ, γ, and β are statistics learned during the model's training process. The batch normalization computation for the output M_i of the convolutional layer can be expressed as:

$$bn(M_1, \mu, \sigma, \gamma, \beta) = (M_1 - \mu_i) \cdot \frac{\gamma_i}{\sigma_i} + \beta_i. \tag{7}$$

After splitting the formula, we can integrate the parameters into the convolution kernel and convolution bias, thereby merging the convolutional layer with the normalization layer. The transformed convolutional kernels and biases after merging are as follows:

$$W_i = W_i \frac{\gamma_i}{\sigma_i}, \quad b_i = \beta_i - \frac{\mu_i \gamma_i}{\sigma_i}. \tag{8}$$

We apply the same transformation to the other two branches, converting the 1×1 convolution and residual branch into an equivalent 3×3 convolution using zero padding, and then merge them with the normalization layer. After merging, all branches of the model become 3×3 convolutional layers. We combine convolution kernel and bias separately:

$$I = I \otimes (k_1 + k_2 + k_3) + (b_1 + b_2 + b_3). \tag{9}$$

3.4 High-Resolution Image Autoencoder

Hiding capacity is a crucial metric for evaluating a steganography scheme. For quite some time, embedding a high-resolution image in a low-resolution image is a central focus of research in the field of image steganography. The complex network structure and model depth of invertible neural network make it have better representation ability than other steganography schemes. Based on these characteristics, we design an autoencoder for high-resolution images, combining it with invertible neural network to achieve high-capacity steganography.

Referring to the vector quantize technique [31], we pre-trained the autoencoders to achieve discrete feature representation of images. The training process is shown in Fig. 4. We construct a learnable feature codebook. The encoder E encodes the secret image into a feature map z_{sec} of size $h \times w \times D$, which contains $h \times w$ D-dimensional vectors. These vectors are indexed in the codebook to find the closest vector and replace it to achieve a discrete representation of the feature map. The quantized feature map z_q is input to the decoder to reconstruct the image. The trainable parameters are encoder, decoder, and codebook. We optimize the encoder and decoder using a simple image reconstruction loss $logp(x|z_q(x))$, and use vector quantization to reduce the difference between the codebook vector and the code vector $\| sg[z_e(x)] - e \|_2^2$. To ensure the image latent vectors are as close as possible to the vectors in the codebook, we introduce a regularization term to constrain the training of encoder parameters. The final loss for the autoencoder is as follows:

$$L = logp(x|z_q(x)) + \| sg[z_e(x)] - e \|_2^2 + \beta \| z_e(x) - sg[e] \|_2^2, \tag{10}$$

where sg is the stop-gradient operation and the partial derivative is set to 0 during the backpropagation of the model.

After pre-training, we retain the encoder and decoder. Unlike typical autoencoders that process only one feature vector for an image, our autoencoder extracts the two-dimensional combination of multiple feature vectors. This type of image feature is relatively discrete. In high-capacity steganography tasks, these low-correlation feature vectors help find reasonable embedding positions and reduce the difficulty of the decoder restoring the secret image. We set the encoder before the forward process of the backbone network and the decoder after the reverse process to realize high-capacity steganography in the invertible neural network.

3.5 Loss Function

Unlike most image steganography schemes that use MSE Loss, this paper employs the Huber Loss function to represent the error between two images.

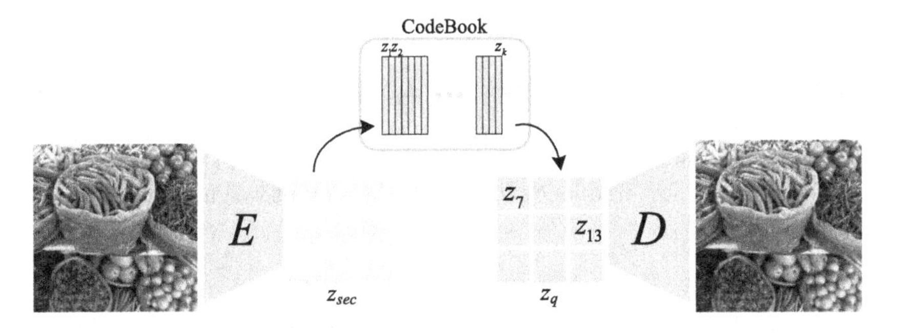

Fig. 4. Pre-training process of high-resolution image autoencoder.

Compared to MSE loss, Huber Loss is less sensitive to outliers in the data, exhibiting better robustness. The calculation of Huber loss function is as follows:

$$L_\delta(y, f(x)) = \begin{cases} \frac{1}{2}(y - f(x))^2 & |y - f(x)| \leqslant \delta \\ \delta|y - f(x)| - \frac{1}{2}\delta^2 & otherwise \end{cases}. \tag{11}$$

The loss function not only includes common steganography loss and extraction loss but also incorporates low-frequency wavelet loss of the stego image. Low-frequency wavelet loss allows for more embedding of secret image features into high-frequency components, to improve the anti-analysis ability of stego image.

The purpose of the forward process is to hide the x_{sec} into the x_{cov}, generating a stego image that matches the visual effect of the cover image. Therefore, the stego loss calculation formula is as follows:

$$L_{stego} = \sum_{n=1}^{N} \ell(x_{cov}^{(n)}, x_{ste}^{(n)}), \tag{12}$$

where N is the number of training samples, ℓ represents the Huber loss between images.

The purpose of the reverse process is to use an auxiliary variable z that conforms to a Gaussian distribution to obtain the extracted image x_{ex} from the stego image:

$$L_{rec} = \sum_{n=1}^{N} E_{z \sim p(z)}[\ell(x_{sec}^{(n)}, x_{ex}^{(n)})]. \tag{13}$$

To hide the secret information in the high-frequency components as much as possible, he low-frequency components should undergo small changes. In the loss function, $\mathcal{H}(\cdot)LL$ represents the decomposed low-frequency sub-band of an image:

$$L_f = \sum_{n=1}^{N} \ell(\mathcal{H}(x_{cov}^{(n)})LL, \mathcal{H}(x_{ste}^{(n)})LL). \tag{14}$$

The training loss function comprises a weighted sum of steganography loss, extraction loss, and low-frequency loss, with the sum of weights equaling 1. The formula is as follows:

$$L = \lambda_s L_{stego} + \lambda_r L_{rec} + \lambda_f L_f. \tag{15}$$

4 Experiment

We train and experimentally validate our model on the DIV2K dataset and COCO dataset, implementing it on an RTX 3090 using the PyTorch framework. In experiments with the same resolution, we use images from the COCO dataset and adjust the resolution of cover images and secret images to 256×256. In high-capacity steganography experiments, we keep the resolution of cover images and adjust the resolution of secret images to 512×512. During the model training process, we observe that maintaining a constant learning rate results in loss oscillation. We set the initial learning rate to 0.0001 and employ a cosine annealing strategy to gradually decay the learning rate, ensuring training stability. Training stops automatically when the average training loss no longer improves. We train the model for 1500 epochs with a batch size of 16.

4.1 Visual Effects

Visual effects are the first impression people have of digital images, so a steganography scheme must first avoid visual differences in images. The stego image x_{ste} generated during the embedding phase must minimize differences with the cover image x_{cov} while ensuring authenticity. The extracted image x_{ex} in the extraction phase must accurately convey the information of the secret image x_{sec}. In this experiment, we choose the pixel distribution histogram as a comparison scheme, which directly compares the difference in pixel frequency between two images. Figure 5 shows the comparison after steganography and extraction of images in our dataset. It is evident that, regardless of the resolution or high-capacity steganography scenarios, third parties cannot visually distinguish differences between images and histograms. This demonstrates that our approach appropriately selects pixel positions to hide x_{sec} features during the steganography process.

4.2 Capacity Analysis

The hidden capacity of data information is a crucial evaluation metric for image steganography schemes. In the case of the same cover image resolution, larger amounts of secret information necessitate higher capabilities from the model in

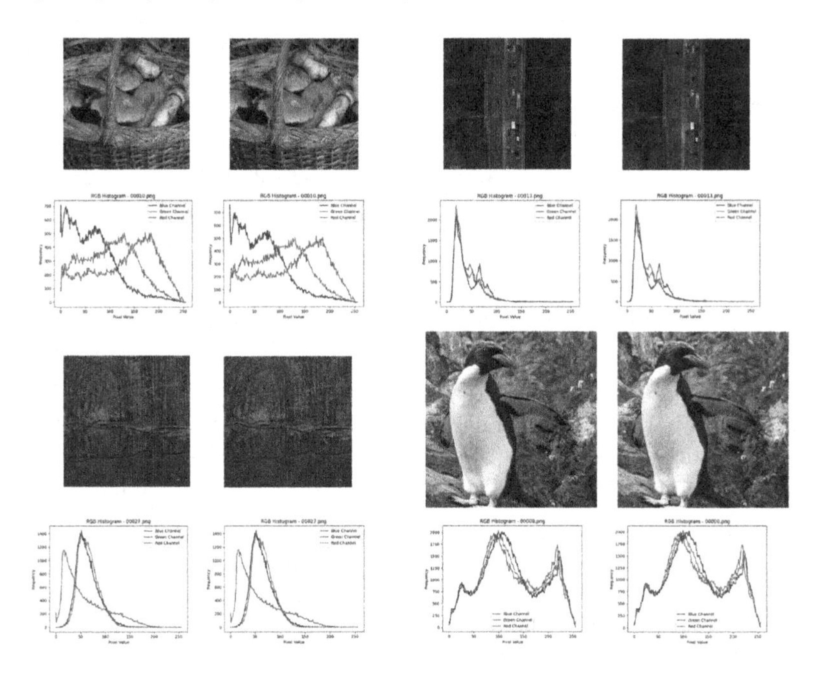

Fig. 5. Experimental results and pixel histograms.

generating stego images. Similarly, increased data volume poses challenges to the accuracy of the extraction model. Traditional image steganography schemes mainly hide a small amount of binary data, and most common deep-learning steganography schemes are limited to steganography at the same resolution. In the capacity analysis, the effective capacity (EC) was chosen to assess the hidden capability of the model. The calculation formula is as follows:

$$EC = NS/NC, \tag{16}$$

where NS represents the quantity of secret information to be hidden (unit: bits), and NC is the number of pixels in the cover image.

Table 1 lists several common traditional image steganography schemes and recent deep-learning steganography schemes for comparison with our proposed scheme. It is evident that our approach far exceeds traditional methods in terms of capacity. Compared to the latest deep-learning methods, our approach can hide high-resolution images with a capacity that is four times greater than theirs. This also demonstrates the effectiveness of our discrete encoding approach for secret images.

4.3 Comparison Against SOTA Methods

To effectively evaluate the invisibility of the proposed steganography scheme, we use the Peak Signal-to-Noise Ratio (PSNR) and Structural Similarity Index

Table 1. Comparison of information capacity between proposed scheme and others.

Type	Schemes	Cover image size	Secret data size	EC(bpp)
Traditional	Ma [32]	512 × 512(gray)	1741(bit)	<0.25
	Liu [33]	512 × 512(RGB)	32 × 32(RGB)	<0.25
Deep learning	Shang [25]	256 × 256(RGB)	256 × 256(RGB)	8
	Yang [26]	224 × 224(RGB)	224 × 224(RGB)	8
	Ours	256 × 256(RGB)	512 × 512(RGB)	32

(SSIM) to compare image similarity. PSNR is widely used for image quality assessment, evaluating image similarity by comparing pixel differences between images. Generally, a PSNR value above 37 indicates that the differences between images are considered imperceptible. The formula is defined as follows:

$$\mathrm{PSNR} = 10 \cdot \log_{10}\left(\frac{(2^n-1)^2}{MSE(I, I_a)}\right). \tag{17}$$

SSIM is an image quality evaluation index similar to human visual perception. The formula is defined as follows:

$$\mathrm{SSIM}(I, I_a) = \frac{(2\mu_I\mu_{I_a} + C_1)(2cov + C_2)}{(\mu_I^2 + \mu_{I_a}^2 + C_1)(\sigma_I^2 + \sigma_{I_a}^2 + C_2)}, \tag{18}$$

where μ_I and μ_{I_a} are the average values of I and I_a, σ_I^2 and $\sigma_{I_a}^2$ are the variances of I and I_a, cov is the covariance of I and I_a, C_1 and C_2 are used to avoid the denominator being zero.

In Table 2, we compare the evaluation results of our scheme with several mainstream schemes, all of which use images of the same resolution for steganography.

Table 2. Comparison of Image Similarity on DIV2K Dataset

Schemes	Index	Stego image	Recover image
4bit-LSB	PSRN	33.19	30.18
	SSIM	0.9453	0.9020
HiDDeN [34]	PSRN	35.21	36.43
	SSIM	0.9691	0.9696
Feng [24]	PSRN	34.32	44.05
	SSIM	0.9350	0.9911
HiNet [9]	PSRN	48.99	52.86
	SSIM	0.997	0.9992
Ours	PSRN	50.72	52.6
	SSIM	0.9980	0.9976
Ours (High-capacity)	PSRN	40.63	40.19
	SSIM	0.9861	0.9936

Our scheme outperforms other schemes on the DIV2K dataset and shows performance advantages over those that also use invertible neural network structures. Additionally, our scheme demonstrates superior capacity and image quality when hiding high-resolution images compared to some mainstream solutions.

4.4 Steganographic Analysis

Third parties use various methods to determine if an image contains secret information. If identified as a stego image, third parties can intercept, tamper with, and interfere with it. Common steganalysis methods include residual image analysis, traditional statistical analysis approaches, and methods based on deep learning. Residual image analysis is the simplest and most common method in steganalysis. The texture information of the secret image is revealed in the residual image. In Fig. 6, we perform residual analysis on the central part of the image data and enhance the residual image by 20 times. It can be observed that the stego images generated by our scheme pass the residual analysis test effectively. Even in the enhanced residual images, the texture information of the secret image cannot be observed.

Fig. 6. Experimental results and residual images.

In statistical analysis, we use the open-source steganalysis tool StegExpose [35] to test 1000 randomly selected stego images and cover images from the

Table 3. The detection results of SRNet.

Methods	Accuracy(%)
4bit-LSB	99.96 ± 0.06
Baluja [37]	99.67 ± 0.01
HiDDeN [34]	76.49 ± 0.11
Proposed	58.42 ± 0.59
Proposed(high-capacity)	72.13 ± 0.21

dataset. We test steganography schemes at the different resolution and plot the Receiver Operating Characteristic (ROC) curves as shown in Fig. 7. We can see that in both cases, the area value under our ROC curve is close to 0.5, and the detection accuracy is close to random guessing. Our approach can effectively deceive open-source platform tools in statistical metric analysis, demonstrating a higher level of security.

For deep learning-based analysis, we choose the SRNet [36] model as the analysis network to distinguish stego images from cover images. The test results of SRNet for different steganography schemes are given in Table 3. The closer the detection results are to the 50% random guessing probability, the better the steganographic effectiveness of the model. Our designed scheme shows better anti-analysis ability than other common steganography schemes in both resolution cases, effectively avoiding detection by deep learning analysis methods.

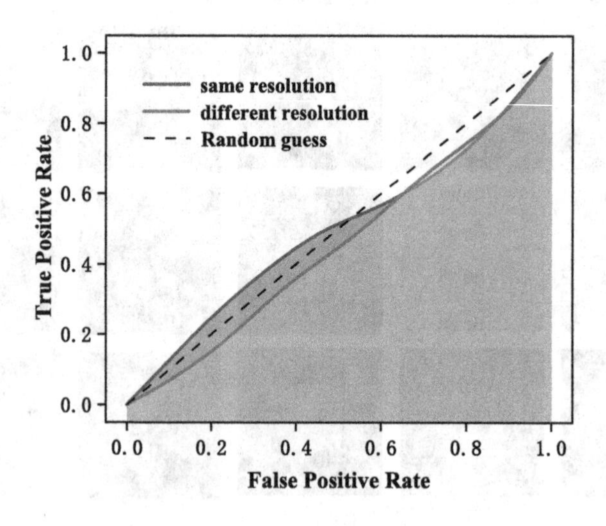

Fig. 7. ROC curve of StegExpose

4.5 Parameter Analysis

A series of steganography schemes based on invertible neural networks adopt the HiNet network or its invertible submodule as the backbone structure of the model. DenseBlock in these schemes has good representation ability, but the need for multiple concat operations increases the difficulty of model calculation. This makes training the model challenging and slows the inference speed, which is a common drawback of such schemes. Our invertible module adopts multi-scale convolutions. We add the feature vector output from different branches, reducing the burden of convolution processing while effectively preserving semantic information. Table 4 shows the calculation results of our model and the HiNet model in terms of inference speed and trainable parameters. Our model has approximately 60% of the trainable parameters of the HiNet model, and after reparameterization, the inference parameters of our model are even fewer. We set the image batch size to 4 and measure the inference time of the model for 10 epochs. Our model exhibits faster inference speed after compression.

Table 4. Model inference time and parameter comparison.

Model	HiNet [9]	Proposed(non-compressed)	Proposed(compressed)
Time(S)	5.38	3.39	3.24
Parameters	4051392	2247864	1954080

5 Conclusion

This paper presents a high-capacity, compressible image steganography scheme predicated on an invertible neural network, which is adept at concealing images at both equivalent and superior resolutions. The core of our invertible module is a multi-scale attention mechanism, which proficiently identifies high-frequency pixel positions within cover images to embed the features of secret images. To facilitate high-capacity steganography, we implement a pre-training strategy involving autoencoders trained with vector quantization techniques, enabling the discrete encoding of high-resolution images. This methodology integrates an invertible neural network to improve the rationality of embedding point selection. Following model training, we employ structural reparameterization techniques to compress the invertible module, thereby significantly enhancing inference speed without sacrificing performance. Future research will incorporate considerations for compression and noise phenomena that may arise during image transmission over network channels. Additionally, we will further optimize the architecture of the invertible neural network and bolster the anti-attack robustness of steganographic models.

Acknowledgements. This research was funded by the National Natural Science Foundation of China under Grant 62462012, and Science and Technology Program of Hebei under Grant 22567606H.

References

1. Mandal, P.C.: Digital image steganography: a literature survey. Inform. Sci. **609**, 1451–1488 (2022)
2. Katzenbeisser, S., Petitcolas, F.A.P.: Defining security in steganographic systems. In: Security and Watermarking of Multimedia Contents IV, pp. 50–56 (2002)
3. Johnson, N.F.: Information hiding: steganography and watermarking-attacks and countermeasures. J. Elect. Imag. **10**(3), 825 (2001)
4. Aghababaiyan, K.: Novel distortion free and histogram based data hiding scheme. IET Image Proc. **14**(9), 1716–1725 (2020)
5. Mielikainen, J.: LSB matching revisited. IEEE Sign. Proc. Let. **13**(5), 285–287 (2006)
6. Holub, V., Fridrich, J.: Designing steganographic distortion using directional filters. In: 2012 IEEE International Workshop on Information Forensics and Security (WIFS), pp. 234–239 (2012)
7. Holub, V., Fridrich, J., Denemark, T.: Universal distortion function for steganography in an arbitrary domain. EURASIP J. Inf. Secur. **2014**(1), 1–13 (2014). https://doi.org/10.1186/1687-417X-2014-1
8. Baluja, S.: Hiding images in plain sight: deep steganography. In: Proceedings of Advances in Neural Information Processing Systems (NIPS), pp. 2069–2079 (2017)
9. Jing, J., Deng, X.: HiNet: deep image hiding by invertible network. In: Proceedings of the IEEE/CVF International Conference on Computer Vision, pp. 4733–4742 (2021)
10. Guo, H., Xue, J.: The analysis of watermarking capacity of packing model and bits replacement model. In: 2016 12th World Congress on Intelligent Control and Automation (WCICA), pp. 2603–2607 (2016)
11. Duan, X.: Reversible image steganography scheme based on a U-net structure. IEEE Access **7**, 9314–9323 (2019)
12. Duan, X.: DUIANet: a double layer U-Net image hiding method based on improved inception module and attention mechanism. J. Vis. Comm. Image Rep. **98**, 104035 (2023)
13. Duan, X.: DHU-Net: high-capacity binary data hiding network based on improved U-Net. Neurocomputing **576**(1), 127314 (2024)
14. Kumar, A.: Encoder-Decoder architecture for image steganography using skip connections. Proc. Comp. Sci. **218**(4), 1122–1131 (2023)
15. Yao, Y.: High invisibility image steganography with wavelet transform and generative adversarial network. Exp. Syst. Appl. **249**, 123540 (2024)
16. Bui, T., Agarwal, S.: RoSteALS: robust steganography using autoencoder latent space. In: Proceedings of the IEEE/CVF International Conference on Computer Vision, pp. 933–942 (2023)
17. Zhang, L.: Joint adjustment image steganography networks. Signal Process. Image Comm. **118**, 117022 (2023)
18. Li, Z.: Adversarial feature hybrid framework for steganography with shifted window local loss. Neur. Netw. **165**, 358–369 (2023)
19. Wang, Z.: Deep image steganography using Transformer and recursive permutation. Entropy **24**(7), 878 (2022)

20. Dinh, L.: Nice: non-linear independent components estimation. arXiv preprint arXiv:1410.8516 (2014)
21. Liu, H.: Unpaired image super-resolution using a lightweight invertible neural network. Pattern Recogn. **144**, 109822 (2023)
22. Ardizzone, L.: Guided image generation with conditional invertible neural networks. arXiv preprint arXiv:1907.02392 (2019)
23. Liu, L.: Lossless image steganography based on invertible neural networks. Entropy **24**(12), 1762 (2022)
24. Feng, Y., Liu, Y.: Image hide with invertible network and Swin Transformer. In: International Conference on Data Mining and Big Data, pp. 385–394 (2022)
25. Shang, F.: Robust data hiding for JPEG images with invertible neural network. Neur. Netw. **163**, 219–232 (2023)
26. Yang, H.: PRIS: practical robust invertible network for image steganography. Eng. Appl. Art. Intell. **133**, 108419 (2024)
27. Ding, X., Guo, Y.: ACNet: strengthening the kernel skeletons for powerful CNN via asymmetric convolution blocks. In: Proceedings of the IEEE/CVF International Conference on Computer Vision, pp. 1911–1920 (2019)
28. Ding, X., Guo, Y.: RepVGG: making VGG-style convnets great again. In: Proceedings of the IEEE/CVF Conference on Computer Vision and Pattern Recognition, pp. 13733–13742 (2021)
29. Mallat, S.G.: A theory for multiresolution signal decomposition: the wavelet representation. IEEE Trans. Patt. Anal. Mach. Intell. **11**(7), 674–693 (1989)
30. Woo, S., Park, J.: CBAM: convolutional block attention module. In: Proceedings of the European Conference on Computer Vision (ECCV), pp. 3–19 (2018)
31. Van Den Oord, A., Vinyals, O.: Neural discrete representation learning. In: Proceedings of Advances in Neural Information Processing Systems (NIPS), p. 30 (2017)
32. Ma, K.: Reversible data hiding in encrypted images by reserving room before encryption. IEEE Trans. Inform. Foren. Secur. **8**(3), 553–562 (2013)
33. Liu, D.: A fusion-domain color image watermarking based on Haar transform and image correction. Expert Syst. Appl. **170**, 114540 (2021)
34. Zhu, J., Kaplan, R.: Hidden: hiding data with deep networks. In: Proceedings of the European Conference on Computer Vision (ECCV), pp. 657–672 (2018)
35. Boehm, B.: Stegexpose - a tool for detecting LSB steganography. arXiv preprint arXiv:1410.6656 (2014)
36. Boroumand, M.: Deep residual network for steganalysis of digital images. IEEE Trans. Inform. Foren. Secur. **14**(5), 1181–1193 (2018)
37. Baluja, S.: Hiding images within images. IEEE Trans. Patt. Anal. Mach. Intell. **42**(7), 1685–1697 (2019)

Crop Classification Methods Based on Siamese CBMM-CNN Architectures Using Hyperspectral Remote Sensing Data

Bin Xie[1,2,3](✉), Jiahao Zhang[1,2], Yuling Li[1,2], Yusong Li[1,2], and Xinyu Dong[1,2]

[1] College of Computer and Cyber Security, Hebei Normal University, Shijiazhuang, Hebei, China
xiebin_hebtu@126.com

[2] Hebei Provincial Key Laboratory of Network and Information Security, Hebei Normal University, Shijiazhuang, Hebei, China

[3] Hebei Provincial Engineering Research Center for Supply Chain Big Data Analytics and Data Security, Hebei Normal University, Shijiazhuang, Hebei, China

Abstract. Amidst global population growth and escalating food demands, real-time agricultural monitoring is crucial for ensuring food security. During the initial stages of crop growth, however, it faces significant challenges in obtaining accurate ground truth data via remote sensing, thereby impeding effective crop classification. Traditional Convolutional Neural Network (CNN) models, although effective in image processing and feature extraction within computer vision, are notably limited in their ability to handle the high dimensionality and complex spectral information inherent in hyperspectral non-image remote sensing data. To address this problem, we designed an enhanced CNN model, CBM-CNN. The CBM-CNN incorporates additional convolutional layers (C), batch normalization (B), and MLP(M) to convert one-dimensional data into two-dimensional data for classification, thus improving the processing efficiency of highly nonlinear data features. Building on CBM-CNN, we add multiple attention mechanisms (CBMM-CNN) to enhance feature extraction capabilities. The Siamese CBMM-CNN (SCBMM-CNN) is employed for in-depth learning of data features, with supervised learning precisely adjusting the model to improve classification performance. Experimental results on hyperspectral remote sensing datasets demonstrate that SCBMM-CNN outperforms traditional machine learning models (PCA-RF, GBDT, GMO-SVM, Random Forest, Decision Tree) and deep learning models (CNN, CNN-LSTM) in terms of accuracy, recall rate, F1 score, and kappa coefficient. These results confirm that SCBMM-CNN significantly improves crop classification accuracy and provides a novel method for early crop growth classification.

Keywords: Crop classification · Siamese convolutional neural network · Hyperspectral non-image remote sensing data · Multi-Head Attention

T. Zhu et al. (Eds.): ICA3PP 2024, LNCS 15252, pp. 54–74, 2025.
https://doi.org/10.1007/978-981-96-1528-5_4

1 Introduction

Agriculture, as a cornerstone of human civilization, holds a central position in the global economy and ecosystems [1]. Over the past few decades, the evolution of remote sensing technology has spurred innovation in crop classification methods [2]. As remote sensing technologies advance, the demand for data processing methods has escalated, particularly for extensive high-spectral remote sensing data. Traditional Convolutional Neural Networks (CNN) excel in computer vision and image processing, particularly in feature extraction and image classification [3–6]. However, their application to hyperspectral remote sensing data processing is less common due to their limitations in effectively extracting features from highly complex and high-dimensional data, where interrelated features across multiple dimensions are not as intuitive or localized as spatial correlations [7].

Classifying crops at early stages using remote sensing imagery presents challenges, necessitating the use of hyperspectral remote sensing data. While CNN are predominantly applied in image processing, their potential in processing hyperspectral remote sensing data remains underexplored. Recent research has focused on effective methods to extract features from high-dimensional data [8–11]. For instance, Zhang et al. [12] proposed the GWO-SVM method using hyperspectral data to detect the severity of Verticillium wilt in cotton, enhancing crop classification accuracy by extracting temporal-spectral features. However, this method faces challenges in handling massive time-series data and high data dimensions. To address these challenges, some researchers have turned to dimensionality reduction techniques. For example, Arunanth TS et al. conducted a study on PCA-based dimensional data reduction and segmentation of DICOM images [13]. This approach simplifies datasets by reducing feature dimensions but may entail information loss, potentially neglecting critical secondary variables for accurate classification. Hyperspectral remote sensing data often exhibit nonlinear relationships, and recent studies by Zhang et al. [14] confirmed the effectiveness of the Gradient Boosting Decision Tree (GBDT) method in handling such relationships and complex classification tasks, albeit with a risk of overfitting.

To better handle high-dimensional and nonlinear features in hyperspectral remote sensing data, we propose an enhanced CNN model, SCBMM-CNN. Traditional neural networks often struggle to effectively extract high-order features from such data, leading to reduced sensitivity in classification tasks. The CBM-CNN model addresses this by incorporating additional convolutional layers for deeper feature extraction, optimizing training through batch normalization, and enhancing classification performance via an integrated MLP. This design improves the model's capacity to process high-dimensional, nonlinear features, boosts generalization, and significantly increases classification accuracy.

Given the inherent imbalance and similarity in agricultural data, we integrate a multi-head self-attention mechanism into the CBM-CNN model to assign higher weights to key features and lower weights to less relevant ones, achieving a more balanced data representation. The model structure is further enhanced using two identical CBM-CNN models with shared weights, inspired by the SiameseNet concept, to capture subtle differences between similar crop samples. This setup allows SCBMM-CNN to effectively distinguish crops with closely related characteristics.

Empirical results indicate that SCBMM-CNN outperforms traditional deep learning methods, significantly enhancing crop classification accuracy and detection. This study not only tackles computational challenges in hyperspectral remote sensing data classification but also demonstrates the potential of CNN-based approaches in processing complex agricultural data, promoting advancements in precision agriculture and sustainable agricultural development.

2 Method

2.1 CNN

The traditional CNN architecture [15] comprises convolutional layers [16], flattening [17], ReLU [18] activation functions, fully connected layers [19], and the Softmax function [20]. Convolutional layers are primarily responsible for feature extraction. Following this, flattening transforms the convolutional layer output into a one-dimensional vector, preparing it for the Multilayer Perceptron (MLP) [21]. Within the MLP, the ReLU (Rectified Linear Unit) activation function processes the features and aids in classification. Subsequently, fully connected layers integrate and categorize these features, while the Softmax function, together with the cross-entropy loss function, converts the MLP output into a probability distribution for classification decisions. Figure 1 illustrates the CNN model process.

The CNN model employs the cross-entropy loss function for crop type classification, which is effective for multi-category classification problems as it quantifies the discrepancy between the model's predicted probability distribution and actual labels. The output from the final fully connected layer is processed by the Softmax function, converting it into predicted probabilities for each category. In the context of crop classification, this means the network predicts the probability of given input data belonging to a specific crop category. The Softmax function is detailed as follows:

$$\bar{y}_i = \frac{exp(a_i)}{\sum_{j=1}^{C} exp(a_j)}, \tag{1}$$

where a_i is the ith element of the fully connected layer's output, and C is the total number of classes.

The probability distributions output by the Softmax function are used to compute the cross-entropy loss, calculated as follows:

$$L(y, \bar{y}) = -\sum_{i=1}^{C} y_i \log(\bar{y}_i), \tag{2}$$

where C is the total number of classes, y is the one-hot encoded representation of the actual labels, $\bar{y}$ is the model's predicted probability distribution outputted by the softmax function, and log is the natural logarithm. Traditional CNN models, although effective in image processing and feature extraction within computer vision, are notably limited in their ability to handle the high dimensionality and complex spectral information inherent in hyperspectral non-image remote sensing data [23]. This limitation arises because standard CNN architectures are primarily designed for processing two-dimensional image data, which constrains their performance in capturing and distinguishing the subtle and intricate spectral features required for accurate hyperspectral data analysis. Figure 2 depicts the overall process of the CNN model handling crop classification.

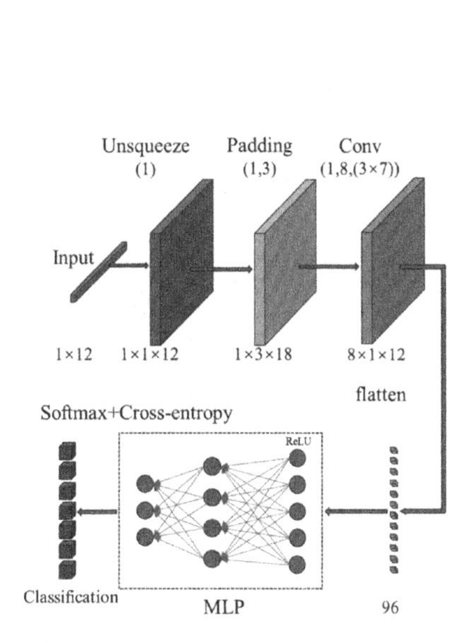

Fig. 1. CNN Model Flowchart.

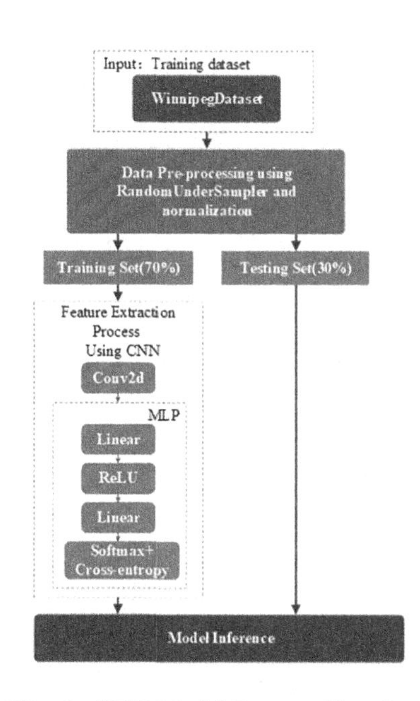

Fig. 2. CNN Model Process Flowchart for Crop Classification.

To address these issues, we propose optimizing CNN models with convolutional layers, batch normalization layers [24], the MLP, multi-head attention, and Siamese CNN. These optimizations aim to enhance crop classification using hyperspectral remote sensing data.

3 Crop Classification Method Based on SCBMM-CNN Architecture

3.1 CBM-CNN

The CBM-CNN model enhances the basic CNN architecture by incorporating additional convolutional layers. Each layer includes padding structures [25], and the convolution kernels are modified from the standard 3×3 to 3×7, allowing for more effective processing of high-dimensional data. Batch normalization layers are added after each convolutional layer to standardize the data, enhancing stability and learning efficiency. Additionally, optimizations and substitutions have been made to the activation functions within the MLP to improve the model's ability to process hyperspectral remote sensing data effectively. The overall framework of the model is illustrated in Fig. 3:

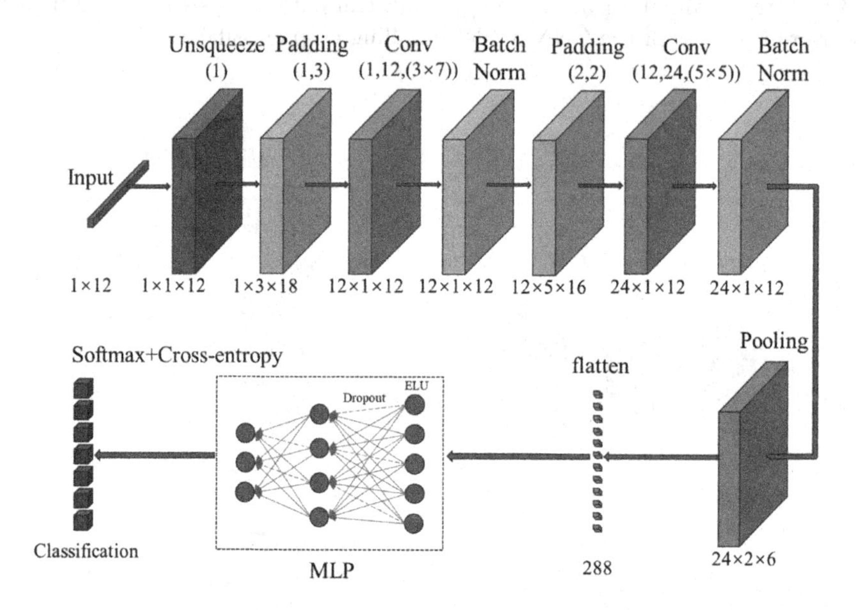

Fig. 3. CBM-CNN Framework.

The CBM-CNN model addresses a common limitation in standard CNNs, where convolution operations without padding gradually reduce the dimensions of feature maps, leading to the loss of critical edge information. To mitigate this issue, the CBM-CNN model incorporates additional convolutional layers, adopted an asymmetric padding strategy that significantly enhances the flexibility and adaptability of CNNs in processing high-dimensional data. By implementing differentiated padding in various directions, the model can more precisely capture and preserve directional features. This approach is particularly effective in refining the model's feature extraction capabilities, especially when handling complex data such as remote sensing images, thereby improving overall performance. The detailed padding structure is as follows:

$$O = \frac{W - K + 2P}{S} + 1, \tag{3}$$

where O is the output size. W is the input size. K is the size of the convolution kernel (kernel). P is the number of pixels for padding. S is the stride (the number of pixels the kernel moves)

In the CBM-CNN model, batch normalization layers are added after each convolutional layer to further enhance the normalization of features, effectively preventing gradient-related issues [26] and improving the model's robustness to minor variations in input data [27]. These layers address the challenges posed by hyperspectral data and boost the overall performance of the model, providing a deeper feature foundation for subsequent classification tasks. The batch normalization process is detailed as follows:

$$\begin{aligned} \bar{x}^{(k)} &= \frac{x^{(k)} - \mu_B}{\sqrt{\sigma_B^2 + \epsilon}}, \\ y^{(k)} &= \gamma^{(k)} \bar{x}^{(k)} + \beta^{(k)}, \end{aligned} \tag{4}$$

where $x^{(k)}$ represents the kth feature,$\mu_{\mathcal{B}}$ and $\sigma_{\mathcal{B}}^2$ are the mean and variance of the features within batch $\mathcal{B}$, ϵ is a small constant added for numerical stability, and $\gamma^{(k)}$ and $\beta^{(k)}$are learnable normalization parameters.

An adaptive average pooling layer [28] is integrated after the final convolutional and batch normalization layers in the CBM-CNN model. This layer adjusts the multi-dimensional features to a format suitable for classification, effectively modifying the size of the feature maps outputted by the convolutional layers to meet the input requirements of the subsequent fully connected layer. This layer provides a valuable summary of information, facilitating efficient classification decisions and ensuring the accuracy and efficiency of the final classification task.

3.2 CBMM-CNN

Building upon the CBM-CNN model, the CBMM-CNN model incorporates a novel multi-head self-attention [29] mechanism to further enhance classification accuracy. Initially, the CBM-CNN model transforms the acquired and processed data features into vector form. The dimensionality is then increased to convert the original vector into a two-dimensional format, thereby expanding the feature set and enabling the model to learn a broader range of crop characteristics. Subsequently, a multi-head self-attention mechanism is integrated after the convolution operation in the CBM-CNN, resulting in the optimized CBMM-CNN model.

The proposed multi-head self-attention mechanism differs from traditional attention methods in several key aspects. Traditional attention mechanisms typically assign a uniform attention weight across all features or rely on a single attention head to capture relationships between features. In contrast, our approach involves multiple attention heads that operate in parallel, each focusing on different aspects of the feature space. This allows the model to capture and emphasize diverse and potentially subtle patterns within the hyperspectral remote sensing data. Moreover, the multi-head structure in our mechanism not only assigns varying importance to features by calculating their encoded weights but also leverages the distinct perspectives offered by each head to aggregate a more comprehensive understanding of the data. The detailed steps of the multi-head self-attention mechanism model are as follows, and its overall process is shown in Fig. 4.

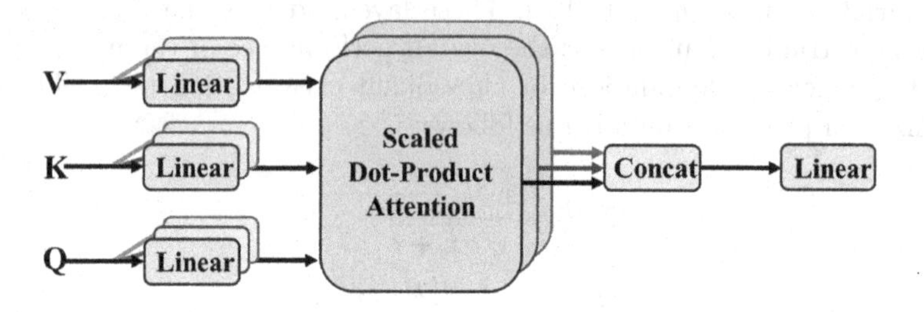

Fig. 4. Multi-head Self-attention Mechanism Model

After the data passes through the convolutional layer, it is transferred to a multi-head self-attention mechanism model. This operation assigns higher weights to important features and lower weights to other features. The multi-head self-attention mechanism model is represented as $S = (a_1, a_2, \ldots, a_i)$, where S is multiplied by the weight data of the attention mechanism to obtain Q, K, and V, specifically $q_i = SW_j^q, k_i = SW_j^k, and v_i = SW_j^v$ with j ranging from 1 to 3. The weight matrices W^q, W^k, and W^v are continually trained through learning, further improving the model's fitting capability. By multiplying Q and K^T, a similarity matrix for different features is obtained, representing the similarity relationships between various characteristics. Subsequently, the similarity is normalized using the softmax function, which reduces the computational load to a certain extent. Finally, as shown in Eqs. (5) and (6), multiplying the results by V produces output data of the same dimension as the input. In Eq. (7), d_K represents the dimension of the K matrix. Ultimately, the final result can be obtained as shown in Eq. (7).

$$\text{head}_j(Q, K, V) = \text{Softmax}\left(\frac{QK^T}{\sqrt{d_K}}\right) V, \tag{5}$$

$$\text{Softmax}(z_i) = \frac{e^{z_i}}{\sum_j e^{z_j}}, \tag{6}$$

$$T = \text{Concat}(\text{head}_1, \text{head}_2, \text{head}_3), \tag{7}$$

where Concat concatenates the attention outputs of different heads, and T is the multi-head attention output result.

3.3 SCBMM-CNN

In the process of crop classification based on hyperspectral remote sensing data, it is difficult to distinguish between samples of different crop categories because of high similarity. In order to capture subtle differences between samples, the idea of

SiameseNet [30] was adopted and two twin CBMM-CNN were used, sharing the same weight, and the Siamese CBMM-CNN model was named SCBMM-CNN, as shown in Fig. 5.

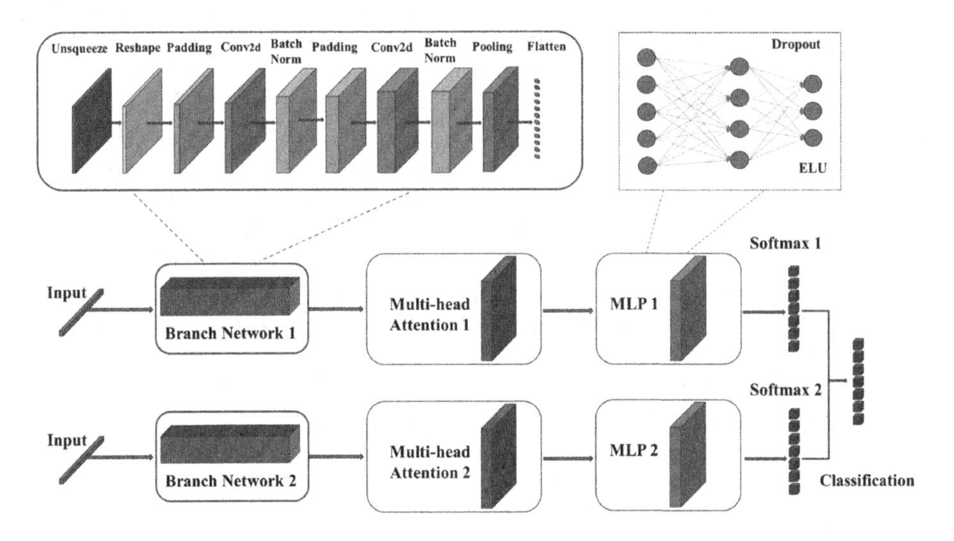

Fig. 5. Framework of SCBMM-CNN

The deep Siamese CNN architecture is a variant of the neural network originally designed to solve the signature verification problem in image matching. It has been used for single-time learning of image classification, facial verification, and dimensionality reduction in the case of unknown categories. The SNN consists of two identical symmetric CNN subnetworks that share the same weight. The similarity between the data is calculated by Euclidean distance, as shown in Eq. (8), while the contrast loss is used to define the following loss function, as shown in Eq. (9):

$$D = \|f(I_1) - f(I_2)\|_2 , \tag{8}$$

$$C(W, I_1, I_2) = (1 - L)\frac{1}{2}D^2 + L\frac{1}{2}\left(\max(0, \text{margin} - D)\right)^2 , \tag{9}$$

where D represents the Euclidean distance. $f(I_1)$ and $f(I_2)$ are the latent representation vectors of I_1 and I_2, respectively. I_1 and I_2 are a pair of data inputs to two identical CNN. L is an indicator function that denotes whether the two input data have the same label. $L = 0$ indicates the same label, while $L = 1$ indicates the opposite. margin is the set safety distance. When the distance between I_1 and I_2 is less than margin, C will become 0. This ensures that I_1 and I_2 are similar rather than identical, enhancing the algorithm's generalization ability.

3.4 Crop Classification Based on SCBMM-CNN Architecture

For hyperspectral remote sensing data in crop classification, converting one-dimensional crop data into two-dimensional format enriches feature representation, enabling the model to capture more detailed crop information. The data, after passing through the convolutional layers, is fed into a multi-head self-attention mechanism, which assigns higher weights to critical features and lower weights to less important ones, enhancing feature extraction. Subsequently, the data is processed by the MLP, which benefits from key optimizations in the SCBMM-CNN model, significantly boosting classification accuracy and efficiency. Given the subtle differences between crop features, a Siamese network architecture is employed to parallelize processing, further improving classification performance. The overall process of SCBMM-CNN for crop classification based on hyperspectral remote sensing data is illustrated in Fig. 6.

In this framework, processed features are converted into feature vectors via adaptive average pooling and then classified using an enhanced MLP. This MLP incorporates fully connected layers with ELU (Exponential Linear Unit) activation and Dropout [31] regularization to mitigate overfitting, as described in Eq. (2). After feature extraction and normalization via convolutional and batch normalization layers, the integrated features are classified through the fully connected layers.

Fully connected layers map locally extracted features from convolutional layers into a higher-dimensional space, facilitating decision-making for the final classification or regression tasks. After processing through the fully connected layer, the ELU activation function captures more complex patterns and relationships within the input data. This function provides a small, non-zero output for negative input values, aiding in the mitigation of the vanishing gradient problem. The ELU activation function is detailed as follows:

$$\mathrm{ELU}: f(x) = \begin{cases} x & \text{if } x > 0 \\ \alpha(\exp(x) - 1) & \text{if } x \leq 0 \end{cases}. \tag{10}$$

After the activation function, Dropout regularization is introduced to enhance the network's generalization ability and prevent overfitting. Dropout works by randomly omitting a subset of neurons during the training process, helping prevent the model from overfitting to the training data. The application of Dropout regularization is mathematically represented as follows:

$$a_{\text{dropout}}^{(l+1)} = a^{(l+1)} \odot r, \tag{11}$$

where $a_{\text{dropout}}^{(l+1)}$ represents the activation output after applying Dropout, $a^{(l+1)}$ is the output after the activation function, and r is a random vector, where each element is independently sampled from a Bernoulli distribution. This distribution has a probability p of producing 1 (keeping the neuron) and a probability $1 - p$ of producing 0 (dropping the neuron).

In the SCBMM-CNN architecture, the MLP classification component is optimized for hyperspectral data by adjusting network layers and enhancing feature

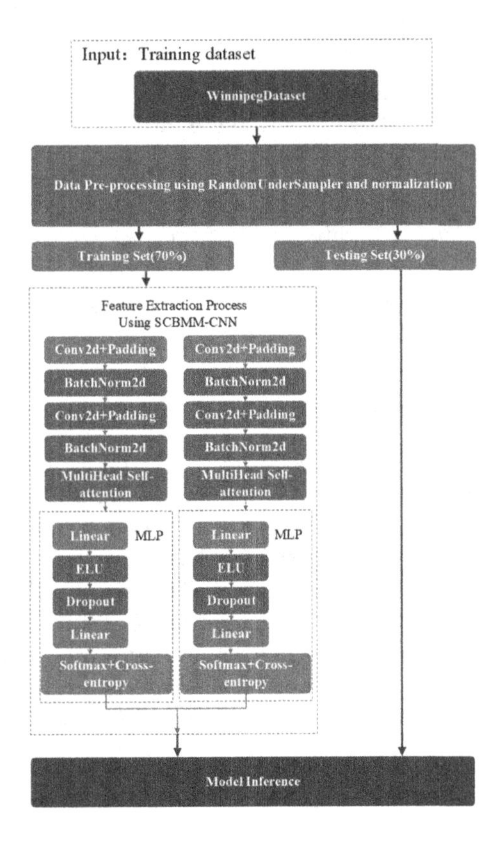

Fig. 6. Flowchart of the SCBMM-CNN Model for Agricultural Crop Classification.

fusion, improving its ability to handle high-dimensional features from the convolutional layers.

The model's effective multi-category classification is driven by robust feature extraction and the use of the cross-entropy loss function combined with softmax activation. This approach minimizes the difference between predicted and actual labels, ensuring accurate and reliable classification of hyperspectral data.

4 Experimental Validation

4.1 Experimental Environment

The proposed model is implemented using the Python 3.6.5 environment. Key parameters include a learning rate of 0.002, a dropout ratio of 0.05, a batch size of 64, and an epoch count of 50. During each training round, the dataset is divided into multiple batches. For each batch, the model executes the following steps:

1. Perform forward propagation to obtain the prediction results [32].

2. Calculate the loss using the cross-entropy loss function.
3. Execute backward propagation to compute gradients.
4. Update the model weights using the Adam optimizer [33].
5. Employ ELU as the activation function.

The selection of these parameters aims to optimize model performance, achieving an efficient and accurate learning process. By carefully tuning these parameters, the study ensures that the model effectively learns from the data, reducing overfitting while maintaining the ability to generalize to new, unseen data.

4.2 Experimental Dataset

To evaluate the performance of the SCBMM-CNN model, the Winnipeg dataset was selected. This dataset contains 325,834 samples across seven categories, with 175-dimensional features divided into subsets representing data characteristics from specific dates, such as July 5, 2012, and July 14, 2012, as follows:
Features f1 to f49 represent data characteristics from July 5, 2012.
Features f50 to f98 represent data characteristics from July 14, 2012.
Features f99 to f136 represent data characteristics from July 5, 2012.
Features f137 to f174 represent optical characteristics from July 14, 2012.
The specific data is shown in Table 1:

Table 1. Display of data from the "WinnipegDataset"

Crops	Corn	Pea	Rapeseed	Soybean	Oats	Wheat	Broadleaf
Num	39162	3598	75673	74067	47117	85074	1143

This dataset provides a comprehensive set of features for testing the SCBMM-CNN model's performance in multi-class classification.

The main challenge of this dataset is the imbalance in the number of samples between categories. For example, categories like corn, rapeseed, wheat, and soybean have significantly more samples compared to categories like pea, oats, and broadleaf plants. To address this, we employed the RandomUnderSampler method to balance the dataset by randomly reducing the number of samples in the majority classes. This approach mitigated the model's bias towards the majority classes and improved its ability to recognize minority classes, ensuring stronger generalization across all categories.

Additionally, we utilized the train-test-split method to divide the dataset into training and testing sets, ensuring the model could generalize to unseen data. The data was also subjected to min-max normalization to reduce the impact of numerical variations on the model's performance during training. The specific process is outlined as follows:

$$X_{norm} = (X - X_{min})/(X_{max} - X_{min}), \quad (12)$$

where X_{max} and X_{min} represent the maximum and minimum values of each feature in the dataset, respectively.

These experimental steps and data preprocessing methods ensure that the SCBMM-CNN model achieves optimal classification performance when dealing with complex and imbalanced hyperspectral datasets.

4.3 Performance Evaluation Indicators for Crop Classification

Evaluating model performance is crucial to ensure its reliability and effectiveness. Based on the confusion matrix analysis, we employed a series of classification metrics to comprehensively evaluate model performance. These metrics include accuracy, precision, recall, specificity, F1 score, and Cohen's Kappa coefficient. These comprehensive metrics thoroughly evaluate our model's theoretical and practical performance in crop classification tasks.

4.4 Experimental Results and Analysis

Results and Analysis for Single Crop Experiments. Figure 7 illustrates the confusion matrix of the SCBMM-CNN model on the Winnipeg dataset, which contains 30% test data. The confusion matrix provides a clear view of the model's classification results for different crop categories, highlighting the high accuracy of the SCBMM-CNN model in various categories, such as corn, pea, rapeseed, soybean, and more.

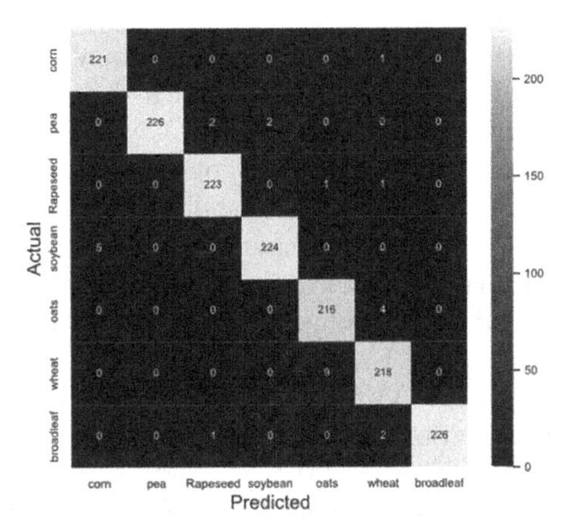

Fig. 7. Confusion Matrix of SCBMM-CNN Technology on the WinnipegDataset

Table 2 compares the performance of the SCBMM-CNN model with other common models on the Winnipeg dataset. This comparative analysis vividly demonstrates the superiority of the SCBMM-CNN model in terms of precision. It outperforms other methods in various crop classification tasks, such as CNN, SVM, decision trees, random forests, and Softmax models.

Table 2. Comparative Analysis of Precision (%) of SCBMM-CNN and Other Methods on the Winnipeg Dataset

Methods	Crops						
	Corn	Pea	Rapeseed	Soybean	Oats	Wheat	Broadleaf
CNN	**98.35**	99.59	99.59	93.85	90.98	95.47	99.18
SVM	95.89	96.95	97.39	99.09	91.66	91.47	95.67
Decision-tree	96.03	99.11	96.92	96.05	92.98	92.14	96.92
Random-forest	94.65	97.53	97.94	94.67	81.14	90.12	99.54
softmax	96.91	**100.0**	99.12	94.73	96.05	85.08	96.92
CNN-LSTM	80.66	99.18	**100.0**	96.72	88.52	95.88	99.18
GWO-SVM	96.02	97.79	96.02	91.15	82.30	79.20	98.23
GBDT	95.15	**100.0**	95.61	94.74	91.67	98.68	98.68
PCA-RF	94.96	98.90	96.03	93.71	83.38	84.16	98.80
Logit-softmax	96.91	99.11	98.68	96.92	91.66	92.98	96.92
SCBMM-CNN	97.79	**100.0**	**100.0**	**98.25**	**98.25**	**96.49**	**99.56**

From the experimental results, traditional classification models exhibit relative disadvantages in the classification accuracy of soybeans, oats, and wheat. Various models, including CNN-LSTM, GWO-SVM, GBDT, PCA-RF, and Logit-softmax, demonstrate a certain level of accuracy in the classification of seven crop types. However, they often do not achieve satisfactory accuracy in one or two crop types. In contrast, the SCBMM-CNN model shows the optimal performance in crop classification tasks, except for being slightly inferior in the classification accuracy of corn and soybeans. This result underscores the reliability of the SCBMM-CNN model in tackling complex classification problems.

Tables 3, 4 and 5 provide a detailed comparison of the performance of the SCBMM-CNN model and other common methods on the WinnipegDataset, including recall, specificity, and F1 scores.

In terms of recall, the SCBMM-CNN demonstrated efficient performance across various crop classifications. Notably, for categories such as corn, peas, and broadleaf, the recall rates for SCBMM-CNN were higher than those achieved by other methods, including traditional CNN, SVM, and decision trees. This result highlights SCBMM-CNN's capability in correctly identifying positive class samples.

Specificity analysis further underscores SCBMM-CNN's superiority in correctly identifying negative class samples. In categories like canola, soybeans, and wheat, SCBMM-CNN's specificity scores were either higher than or close to those

Table 3. Comparative Analysis of SCBMM-CNN vs Other Methods on Winnipeg Dataset for Recall (%)

Methods	Crops						
	Corn	Pea	Rapeseed	Soybean	Oats	Wheat	Broadleaf
CNN	98.14	99.37	96.42	96.85	98.07	94.57	98.78
SVM	92.92	98.67	99.11	96.90	92.47	90.26	97.78
Decision-tree	94.34	99.55	98.23	94.42	91.01	92.44	99.09
Random-forest	93.87	99.16	92.60	95.06	90.41	86.22	98.38
Softmax	95.65	99.12	99.12	97.35	97.07	89.20	98.66
CNN-LSTM	**100.0**	94.14	95.29	93.28	96.42	92.46	90.29
GWO-SVM	96.02	94.85	90.80	96.26	82.30	81.00	99.55
GBDT	98.18	97.42	93.97	94.32	95.87	91.06	99.12
PCA-RF	96.51	82.23	95.44	96.04	77.02	88.06	49.10
Logit-softmax	95.65	97.83	97.83	95.67	93.72	93.39	99.10
SCBMM-CNN	99.10	**99.56**	**98.70**	**97.81**	**96.55**	**99.09**	**99.56**

obtained with other methods, indicating SCBMM-CNN's efficiency in reducing false positives.

Comparative analysis of F1 scores provides a more comprehensive view of the SCBMM-CNN model's balance and consistency. Across various categories, including corn, soybeans, and wheat, SCBMM-CNN's F1 scores surpassed those of other traditional models like CNN, SVM, and random forests. These results indicate that SCBMM-CNN achieves a good balance between precision and recall, ensuring the overall quality of classification results.

Overall, SCBMM-CNN exhibited outstanding performance on the Winnipeg-Dataset, particularly in terms of F1 score, consistently surpassing other traditional classification methods. Figure 8 further validates the accuracy of the SCBMM-CNN model on the WinnipegDataset. The figure illustrates that the SCBMM-CNN model not only performed exceptionally well during the training phase but also maintained high accuracy during validation, demonstrating the model's strong generalization capability.

Figure 9 illustrates the loss analysis of the SCBMM-CNN model during the training and validation phases. The results indicate that throughout the training process, the loss values of the SCBMM-CNN model continuously decreased, reaching a minimum value. This decline in loss reflects the stability and effectiveness of the model. Such a trend demonstrates that the SCBMM-CNN model successfully learned from the training data, optimizing its parameters effectively to minimize the discrepancy between the predicted outcomes and the actual labels, thereby indicating strong convergence and robustness of the model during the learning process.

Table 4. Comparative Analysis of SCBMM-CNN vs Other Methods on Winnipeg Dataset for Specificity (%)

Methods	Crops						
	Corn	Pea	Rapeseed	Soybean	Oats	Wheat	Broadleaf
CNN	99.69	99.90	99.38	99.49	**99.69**	99.08	99.79
SVM	99.34	99.48	99.56	**99.85**	98.60	98.60	99.26
Decision-tree	99.05	99.85	99.78	98.90	98.61	99.05	99.85
Random-forest	98.97	99.86	98.69	99.17	98.56	97.60	99.72
Softmax	98.39	99.85	**99.85**	99.56	99.12	98.76	99.85
CNN-LSTM	99.83	98.97	99.18	98.84	99.45	98.70	98.22
GWO-SVM	99.34	99.12	98.38	99.41	97.05	96.90	99.91
GBDT	99.71	99.56	98.98	99.05	99.34	98.46	99.85
PCA-RF	99.53	99.76	98.61	98.86	95.82	95.97	99.65
Logit-softmax	99.27	99.63	99.63	99.27	98.98	98.90	99.85
SCBMM-CNN	**99.85**	**99.92**	99.78	99.63	99.41	**99.85**	**99.92**

Table 5. Comparative Analysis of SCBMM-CNN vs Other Methods on Winnipeg Dataset for F1-Score (%)

Methods	Crops						
	Corn	Pea	Rapeseed	Soybean	Oats	Wheat	Broadleaf
CNN	98.15	99.07	98.18	95.65	96.23	95.73	99.08
SVM	94.38	97.81	98.25	97.99	92.07	90.87	96.72
Decision-tree	94.97	99.12	97.80	95.44	92.21	91.83	97.55
Random-forest	94.26	98.34	95.20	94.86	85.52	88.12	98.98
Softmax	96.28	99.34	**99.34**	97.14	91.92	93.31	98.01
CNN-LSTM	89.29	96.59	97.59	94.97	92.31	94.14	94.53
GWO-SVM	96.02	96.30	93.33	93.64	82.30	80.09	98.89
GBDT	96.64	98.70	94.78	94.53	93.72	92.44	98.90
PCA-RF	95.73	89.80	95.73	94.86	80.07	86.07	65.60
Logit-softmax	96.28	98.47	98.25	96.30	92.68	93.19	98.00
SCBMM-CNN	**98.44**	**99.78**	**99.34**	**98.03**	**97.39**	**97.77**	**99.56**

Model Experimental Results and Analysis. To comprehensively evaluate the performance of the SCBMM-CNN model in crop classification tasks, a comparative analysis of the model's average accuracy and Kappa coefficient on the WinnipegDataset, encompassing seven types of crops, was conducted, with the results summarized in Fig. 10. This table reveals the significant advantages of the SCBMM-CNN model in terms of two key performance metrics, accuracy,

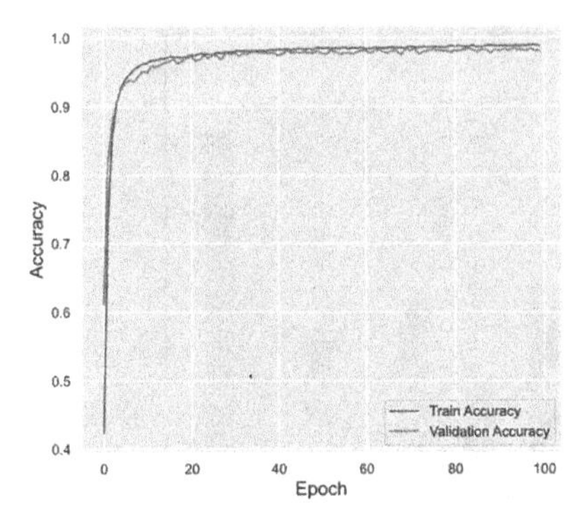

Fig. 8. Loss Analysis of SCBMM-CNN Technology on the WinnipegDataset

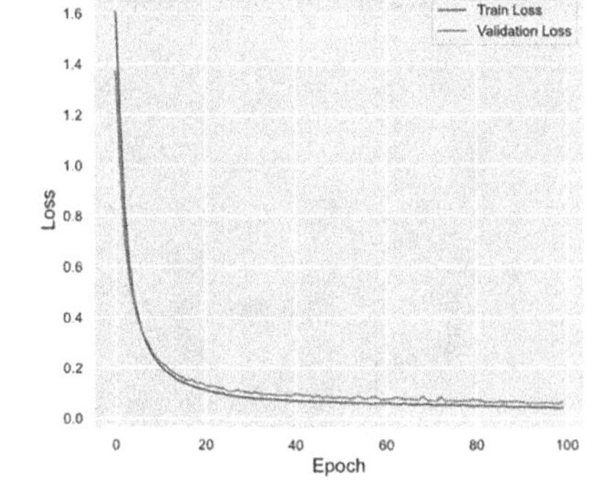

Fig. 9. Accuracy Analysis of SCBMM-CNN Technology on the Winnipeg-Dataset

and Kappa value, compared to conventional classification algorithms (including CNN, SVM, decision trees, and random forests).

The comprehensive data demonstrates that the SCBMM-CNN model surpasses traditional classification techniques not only on individual metrics but also in terms of overall performance, showcasing its exceptional generalization capability and efficiency. These experimental results robustly validate the effectiveness of the SCBMM-CNN model as an efficient and reliable tool for crop classification, underlining its significant implications for the analysis and application of hyperspectral remote sensing data. The model's strong performance highlights its potential to contribute valuable insights in agricultural monitoring, precision farming, and environmental assessment, leveraging the rich information content of hyperspectral data for detailed and accurate classification tasks.

4.5 Ablation Experiment

To further validate the effectiveness of each improvement module, this section presents five variants of the CNN algorithm:

1. CNN-1 algorithm: Introduces extra convolutional layers and an adaptive average pooling layer to maintain feature map dimensions and mitigate overfitting, optimizing the convolutional layers and forward propagation.
2. CNN-2 algorithm: Replaces the activation function with ELU and adds dropout to address gradient disappearance and prevent overfitting, further improving model stability.
3. CNN-3 algorithm: Adds a batch normalization layer after each convolutional layer in the CNN-2 algorithm, accelerating the training process and enhancing model stability and learning efficiency. This normalization reduces internal

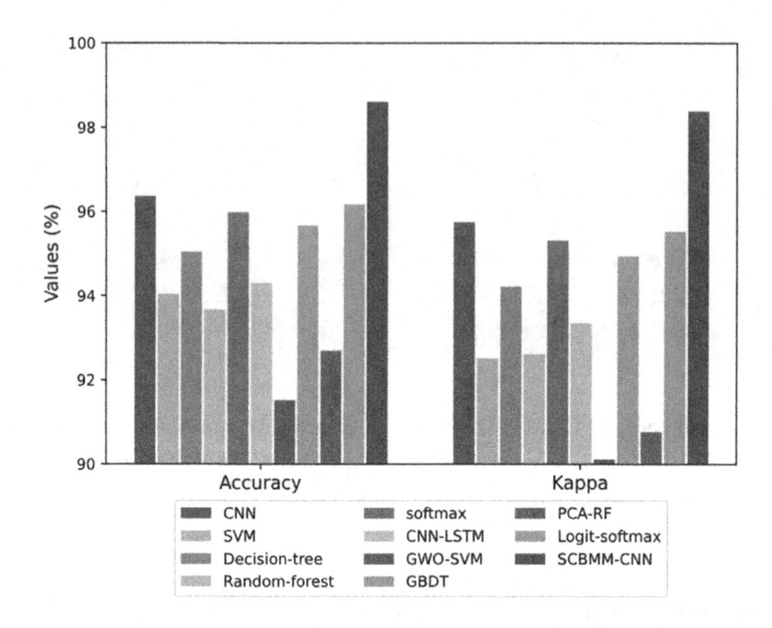

Fig. 10. Comparative analysis of SCBMM-CNN technology on the WinnipegDataset dataset.

covariate shift by standardizing the inputs to the layer, aiding in faster convergence. The model algorithm continues to enhance overall optimization by adding batch normalization layers.

4. CNN-4 algorithm: Based on CNN-3, adds the multi-head attention mechanism to allocate more weight to more important features and improve the detection accuracy of the model.
5. CNN-5 algorithm: Parallelizes two CNN-4 models for crop classification. The multi-head attention enhances feature extraction ability, and the Siamese network enhances model learning ability, making the model more effective for hyperspectral remote sensing crop classification.

Table 6. The Accuracy and Kappa (%) Values for Five Variants of the CNN Algorithm

Methods	Accuracy	Kappa
CNN	96.61	96.04
CNN-1	96.65	96.09
CNN-2	97.27	96.81
CNN-3	97.71	97.33
CNN-4	97.86	98.02
CNN-5	**98.61**	**98.39**

Table 6 present the accuracy and Kappa coefficient values for five CNN algorithm variants.

5 Conclusion

This study introduces the SCBMM-CNN model for hyperspectral data processing in crop classification, significantly enhancing feature extraction and processing through integrated convolutional layers, adaptive pooling, batch normalization, dropout techniques, the ELU activation function, Multi-Head Attention, and a Siamese CNN structure. Extensive experiments on the WinnipegDataset demonstrate that SCBMM-CNN consistently outperforms traditional methods, achieving superior classification accuracy and efficiency. Ablation studies further validate the effectiveness of each model component and the optimal network configuration.

In summary, this research presents a novel and effective method for hyperspectral remote sensing data processing, which supports precision agriculture and promotes sustainable practices. The SCBMM-CNN model's robustness and broad applicability offer valuable insights for future research. Potential adaptations could include extending the model to other types of remote sensing data or agricultural challenges. Additionally, future studies could explore integrating this model with advanced technologies, such as deep learning algorithms or IoT devices, to further enhance the precision and timeliness of agricultural interventions, laying a solid foundation for innovations in agricultural science.

Acknowledgments. This work was supported by the Central Guidance on Local Science and Technology Development Fund of Hebei Province (no. 236Z0104G).

Disclosure of Interests. The authors have no competing interests to declare that are relevant to the content of this article.

References

1. Vishnoi, S., Goel, R.K.: Climate smart agriculture for sustainable productivity and healthy landscapes. Environ. Sci. Policy **151**, 103600 (2024). https://doi.org/10.1016/j.envsci.2023.103600
2. Omia, E., Bae, H., Park, E., et al.: Remote sensing in field crop monitoring: a comprehensive review of sensor systems, data analyses and recent advances. Remote Sens. **15**(2), 354 (2023). https://doi.org/10.3390/rs15020354
3. Li, Z., Liu, F., Yang, W., et al.: A survey of convolutional neural networks: analysis, applications, and prospects. IEEE Trans. Neural Netw. Learn. Syst. **33**(12), 6999–7019 (2021). https://doi.org/10.1109/TNNLS.2021.3084827
4. Kamilaris, A., Prenafeta-Boldú, F.X.: A review of the use of convolutional neural networks in agriculture. J. Agric. Sci. **156**(3), 312–322 (2018). https://doi.org/10.1017/S0021859618000436
5. Khan, A., Sohail, A., Zahoora, U., Qureshi, A.S.: A survey of the recent architectures of deep convolutional neural networks. Artif. Intell. Rev. **53**(8), 5455–5516 (2020). https://doi.org/10.1007/s10462-020-09825-6

6. Krichen, M.: Convolutional neural networks: a survey. Computers **12**(8), 151 (2023). https://doi.org/10.3390/computers12080151
7. Song, L., Xia, M., Weng, L., et al.: Axial cross attention meets CNN: Bibranch fusion network for change detection. IEEE J. Sel. Top. Appl. Earth Obs. Remote Sens. **16**, 21–32 (2022). https://doi.org/10.1109/JSTARS.2022.3224081
8. Ahadzadeh, B., Abdar, M., Safara, F., et al.: SFE: a simple, fast and efficient feature selection algorithm for high-dimensional data. IEEE Trans. Evol. Comput. (2023). https://doi.org/10.1109/TEVC.2023.3238420
9. Jia, W., Sun, M., Lian, J., et al.: Feature dimensionality reduction: a review. Complex Intell. Syst. **8**(3), 2663–2693 (2022). https://doi.org/10.1007/s40747-021-00637-x
10. Song, D., Yuan, X., Li, Q., et al.: Intrusion detection model using gene expression programming to optimize parameters of convolutional neural network for energy internet. Appl. Soft Comput. **134**, 109960 (2023). https://doi.org/10.1016/j.asoc.2022.109960
11. Feng, S., Zhao, L., Shi, H., et al.: One-dimensional VGGNet for high-dimensional data. Appl. Soft Comput. **135**, 110035 (2023). https://doi.org/10.1016/j.asoc.2023.110035
12. Zhang, N., Zhang, X., Shang, P., et al.: Detection of cotton verticillium wilt disease severity based on hyperspectrum and GWO-SVM. Remote Sens. **15**(13), 3373 (2023). https://doi.org/10.3390/rs15133373
13. Arulananth, T.S., Balaji, L., Baskar, M., et al.: PCA based dimensional data reduction and segmentation for DICOM images. Neural Process. Lett. **55**(1), 3–17 (2023). https://doi.org/10.1007/s11063-020-10391-9
14. Zhang, T., Huang, Y., Liao, H., et al.: A hybrid electric vehicle load classification and forecasting approach based on GBDT algorithm and temporal convolutional network. Appl. Energy **351**, 121768 (2023). https://doi.org/10.1016/j.apenergy.2023.121768
15. Hershey, S., Chaudhuri, S., Ellis, D.P.W., et al.: CNN architectures for large-scale audio classification. In: 2017 IEEE International Conference on Acoustics, Speech and Signal Processing (ICASSP), pp. 131–135. IEEE (2017). https://doi.org/10.1109/ICASSP.2017.7952132
16. Zhang, Q., Xiao, J., Tian, C., et al.: A robust deformed convolutional neural network (CNN) for image denoising. CAAI Trans. Intell. Technol. **8**(2), 331–342 (2023). https://doi.org/10.1049/cit2.12110
17. Shastri, S.S., Bhadrashetty, A., Kulkarni, S.: Detection and classification of Alzheimer's disease by employing CNN. Int. J. Intell. Syst. Appl. **15**, 14–22 (2023). https://doi.org/10.5815/ijisa.2023.02.02
18. Jenifer, A.L.L., Indumathi, B.K., Mahalakshmi, C.P.: Detection of female anopheles mosquito-infected cells: exploring CNN, ReLU, and sigmoid activation methods. EAI Endorsed Trans. Pervasive Health Technol. **10** (2024). https://doi.org/10.4108/eetpht.10.5269
19. Hirata, D., Takahashi, N.: Ensemble learning in CNN augmented with fully connected subnetworks. IEICE Trans. Inf. Syst. **106**(7), 1258–1261 (2023). https://doi.org/10.1587/transinf.2022EDL8098
20. Mehra, S., Raut, G., Purkayastha, R.D., et al.: An empirical evaluation of enhanced performance softmax function in deep learning. IEEE Access **11**, 34912–34924 (2023). https://doi.org/10.1109/ACCESS.2023.3265327
21. Ghate, V.N., Dudul, S.V.: Optimal MLP neural network classifier for fault detection of three phase induction motor. Expert Syst. Appl. **37**(4), 3468–3481 (2010). https://doi.org/10.1016/j.eswa.2009.10.041

22. Ghosh, J., Gupta, S.: ADAM optimizer and CATEGORICAL CROSSENTROPY loss function-based CNN method for diagnosing colorectal cancer. In: 2023 International Conference on Computational Intelligence and Sustainable Engineering Solutions (CISES), pp. 470–474. IEEE (2023). https://doi.org/10.1109/CISES58720.2023.10183491
23. Yuan, X., et al.: Variable correlation analysis-based convolutional neural network for far topological feature extraction and industrial predictive modeling. IEEE Trans. Instrum. Meas. (2024). https://doi.org/10.1109/TIM.2024.3373085
24. Wang, L., Li, H.: Hmcnet: hybrid efficient remote sensing images change detection network based on cross-axis attention MLP and CNN. IEEE Trans. Geosci. Remote Sens. **60**, 1–14 (2022). https://doi.org/10.1109/TGRS.2022.3215244
25. Tummalapalli, S., Kumar, L., Bhanu Murthy, N.L.: Web service anti-patterns detection using CNN with varying sequence padding size. In: Mobile Application Development: Practice and Experience: 12th Industry Symposium in Conjunction with 18th ICDCIT 2022, pp. 153–165. Springer, Singapore (2023). https://doi.org/10.1007/978-981-19-6893-8_13
26. Elbaz, K., Shaban, W.M., Zhou, A., et al.: Real time image-based air quality forecasts using a 3D-CNN approach with an attention mechanism. Chemosphere **333**, 138867 (2023). https://doi.org/10.1016/j.chemosphere.2023.138867
27. Chen, X., Long, G., Tao, C., et al.: Improving the robustness of summarization systems with dual augmentation. arXiv preprint arXiv:2306.01090 (2023). https://doi.org/10.48550/arXiv.2306.01090
28. Zhao, L., Jiao, J., Yang, L., et al.: A CNN-based layer-adaptive GCPs extraction method for TIR remote sensing images. Remote Sens. **15**(10), 2628 (2023). https://doi.org/10.3390/rs15102628
29. Zhao, L., et al.: SSIR: spatial shuffle multi-head self-attention for single image super-resolution. Pattern Recognit. **148**, 110195 (2024). https://doi.org/10.1016/j.patcog.2023.110195
30. Seong, S., et al.: Crop classification in South Korea for multitemporal PlanetScope imagery using SFC-DenseNet-AM. Int. J. Appl. Earth Obs. Geoinformation **126**, 103619 (2024). https://doi.org/10.1016/j.jag.2023.103619
31. Talebi, K., Torabi, Z., Daneshpour, N.: Ensemble models based on CNN and LSTM for dropout prediction in MOOC. Expert Syst. Appl. **235**, 121187 (2024). https://doi.org/10.1016/j.eswa.2023.121187
32. Robertson, B., Gharabaghi, B., Hall, K.: Prediction of incipient breaking wave-heights using artificial neural networks and empirical relationships. Coast. Eng. J. **57**(04), 1550018 (2015). https://doi.org/10.1142/S0578563415500187
33. Singh, R., Sharma, A., Sharma, N., et al.: Impact of adam, adadelta, SGD on CNN for white blood cell classification. In: 2023 5th International Conference on Smart Systems and Inventive Technology (ICSSIT), pp. 1702–1709. IEEE (2023). https://doi.org/10.1109/ICSSIT55814.2023.10061068
34. Lee, W., Sim, D., Oh, S.J.: A CNN-based high-accuracy registration for remote sensing images. Remote Sens. **13**(8), 1482 (2021). https://doi.org/10.3390/rs13081482
35. Xie, J., Hu, K., Li, G., et al.: CNN-based driving maneuver classification using multi-sliding window fusion. Expert Syst. Appl. **169**, 114442 (2021). https://doi.org/10.1016/j.eswa.2020.114442
36. Zhao, W., Jiao, L., Ma, W., et al.: Superpixel-based multiple local CNN for panchromatic and multispectral image classification. IEEE Trans. Geosci. Remote Sens. **55**(7), 4141–4156 (2017). https://doi.org/10.1109/TGRS.2017.2689018

37. Banerjee, D., Kukreja, V., Hariharan, S., et al.: Hybrid CNN-SVM approach with regularization for accurate classification of images: a case study on rudraksha beads. In: 2023 International Conference on Disruptive Technologies (ICDT), pp. 676–680. IEEE (2023). https://doi.org/10.1109/ICDT57929.2023.10151056
38. Sui, X., Lv, Q., Zhi, L., et al.: A hardware-friendly high-precision CNN pruning method and its FPGA implementation. Sensors **23**(2), 824 (2023). https://doi.org/10.3390/s23020824

Secure and Revocable Multi-authority CP-ABE for Mobile Cloud Computing

Junyang Li[1], Hongyang Yan[1](✉), Arthur Sandor Voundi Koe[2], Weichu Deng[1], and Zhengxi Zhong[1]

[1] Guangdong Key Laboratory of Blockchain Security, Institute of Artificial Intelligence, Guangzhou University, Guangzhou, China
{jyang_li,2112106114,2112206130}@e.gzhu.edu.cn, hyang_yan@gzhu.edu.cn
[2] State Key Laboratory of Integrated Service Networks, Xidian University, Xi'an, China

Abstract. Mobile cloud computing (MCC) enhances mobile device capabilities with cloud-based resources, allowing mobile users to outsource more data to the cloud to enjoy services of interest. It uses ciphertext policy attribute-based encryption (CP-ABE) to ensure the security and privacy of outsourced data and to implement fine-grained access control of data. However, the existing CP-ABE schemes exhibit the key escrow problem on the central authority and fail to grant data users full control over their identity which may expose user privacy. This paper proposes a multi-authority CP-ABE scheme to solve the key escrow problem and lack of user attribute privacy in the current literature. We design a novel SSI-aware communication model and seamlessly integrate it with our multi-authority CP-ABE to protect user identity privacy and strengthen the user's sovereignty over their identity. We show how to conduct attribute revocation in our system. We prove security under the Decisional Bilinear Diffie-Hellman assumption. This research contributes to the existing body of knowledge on secure and privacy-preserving data sharing in the mobile cloud.

Keywords: CP-ABE · self-sovereign identity · mobile cloud computing · attribute revocation · key-escrow free

1 Introduction

Cloud services have become a significant strategic deployment for many organizations and institutions, which has led to an increasing market demand. According to statistics from the well-known Statista platform, the market for public cloud services in 2023 has doubled compared to five years ago. Such increasing tendencies have put mobile cloud computing (MCC) under the spotlight and recent research advocates the study of MCC-based techniques to enhance user experience. MCC can share, process and analyze data on cloud resources through mobile applications, allowing users to monitor their digital resources at

T. Zhu et al. (Eds.): ICA3PP 2024, LNCS 15252, pp. 75–84, 2025.
https://doi.org/10.1007/978-981-96-1528-5_5

any time. Most MCC service providers only provide basic security mechanisms such as firewalls, anti-virus, and logging, while fine-grained access control for data is excluded from the security mechanisms [9].

Cipher Policy Attribute-base Encryption (CP-ABE) [1,10] is a well-known public key cryptography primitive that can achieve fine-grained access control of data. It is suitable for dynamic environments with complex access policies, such as MCC. However, existing CP-ABE schemes have some shortcomings. On the one hand, users need to hand over all attributes to attribute authorities to generate and update attribute keys, which is not flexible enough for attribute private key updates and also infringes on the user's attribute sovereignty. On the other hand, existing CP-ABE schemes have the key-escrow problem. Attribute authorities can know the attribute keys generated for users, resulting in user privacy leaks.

In this paper, we propose an MA-CP-ABE scheme with key-escrow free and user attribute sovereignty, that tailors mobile cloud computing scenarios. The contributions of this paper are as follows: 1. We achieve seamless integration by leveraging MA-CP-ABE and SSI, which realize user attribute sovereignty; 2. We design a novel attribute revocation mechanism in the SSI environment to achieve flexible updates of user attribute private keys; 3. We design a novel key-escrow free MA-CP-ABE scheme to ensure that only data users know their complete attribute keys.

2 Related Work and Preliminary

2.1 Related Work

Existing CP-ABE [1] schemes have three main issues: key-escrow [7,16], lack of user attribute sovereignty [5,14] and lack of attribute revocation mechanism. Regarding key-escrow, existing solutions manage different attribute domains through multiple attribute authorities to prevent the attribute authority from mastering the user's complete attribute key [3,4,6]. Some solutions [3,4] require a trusted authority to integrate the user's private keys from each attribute authority to get a complete user private key. In terms of attribute sovereignty, some schemes [5,14] do not confer users autonomy over their attributes, resulting in malicious authorities tampering with users' attributes, which leads to an infringement of their access rights [12]. User attributes in the MCC environment will be updated at any time, and real-time attribute revocation is required to maintain privacy. Existing solutions [11,13] are insufficient in this regard. It is necessary to design an attribute revocation scheme for CP-ABE in the MCC scenario to provide more reliable protection for data security in the MCC environment.

2.2 Self-sovereign Identity

Self-sovereign identity (SSI), as a new paradigm for decentralized digital identity management, aims to establish trusted access authorization between entities, thereby enabling secure data exchange. SSI is mainly implemented by using blockchain-based decentralized identifiers (DID) [8] and verifiable credentials (VC). In the system using SSI as the access control model (ACM), each entity has a unique DID, and selects VCs that comply with access policies to access resources which are combined into VPs and sent to resource owners for verification. SSI allows users to disclose their identities as needed, avoiding the risk of centralized digital identity leakage and ensuring privacy and security.

2.3 BLS Signature

The BLS signature scheme [2] includes three algorithms: *KeyGen*, *Sign* and *Verify*. This scheme uses a hash function to construct, $H : \{0,1\}^* \rightarrow \mathbb{G}_0$, where H is viewed as a random oracle in the security analysis.

- *KeyGen*. The signer chooses a random number $x \in \mathbb{Z}_p$ and computes $v = g^x$. It marks v and x as its public key and secret key.
- *Sign*. The signer inputs a message $m \in \{0,1\}^*$ and computes $\sigma = H(m)^x$. And σ is the signature of m.
- *Verify*. It inputs public key v, message m, and signature σ. The verifier uses $e(\sigma, g) = e(H(m), v)$ to verify the correctness of σ.

2.4 Decisional Bilinear Diffie-Hellman (DBDH) Assumption

Let $e : \mathbb{G}_0 \times \mathbb{G}_0 \rightarrow \mathbb{G}_T$ be a bilinear map, and $\mathbb{G}_0$ be group of prime order p with generator g. Then radomly chooses $a, b, c \in \mathbb{Z}_p$ and computes g^a, g^b, g^c. The DBDH problem takes (g, g^a, g^b, g^c) as input, and computes $e(g,g)^{abc}$. The probabilistic polynomial time (PPT) adversary $\mathcal{A}$ distinguishes the tuples $(g, g^a, g^b, g^c, e(g,g)^{abc})$ from the tuples $(g, g^a, g^b, g^c, e(g,g)^z)$ whether $z = abc$ or z is a random element. The DBDH assumption is that no PPT adversary $\mathcal{A}$ can resolve the DBDH problem with nonnegligible advantage.

3 Framework Design

3.1 System Model

The framework of the system is shown in Fig. 1. The main roles in this framework are Certificate Authority (CA), Attribute Authority (AA), Data Owner (DO), Data User (DU), Decryption Service Provider (DSP), Blockchain(BC) and Inter-Planetary File System (IPFS), totally seven roles in system framework.

3.2 System Setup

The CA executes the CA.Setup function to initialize the system. Let $e : \mathbb{G}_0 \times \mathbb{G}_0 \rightarrow \mathbb{G}_T$ be a bilinear map, where $\mathbb{G}_0$ and $\mathbb{G}_T$ are groups with same prime order p, and g is the generator of $\mathbb{G}_0$. Let $H : \{0,1\}^* \rightarrow \mathbb{G}_0$ be a hash function. Let the lagrange coefficient be $\Delta_{i,s} = \prod_{l \in S, l \neq i}(x-l)/(i-l)$ where $i \in \mathbb{Z}_p$ and

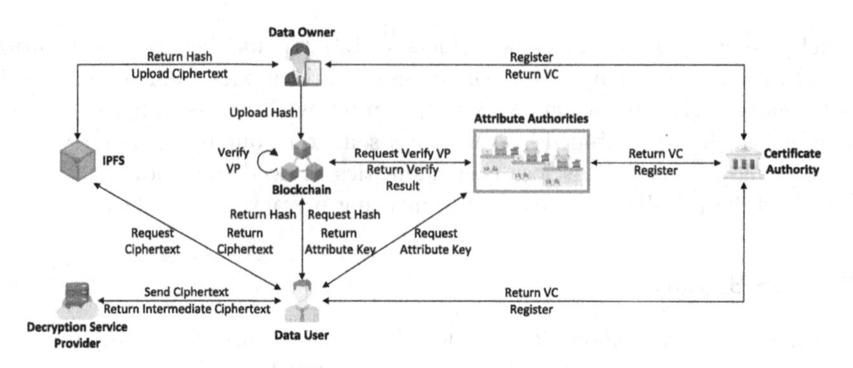

Fig. 1. The system model.

$S = \{s_1, s_2, \ldots, s_m\}$ $(s_i \in \mathbb{Z}_p)$. Let $\mathbb{A} = \{a_1, a_2, \ldots, a_n\}$ be the universe of attribute set, $\mathbb{A}_i$ $(i = \{1, 2, \ldots, k\})$ be the proper subset of $\mathbb{A}$ and the k is the number of AA. The specific form of the function is as follows:

$$\mathsf{CA.Setup}(1^\lambda) \rightarrow params, MSK, SK_{AA} \tag{1}$$

CA inputs the security parameter 1^λ. Then, CA selects randomly $\alpha, \beta, \{\theta_i | i \in \{1, 2, \ldots, k\}\} \in \mathbb{Z}_p$, $h \in \mathbb{G}_0$, and computes g^α, g^β, and $\{g^{\frac{1}{\theta_i}} | i \in \{1, 2, \ldots, k\}\}$. CA generates public parameters $params = \{\mathbb{G}_0, g, h, g^\beta, \{g^{\frac{1}{\theta_i}} | i \in \{1, 2, \ldots, k\}\}, e(g, g)^\alpha, H\}$, master secret key $MSK = \{g^\alpha, \beta\}$ and AAs' secret key $SK_{AA} = \{SK_{AA_i} = \theta_i | i \in \{1, 2, \ldots, k\}\}$. After that, CA broadcasts $params$, stores MSK locally, and sends SK_{AA_i} to the corresponding AA_i.

3.3 Key Generation

DU's $\mathbf{SK_1}$ and $\mathbf{OSK_{AA}}$. First, CA randomly chooses two numbers $r, r^* \in \mathbb{Z}_p$ for each user. Second, CA computes and obtains partial secret key $SK_1 = \{D = g^\alpha g^{\beta r}, \tilde{D} = g^{\beta r} h^{r^*}, \tilde{D}' = g^{r^*}\}$ and the outsourced secret keys $OSK_{AA} = \{OSK_{AA_i} | i \in \{1, 2, \ldots, k\}\}$. For each outsourced secret key $OSK_{AA_i} = g^{\frac{r + r^*}{\theta_i}}$ where $i \in \{1, 2, \ldots, k\}$.

$$\mathsf{CA.KeyGen}(params, MSK) \rightarrow SK_1, OSK_{AA} \tag{2}$$

DU's $\mathbf{SK_{2,i}}$. Before executing function $\mathsf{AA.KeyGen}$, DU needs to fill in the template of AA's verifiable presentation (VP) with the attributes on the verifiable credentials (VC), generate the corresponding VP and send it to AA for verification. After the VP has been verified, attributes within the VP will be extracted to an attribute set S_i, and run the function as follows:

$$\mathsf{AA.KeyGen}(params, S_i, SK_{AA_i}, OSK_{AA_i}) \rightarrow SK_{2,i} \tag{3}$$

In details, $SK_{2,i} = \{\forall j \in S_i \; and \; j \in \mathbb{A}_i, D_j^* = g^{r + r^*} \cdot H(j)^{r_j}, D_j' = g^{\beta r_j}\}$, where, $\mathbb{A}_i$ is the sub-universe of attribute set managed by current AA_i.

DU aggregate $\mathbf{SK_1}$ and $\mathbf{SK_{2,i}}$. DU aggregates SK_1 and $SK_{2,i}$ from CA and AA respectively, then generates the private key SK used to decrypt the ciphertext.

$$\mathsf{DU.Aggregate}(SK_1, \{SK_{2,i} | i \in \{1, 2, \ldots, k\}\}) \rightarrow SK \tag{4}$$

In the above formula, $SK = \{D = g^{\alpha} g^{\beta r}, \tilde{D} = g^{\beta r} h^{r^*}, \tilde{D}' = g^{r^*}, \forall j \in S', D_j = g^r \cdot H(j)^{r_j}, D_j' = g^{\beta r_j}\}$, where $D_j = D_j^* / \tilde{D}' = (g^{r+r^*} \cdot H(j)^{r_j})/g^{r^*} = g^r \cdot H(j)^{r_j}$ and S' is the attribute set which DU has sent to AAs.

3.4 Encryption

DO executes the following function where $CT_1 = \{\tilde{C} = Me(g,g)^{\alpha s}, C_1 = g^s, C_2 = g^{s_2}, C_3 = h^{s_2}, \sigma = H(M)^s\}$ and $CT_2 = \{\forall y \in Y, C_y = g^{\beta q_y(0)}, C_y' = H(att(y))^{q_y(0)}\}$. Specifically, DO selects two random numbers $s, s_1 \in \mathbb{Z}_p$ and computes $s_2 = (s - s_1) \bmod p$. Then DO selects the polynomial q_x for each node x (including the leaf nodes) in $\mathcal{T}$. After that, DO selects the nodes' information of q_x in a top-to-bottom manner randomly. For each node x in $\mathcal{T}$, the degree of the polynomial d_x is set to $cx - 1$, where cx is the threshold value. And let $q_R(0) = s_1$ (R is the root), for each nonroot node x, sets $q_x(0) = q_{parent(x)}(index(x))$ and randomly chooses d_x and other points to completely define q_x. As for $\sigma = H(M)^s$, it is a BLS signature to verify the correctness of the outsourced decryption of ciphertext. After DO completes the encryption of the plaintext M, it uploads the ciphertext $CT = \{CT_1, CT_2\}$ to IPFS, and IPFS will return the DO ciphertext storage address $Hash$. Subsequently, DO initiates a transaction request, invokes smart contacts to store the $Hash$ in the database, and deploys smart contacts to verify whether DU satisfies the access policy.

$$\mathsf{DO.Encrypt}(params, \mathcal{T}, M) \rightarrow CT_1, CT_2 \tag{5}$$

3.5 Decryption

Delegate Key Generation. DU generates the outsourcing key OSK_{DSP} for outsourcing decryption and the delegate secret key $SK_{delegate}$ for decrypting the ciphertext after outsourcing decryption. The keys mentioned above are both generated using the private key SK.

$$\mathsf{DU.Delegate}(SK) \rightarrow OSK_{DSP}, SK_{delegate} \tag{6}$$

DU randomly chooses the number $t \in \mathbb{Z}_p$. Then computes outsourcing key $OSK_{DSP} = \{D' = (g^{\alpha} g^{\beta r})^{\frac{1}{t}}, \tilde{D} = g^{\beta r} h^{r^*}, \tilde{D}' = g^{r^*}, \{\forall j \in S' \ and \ j \in \mathbb{A}, D_j = g^r \cdot H(j)^{r_j}, D_j' = g^{\beta r_j}\}\}$ and delegated secret key $SK_{delegate} = \{t\}$.

Outsourcing Decryption. DSP executes this function to compute intermediate ciphertexts IT_1 and IT_2, where ciphertext $CT_1' = CT_1 \setminus \{\tilde{C}\}$.

$$\mathsf{DSP.Decrypt}(params, CT_1', CT_2, OSK_{DSP}) \rightarrow IT \tag{7}$$

In details, $IT = \{IT_1, IT_2\}$, and IT_1, IT_2 are as follows:

$$IT_1 = \frac{e(\tilde{D},C_2)}{e(\tilde{D}',C_3)} \cdot \frac{\prod_{i\in S'} e(D_i,C_x)^{\Delta_{i,S'}(0)}}{\prod_{i\in S'} e(D_i',C_x')^{\Delta_{i,S'}(0)}} = e(g,g)^{r\beta s} \tag{8}$$

$$IT_2 = e(D',C_1) = e(g,g)^{\frac{\alpha s}{t}} \cdot e(g,g)^{\frac{r\beta s}{t}} \tag{9}$$

Local Decryption. DU executes DU.Decrypt to get the plaintext M.

$$\mathsf{DU.Decrypt}(IT_1, IT_2, \tilde{C}, \sigma, SK_{delegate}) \rightarrow M \tag{10}$$

In details, M is computed as follows:

$$M = \frac{\tilde{C}}{(IT_2^t)/(IT_1)} = \frac{Me(g,g)^{\alpha s}}{(e(g,g)^{\alpha s} \cdot e(g,g)^{r\beta s})/(e(g,g)^{r\beta s})} \tag{11}$$

After the function finishes the computation of plaintext M, it will compute $e(H(M), g^s)$ and $e(\sigma, g)$ to verify if they are consistent. If not, returns error $\perp$ to DU; else returns plaintext M.

3.6 Attribute Revocation

We assume that the old attribute version is r_{ver}, where the value of r_{ver} before the first time of attribute revocation is $1 \in \mathbb{Z}_p$, and the new attribute version is r_{ver}'.

Update Key Generation. When a specific DU's attribute is revoked, the CA first updates the VC of the DU and then transmits the message to the AA that manages the corresponding attribute. AA uses the public parameters $params$ and its own secret key SK_{AA_i} to generate an update key UK_D for the attribute y to update the secret key for the unrevoked DUs, and an update key UK_C for DO to update corresponding ciphertext.

$$\mathsf{AA.UpdateKeyGen}(params, y) \rightarrow (UK_D, UK_C) \tag{12}$$

AA randomly chooses a number $r_{ver}' \in \mathbb{Z}_p$ for the revoked attribute version update, then computes $UK_D = H(j)^{r_i \cdot r_{ver} \cdot (r_{ver}' - 1)}$ and $UK_C = r_{ver}'$. After generating two update keys, AA transmits the message to DOs and unrevoked DUs that the attribute has been updated.

Ciphertext Update. DO generates VP from its own VC, which is based on the template specified by AA, and then sends the VP to AA. If AA confirms that the VP is valid, AA will send the updated attribute y and the update key UK_C of the ciphertext corresponding to the updated attribute y to DO.

$$\mathsf{DO.UpdateCiphertext}(y, CT_2, UK_C) \rightarrow CT_2' \tag{13}$$

The update key UK_C will not update the entire ciphertext CT_2, only update the ciphertext component $C_y'^{*} = (C_y')^{UK_C} = H(j)^{q_y(0) \cdot r_{ver} \cdot r_{ver}'}$ ($C_y'^{*}$ is the updated C_y') of the corresponding attribute y in the ciphertext CT_2.

DU's Key Update. DU generates VP from its own VC and then sends the VP to AA. If AA confirms that the VP is valid, AA will send the updated attribute y and the update key UK_D of the secret key corresponding to the updated attribute y to DU.

$$\mathsf{DU.UpdateKey}(y, SK, UK_D) \rightarrow SK' \tag{14}$$

The update key UK_D will not update the entire secert key $SK_{2,i}$, only update the secert key component $D_j^* = D_j \cdot UK_D = g^r \cdot H(j)^{r_j \cdot r_{ver} \cdot r'_{ver}}$ (D_j^* is the updated D_j) of the corresponding attribute y in the secert key $SK_{2,i}$.

4 Performance Analysis

In this section, we analyze and compare the efficiency of our proposed scheme with some related works [1,5,15] through efficiency analysis and experiments. We use the symbol $|\mathbb{Z}_p|$, $|\mathbb{G}_0|$ and $|\mathbb{G}_T|$ to represent the size of the element in $\mathbb{Z}_p$, $\mathbb{G}_0$ and $\mathbb{G}_T$. k is the number of AA. N_{DU}, N_{VP} and N_T to represent the number of attributes associated with a DU, the number of attributes provided by DU on VPs, and the number of attributes that satisfies the access policy T. C_0, C_T and C_e to represent the cost of an operation in $\mathbb{G}_0$, an operation in $\mathbb{G}_T$, and a bilinear paring. Since DU provides attributes to AA through VP in this scheme, the number of attributes provided to AA is less than or equal to the number owned by DU, that is, $N_{VP} \leq N_{DU}$.

Table 1 shows the feature comparison between our proposed scheme and related schemes [1,5,15]. We find that only our scheme supports key-escrow freeness, user attribute sovereignty, and attribute revocation at the same time.

Table 1. Feature comparisons.

Schemes	Multi-Authority	Key-Escrow Free	User Attribute Sovereignty	Outsourced Decryption	Attribute Revocation
[1] Bethencourt. et, al.	✗	✗	✗	✗	✗
[15] Yan. et, al.	✗	✗	✗	✓	✓
[5] Huang. et, al.	✓	✗	✗	✓	✓
our	✓	✓	✓	✓	✓

As shown in Table 2 and 3, compared with the same multi-authority scheme [5] in public parameters, DU's private key and ciphertext size, we found our scheme is better than the comparison scheme. In the time cost of generating update keys, our scheme is better than scheme [15] but slightly higher than scheme [5], which is the direction we need to improve in the future. In DU's private key generation, encryption and decryption time cost, our scheme is better than the scheme [5]. Although our scheme private key, ciphertext, and DU's private key size are both bigger than [1,15], our scheme encryption time cost is close to the scheme [1], and the total decryption cost (sum of time cost for partial decryption plus full decryption) close to the scheme [15].

Table 2. Storage and update time cost comparisons.

Schemes	Public Parameters Size	DU's Private Key Size	Cipertext Size	Update Keys Generation	Ciphertext Update	DU's Private Key Update
[1]	$3\|\mathbb{G}_0\| + \|\mathbb{G}_T\|$	$(2 + N_{DU})\|\mathbb{G}_0\|$	$(1 + 2N_T)\|\mathbb{G}_0\| + \|\mathbb{G}_T\|$	-	-	-
[15]	$3\|\mathbb{G}_0\| + \|\mathbb{G}_T\|$	$(2 + N_{DU})\|\mathbb{G}_0\|$	$(1 + N_T)\|\mathbb{G}_0\| + \|\mathbb{G}_T\|$	$10C_0$	C_0	C_0
[5]	$(1 + k)\|\mathbb{G}_0\| + k\|\mathbb{G}_T\|$	$2N_{DU} \cdot \|\mathbb{G}_0\|$	$3N_T \cdot \|\mathbb{G}_0\| + (1 + N_T)\|\mathbb{G}_T\|$	$2C_0$	C_0	C_0
our	$(3 + k)\|\mathbb{G}_0\| + \|\mathbb{G}_T\|$	$(3 + 2N_{DU})\|\mathbb{G}_0\|$	$(4 + 2N_T)\|\mathbb{G}_0\| + \|\mathbb{G}_T\|$	$4C_0$	C_0	C_0

Table 3. DU's private key generation, encryption, decryption time cost comparisions.

Schemes	DU's Private Key Generation	Encryption	Partial Decryption	Full Decryption
[1]	$(3N_{DU} + 3)C_0$	$(2N_T + 1)C_0 + 2C_T$	NaN	$(2N_T - 1)C_T + 2N_T \cdot C_e$
[15]	$(4N_{DU} + 8)C_0$	$(5N_T + 3)C_0 + C_T$	$2N_{DU} \cdot C_T + (2N_{DU} + 1)C_e$	C_T
[5]	$8N_{DU} \cdot M_0$	$(6N_T + 2)C_0 + 4N_T \cdot C_T + (N_T + 1)C_e$	$(4N_{DU} - 4)C_T + (3N_{DU} - 3)C_e$	$C_0 + 3C_T$
our	$(5N_{VP} + 5)C_0$	$(2N_T + 4)C_0 + 2C_T$	$(2N_{VP} + 1)C_T + (2N_{VP} + 3)C_e$	$3C_T$

5 Security Analysis

We assume there are three roles in this game: a PPT adversary $\mathcal{A}$ with the non-negligible advantage ϵ in this game; a simulator $\mathcal{S}im$ with the advantage $\epsilon/2$ in distinguish between tuples of DBDH and random tuples; a DBDH challenger $\mathcal{C}$.

Let $e : \mathbb{G}_0 \times \mathbb{G}_0 \to \mathbb{G}_T$ be a bilinear mapping, where $\mathbb{G}_0$ is a group of prime p order and has generator g. Next, the DBDH challenger $\mathcal{C}$ randomly selects elements $\mu \in \{0, 1\}$ and $a, b, c, z \in \mathbb{Z}_p$. Challenger $\mathcal{C}$ sets a tuple (g, A, B, C, Z), where $A = g^a$, $B = g^b$ and $C = g^c$. As for Z, if $\mu = 0$, $\mathcal{C}$ lets $Z = e(g, g)^z$ and sends tuple $(g, g^a, g^b, g^c, e(g, g)^z)$ to $\mathcal{S}im$. If $\mu = 1$, $\mathcal{C}$ lets $Z = e(g, g)^{abc}$ and sends tuple $(g, g^a, g^b, g^c, e(g, g)^{abc})$ to $\mathcal{S}im$. The simulator $\mathcal{S}im$ stands a challenger in the following DBDH game.

1. **Setup**: The simulator $\mathcal{S}im$ randomly chooses elements $d, \alpha', \{\theta_i' | i \in \{1, 2, \ldots, k\}\} \in \mathbb{Z}_p$, and sets $\alpha = \alpha' + ab$ and $\{\theta_i = \theta_i' | i \in \{1, 2, \ldots, k\}\}$. $\mathcal{S}im$ computes and gets public parameters $params = \{\mathbb{G}_0, g, h = g^d, \{g^{1/\theta_i} | i \in \{1, 2, \ldots, k\}\}, e(g, g)^{\alpha}, H\}$ where $g^{1/\theta_i} = g^{1/\theta_i'}$ and $e(g, g)^{\alpha} = e(g, g)^{\alpha'} \cdot e(g, g)^{ab}$. $\mathcal{S}im$ sends $params$ to the adversary $\mathcal{A}$.
2. **Phase 1**: The adversary $\mathcal{A}$ submits the attribute set $S = \{S_1, S_2, \ldots, S_n\}$ to $\mathcal{S}im$ to get the outsourced secret keys $OSK_{AA} = \{OSK_{AA_i} | i \in \{1, 2, \ldots, k\}\}$. After $\mathcal{S}im$ receives attribute set S, $\mathcal{S}im$ randomly selects $r' \in \mathbb{Z}_p$ and set $r + r^* = r' - a$. $\mathcal{S}im$ computes $\forall i \in \{1, 2, \ldots, k\}, OSK_{AA_i} = g^{(r+r^*)/\theta} = g^{(r'-a)/\theta'}$ and sends OSK_{AA} to $\mathcal{A}$.
3. **Challenge**: The adversary $\mathcal{A}$ submits two messages m_0 and m_1 of equal length and the challenge access structure $\mathcal{T}_C$ (where the outsourced secret keys OSK_{AA} fail to satisfy to $\mathcal{T}_C$) to $\mathcal{C}$. $\mathcal{S}im$ randomly generates a bit $x \in \{0, 1\}$ and encrypts m_x with . $\mathcal{S}im$ chooses random numbers $s, s_1 \in \mathbb{Z}_p$ and lets $s = c$. $\mathcal{S}im$ computes the challenge ciphertext CT_C as follows:

$$CT_C = \begin{pmatrix} \tilde{C} = m_x \cdot e(g, g)^{\alpha'} \cdot Z, C_1 = g^c \\ C_2 = g^{s-s_1} = C/g^{s_1}, C_3 = g^{x \cdot s_2} \end{pmatrix} \tag{15}$$

$\mathcal{S}im$ sends CT_C to $\mathcal{A}$.

4. **Phase 2**: Same as Phase 1.
5. **Guess**: The adversary $\mathcal{A}$ outputs a guess $x^{'} \in \{0, 1\}$ of x. If $x^{'} = x$, $\mathcal{S}im$ outputs 0 to express $Z = e(g, g)^{abc}$. If $x^{'} \neq x$, $\mathcal{S}im$ outputs 1 to indicate that Z is a random group element in $\mathbb{G}_T$. Let $Pr[Z = e(g, g)^{abc}|x^{'} = x]$ be the probability that $\mathcal{S}im$ outputs 0 when $x^{'} = x$:

$$Pr[Z = e(g, g)^{abc}|x^{'} = x] = \frac{1}{2} + \epsilon. \tag{16}$$

If $Z = e(g, g)^z$, $\tilde{C} = m_x \cdot e(g, g)^{\alpha^{'}} \cdot e(g, g)^z$, which means CT_C is completely random for $\mathcal{A}$. Let $Pr[Z = e(g, g)^z|x^{'} = x]$ be the probability that $\mathcal{S}im$ outputs 0 when $x^{'} = x$:

$$Pr[Z = e(g, g)^z|x^{'} = x] = \frac{1}{2}. \tag{17}$$

Let $Pr[x^{'} = x] = Pr[x^{'} \neq x] = \frac{1}{2}$, regardless the distribution of $x \in \{0, 1\}$. We get $Pr[Z = e(g, g)^{abc}] = Pr[Z = e(g, g)^{abc}|x^{'} = x] \cdot Pr[x^{'} = x]$ and $Pr[Z = e(g, g)^z] = Pr[Z = e(g, g)^z|x^{'} = x] \cdot Pr[x^{'} = x]$.The advantage of the simulator in this security game is described as follows:

$$Adv_{\mathcal{S}} = Pr[Z = e(g, g)^{abc}] + Pr[Z = e(g, g)^z] - \tfrac{1}{2} = \tfrac{\epsilon}{2} \tag{18}$$

If the adversary $\mathcal{A}$ wins the security game with a non-negligible advantage ϵ, then the simulator $\mathcal{S}im$ wins the DBDH game with the probability $\epsilon/2$ in the DBDH game. Since there is no polynomial-time algorithm that can solve the DBDH assumption, the adversary $\mathcal{A}$ can not win the game with a non-negligible advantage ϵ. Our scheme exhibits indistinguishable security under the CPA model.

6 Conclusion

In this article, we design an MA-CP-ABE scheme with key-escrow free and user attribute sovereignty. In this scheme, we split the issuance of secret keys between the certificate authority and the attribute authority to achieve key-escrow free. We integrate MA-CP-ABE and SSI to enhance user attribute privacy and key generation efficiency. Furthermore, we design an efficient attribute update mechanism to revoke user attributes, enabling our solution to adapt to the attribute update requirements of MA-CP-ABE in dynamic scenarios. The performance analysis indicates that the proposed scheme has the characteristics of high availability, high security, and fine-grained access control.

Acknowledgments. This work was supported by the Open Research Fund of Guangdong Key Laboratory of Blockchain Security, Guangzhou University, the National Key Research and Development Program Young Scientist Scheme (No. 2022YFB3102400), the City School Joint Funding Project of Guangzhou City (No. 2023A03J0117) and Guangzhou University Research Project under Grant RQ2021013.

References

1. Bethencourt, J., Sahai, A., Waters, B.: Ciphertext-policy attribute-based encryption. In: 2007 IEEE Symposium on Security And Privacy (SP 2007), pp. 321–334. IEEE (2007)
2. Boneh, D., Lynn, B., Shacham, H.: Short signatures from the Weil pairing. In: International Conference on the Theory and Application of Cryptology and Information Security, pp. 514–532. Springer (2001)
3. Chase, M.: Multi-authority attribute based encryption. In: Vadhan, S.P. (ed.) TCC 2007. LNCS, vol. 4392, pp. 515–534. Springer, Heidelberg (2007). https://doi.org/10.1007/978-3-540-70936-7_28
4. He, Z., Chen, Y., Luo, Y., Zhang, L., Tang, Y.: Revocable and traceable undeniable attribute-based encryption in cloud-enabled e-health systems. Entropy **26**(1), 45 (2023)
5. Huang, K.: Secure efficient revocable large universe multi-authority attribute-based encryption for cloud-aided iot. IEEE Access **9**, 53576–53588 (2021)
6. Lewko, A., Waters, B.: Decentralizing attribute-based encryption. In: Annual International Conference on the Theory and Applications of Cryptographic Techniques, pp. 568–588. Springer (2011)
7. Liu, Y., Zhang, Y., Ling, J., Liu, Z.: Secure and fine-grained access control on e-healthcare records in mobile cloud computing. Futur. Gener. Comput. Syst. **78**, 1020–1026 (2018)
8. Reed, D., et al.: Decentralized identifiers (DIDS) v1.0. Tech. rep., W3C (2022). https://www.w3.org/TR/2022/REC-did-core-20220719
9. Saeed, I., Baras, S., Hajjdiab, H.: Security and privacy of AWS S3 and azure blob storage services. In: 2019 IEEE 4th International Conference on Computer and Communication Systems (ICCCS), pp. 388–394. IEEE (2019)
10. Sanchol, P., Fugkeaw, S.: An analytical review of data access control schemes in mobile cloud computing. In: International Conference on Computing and Information Technology, pp. 310–321. Springer (2021)
11. Taha, M.B., Talhi, C., Ould-Slimane, H.: Performance evaluation of CP-ABE schemes under constrained devices. Procedia Comput. Sci. **155**, 425–432 (2019)
12. Xiao, M., Ma, Z., Li, T.: Privacy-preserving and scalable data access control based on self-sovereign identity management in large-scale cloud storage. In: Security, Privacy, and Anonymity in Computation, Communication, and Storage: 13th International Conference, SpaCCS 2020, Nanjing, 18–20 December 2020, Proceedings 13, pp. 1–18. Springer (2021)
13. Xie, M., Ruan, Y., Hong, H., Shao, J.: A CP-ABE scheme based on multi-authority in hybrid clouds for mobile devices. Futur. Gener. Comput. Syst. **121**, 114–122 (2021)
14. Yan, X., Tu, S., Alasmary, H., Huang, F.: Multiauthority ciphertext policy-attribute-based encryption (MA-CP-ABE) with revocation and computation outsourcing for resource-constraint devices. Appl. Sci. **13**(20), 11269 (2023)
15. Yan, X., Chen, Y., Zhai, Y., Ba, Y., Li, X., Jia, H.: An encryption and decryption outsourcing CP-ABE scheme supporting efficient ciphertext evolution. In: Proceedings of the 2020 4th International Conference on Cryptography, Security and Privacy, pp. 116–125 (2020)
16. Zhao, Y., Ren, M., Jiang, S., Zhu, G., Xiong, H.: An efficient and revocable storage CP-ABE scheme in the cloud computing. Computing **101**, 1041–1065 (2019)

Software Crowdsourcing Allocation Algorithm Based on Task Priority

Ao Mei[1] and Dunhui Yu[1,2(✉)]

[1] School of Computer Science and Information Engineering, Hubei University, Wuhan 430062, China
yumhy@hubu.edu.cn

[2] Engineering and Technical Research Center of Hubei Province in Educational Informatization, Wuhan 430062, China

Abstract. Aiming at the problem of insufficient consideration of skill rarity and worker skill coverage in the task allocation decision of the current software crowdsourcing platform, this paper proposes a task priority-based software crowdsourcing allocation algorithm. First, the skill rarity is calculated based on the tasks to be assigned and the development skills of workers, and the task priority is determined accordingly. Then, the pre-allocation scheme based on the priority is used as the initial input of the allocation algorithm. Finally, the parameters of the algorithm are set based on the global optimization objectives of maximizing the number of tasks to be assigned and minimizing the number of workers to complete the crowdsourcing task allocation. The experimental results show that compared with allocation algorithms such as candidate task worker group algorithm based on allocation utility and preference matching satisfaction maximization, The algorithms in this paper improves the number of successful task allocations and average worker overhead by 15.23% and 12.76% on average.

Keywords: Software crowdsourcing · Skill rarity · Task priority · Task allocation

1 Introduction

With the widespread application of the Internet, crowdsourcing [1] has rapidly emerged as a work mode that gathers a large number of participants to form a huge and diverse workforce. In this context, software crowdsourcing platforms play a vital role as the technical foundation and intermediary for realizing the crowdsourcing model [2, 3]. Such platforms build an online framework and infrastructure aimed at facilitating effective task posting, participant recruitment, task management, and result delivery.

In the crowdsourcing platform, task allocation [4–6] is a crucial link. It involves how to reasonably assign multiple tasks to workers in a certain period of time, so as to ensure that the task dispatching is maximized, earnings are obtained, and quality and fairness of work are maintained at the same time. From the perspective of platforms, current task allocation researches mainly focuses on selecting worker characteristics, on

T. Zhu et al. (Eds.): ICA3PP 2024, LNCS 15252, pp. 85–94, 2025.
https://doi.org/10.1007/978-981-96-1528-5_6

the one hand, after measuring workers' ability by multiple indicators [7], task allocation is carried out; on the other hand, setting up incentive mechanism for workers [8] to improve their enthusiasm and thus improve the efficiency of platform allocation. [9] proposed a dynamic utility-based software crowdsourcing task allocation algorithm, which dynamically updates workers' capabilities and performs task allocation. [10] proposed a software crowdsourcing task allocation method that supports workers' capability fuzzy measure and role collaboration, and selects groups of workers after measuring their comprehensive capabilities. [11] updates the affinity matrix and skills of workers based on the response of workers and task requesters to accomplish high quality assignments. [12, 13] motivated workers to improve their credibility by setting up an integral mechanism to increase user participation and platform revenue, but lacks analysis on task completion quality [14]. [15] proposed a malicious bidding prevention incentive mechanism for multitask crowdsourcing allocation, which maintains fairness and stability of the environment of the crowdsourcing platform, but does not consider the problem of platform revenue. After obtaining the worker measures, task-worker group selection is conducted under the many-to-many type of task allocation. [16–18] adopted a task allocation scheme based on KM (Kuhn-Munkres) algorithm to transform the allocation into a weighted bipartite graph matching problem seeking for an optimal solution with different optimization objectives as weights. [10, 19] adopted an allocation optimization method based on role collaboration [20, 21], setting constraints between workers to improve the efficiency of allocation. [22] formed high-quality worker teams based on worker reputation using particle swarm optimization algorithm.

In software crowdsourcing, there is a lack of literature that conducts in-depth exploration of the combinational optimization framework that views tasks and worker groups as a whole. Meanwhile, existing methods do not consider the impact of skill rarity degree on the order of task allocation when allocating, resulting in some tasks requiring rare skills cannot pick workers with full coverage skills due to late allocation. In response to the above problems, this paper considers calculating task allocation priority based on skill rarity under the same allocation period and optimizes crowdsourcing allocation based on imperialist competition algorithm [23–25], and proposes a software Crowdsourcing Assignment algorithm based on Task Priority (TPAC). Unlike existing methods that determine task posting priority by calculating the task publisher and task weight [26], TPAC algorithm determines task priority by calculating skill rarity.

2 Overall Framework of the Algorithm

The TPAC algorithm first calculates skill rarity according to the frequency of each development skill appearing in the tasks to be assigned and the bidding workers. For each task, select the value with the highest skill rarity as the task rarity and then determine the task priority. The imperialist competition algorithm is used in the task assignment stage. The TPAC first assigns workers to each task in turn according to the priority order, then continuously optimizes the assignment scheme through assimilation strategy, multi-point mutation and jump out of local optimum strategy and other means to complete the crowdsourcing task assignment. The overall framework of the algorithm is shown in Fig. 1.

This work makes the following contributions:

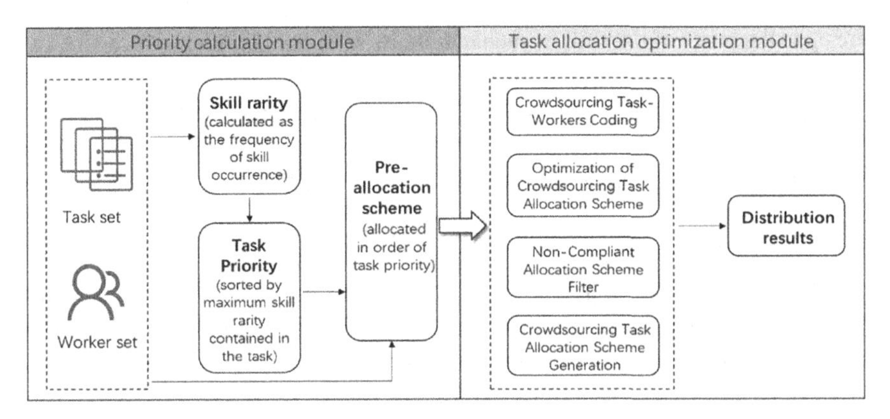

Fig. 1. Algorithm framework

1) We design a task assignment priority method that considers both the task's skill requirements and the bidding workers' skill backgrounds, ensuring accurate reflection of the demand for rare skills and enabling efficient matching.
2) The allocation scheme optimization is driven by setting up proper constraints and assimilation strategies, ensuring that as many tasks as possible are dispatched while minimizing the number of workers.

3 Task Priority Calculation Based on Skill Rarity

3.1 Calculation of Skill Rarity

For software development, skill rarity reflects the supply and demand of required skills. If many tasks require Skill A but only a few workers have it, the platform should prioritize tasks containing Skill A. Otherwise, workers with Skill A might be matched too early, making it hard to find matches for later tasks requiring Skill A, thus reducing successful task assignments and platform revenue.

Based on the above, the development skills of the bidding workers when they register on the platform are extracted to obtain the set of $S = \{Sk_1, Sk_2, \ldots, Sk_x\}$, Sk_x represents the development skills contained in worker x. Similarly, collect the development requirement skills of the tasks to be assigned to get the set $T = \{T_{c1}, T_{c2}, \ldots, T_{ci}\}$, T_{ci} represents the requirement skills contained in task i. Count the frequency s_{xj}, t_{ij} of each skill in the sets S and T, respectively. s_{xj} represents the possession degree of the jth skill of worker x among workers and t_{ij} represents the demand degree of the jth skill of task i in the to-be-assigned task. The rarity δ_{ij} of skill j in task i is defined as follows:

$$\delta_{ij} = t_{ij} \big/ s_{xj} \tag{1}$$

3.2 Calculation of Task Priority

For T_{ci}, the maximum value of skill rarity contained in task i is used as a priority metric. The priority P_{ci} will be calculated by Eq. 2:

$$P_{ci} = \max\{\delta_{ij}\} \tag{2}$$

where δ_{ij} is the rarity of the skill j contained in task i.

4 Task Allocation Based on Imperial Competition Algorithm

4.1 Imperialist Competitive Algorithm

The Imperialist Competitive Algorithm (ICA) is an optimization method inspired by historical imperialist competition, introduced by Atashpaz-Gargari and Lucas in 2007 [23]. In ICA, solutions are represented as empires or colonies, with empires being more optimal. The algorithm iterates through competitive and improvement processes among empires and colonies, using assimilation and mutation to enhance solutions.

4.2 Task Allocation Algorithm Based on ICA

This paper adapts ICA for crowdsourcing task allocation, aiming to maximize task assignments while minimizing worker usage, similar to the Multiple Knapsack Problem [27]. To avoid getting trapped in local optima, the ICA is enhanced with recombination mutation and a local jump-out mechanism, drawing on the idea of literature [28].

Crowdsourcing Task-Workers Coding. The algorithm treats each task and its corresponding worker group as a separate assignment unit and a dictionary data structure is used for encoding, where the task is keyed by its unique number and the corresponding value is a list containing the assigned worker number.

Optimization of Crowdsourcing Task Allocation Scheme. In the optimization process, the algorithm applies an evaluation function to measure the value of each set of task-worker allocation schemes. The evaluation function is described as:

$$F_{ci} = \frac{P_{ci} \cdot I_{ci}}{N_{ci}} \tag{3}$$

where P_{ci} is the task priority, I_{ci} is a boolean variable indicating whether the skills of the worker group completely cover the task requirements. If the skills of the worker group completely cover the task requirements, $I_{ci} = 1$, otherwise $I_{ci} = 0$. N_{ci} is the number of workers assigned to task i.

Non-compliant Allocation Scheme Filter. In the allocation algorithm, it is crucial to ensure that each task must be undertaken by a group of workers with full coverage of its required skills. This is manifested in the variable $I_{ci} = 0$, thus $F_{ci} = 0$.

To fix non-compliant assignment schemes, if the currently assigned worker does not fully cover the task skill, the set of workers is traversed again to find the worker with the highest skill match to the current task.

Crowdsourcing Task Allocation Scheme Generation. In order to avoid the TPAC algorithm from falling into local optimal solutions in the later iterations, a local search strategy, i.e., the reorganization mutation strategy, is designed. That is, one or more new workers are randomly selected from the original worker group and reassigned to the task. Ensuring that the reorganized worker group is assigned differently from the original task ensures the validity of the mutation.

From the experimental results, we know that the global optimum is better when the probability of executing the reorganization variant policy for each task-workers assignment scheme is $P_{reorg} = 0.2$.

Specific Algorithm Implementation. It is first necessary to define the initial allocation unit, i.e., the initial allocation of tasks in order of priority based on skill coverage(see Algorithm 1). Polling allocation is used for unsuccessfully assigned tasks and workers, i.e., workers are assigned to tasks sequentially until all workers have been assigned.

Algorithm 1 Preliminary Allocation

Input: The set of required skills SK_{ci} of T_{ci} , $Work$

Output: Group of assigned workers W_{ci}

```
1   for (each T_ci to be assigned) do:
2       for SK_ci, sort by skill δ in descending order;
3       for (element sk_tx in SK_ci) do:
4           if SK_ci is null: return W_ci;
5           else: select the worker containing sk_tx in the set Work and add it to the set W_ci;
6               SK_ci = SK_ci - SK_xj;
7       Back to step3;
8   return W_ci
```

When the total number of tasks is n, after obtaining the initial allocation units, the top pri_num units with the highest evaluation function values are selected to form the preferred set. Constraints during execution ensure no worker is assigned more than one task simultaneously; if violated, the task allocation is readjusted. The number of iterations M is determined by the task assignment size, and during iterations, reorganization and jumping out strategies are used to prevent premature convergence, leading to the optimal set of task allocations.

The algorithm is described in Algorithm 2.

Algorithm 2 Task Assignment

Input: initial assignment units n , assignment set units pri_num , M

Output: set of assignment schemes W

```
1   Initialize the scheme and priority scheme;
2   For (m from 1 to M) do:
3       for (i from 1 to pri_num) do
4           Searches the solution space by assimilation strategy and optimizes F_cj;
5           Reorganize the variations;
6           Calculate the new F_cj, and if it is larger than the original, then overwrite;
7       End for;
8       JLOA;
9   End for;
10  If schemes are no longer updated: return W;
```

5 Experiment and Result Analysis

The algorithm was implemented in Python 3.7 on a machine with a 2.6 GHz Intel Core i7 processor, 16 GB of RAM, and Windows 10 as the operating system.

5.1 Experimental Data Set

To verify the rationality and effectiveness of the proposed method in this paper, this paper crawls 1500 workers' information and their corresponding historical task information on open source crowdsourcing. After removing duplicates and preprocessing the data, we selected 500 valid tasks. The task information includes release time, task name, and required skills, while worker information includes worker number and development skills. Some of this task information is presented in Table 1.

Table 1. Information on pending assignments

Release time	Task name	Required skills
2024-01-05	WeChat water measurement	['Java', 'PHP', 'Python', 'C++', 'C#', 'HTML5', 'JavaScript']
2023-12-29	Convert methods in cpp file to dll	['C', 'C++', 'C#']
2023-12-27	Automatic photo insertion script	['Python']
2023-12-20	Android webview wrapper	['Java', 'Android']
2023-11-26	Document recognition	['Python', 'Java', 'PHP', 'C#', 'C++', 'C', 'Scala', 'Go']
...	...	...

5.2 Compared Algorithms

In order to verify the effectiveness of TPAC algorithm for task allocation, the allocation utility-based candidate task worker group algorithm [17], the preference matching satisfaction maximization algorithm (MA for short) [29], and NBFA [30], which are also under many-to-many mode of allocation, are selected as comparison algorithms. The Mallowa model is used in MA to generate sequences of preferences for both parties.

5.3 Evaluation Indicators

The number of successful task assignments ($task\_count$) and the average worker overhead of tasks ($task\_\cos t$) are used as evaluation metrics

$task\_\cos t$ is the ratio of the number of workers used ($work\_count$) and the number of tasks successfully assigned:

$$task\_\cos t = \frac{work\_count}{task\_count} \tag{4}$$

5.4 Parameter Determination Experiments

Number of Preferred Allocation Units *pri_num***.** To determine the number of preferred allocation units for initialization of the task allocation phase of the TPAC algorithm, we set $P_{reorg} = 0.2$, the number of tasks is 500, and the number of workers is 1500. The results of the operation of the two evaluation indexes are shown in Fig. 2 for *pri_num* of 25, 50, 75, 100, and 125.

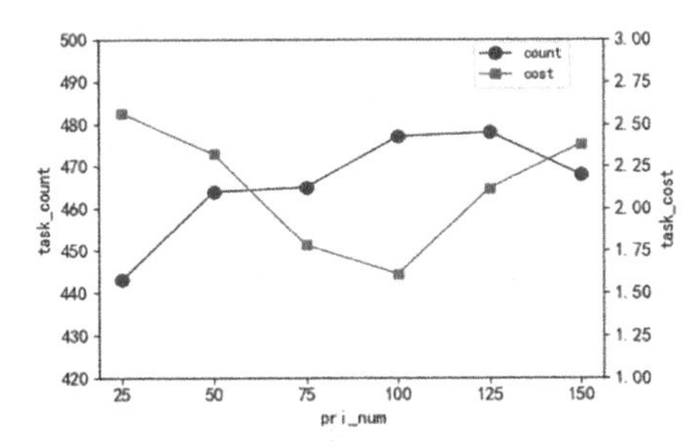

Fig. 2. Changes in evaluation indicators in 5.3 with increasing *pri_num*

From Fig. 2, it can be seen that fewer preferred allocation units limit the algorithm's ability to explore more allocation scenarios and may cause the algorithm to fall into a local optimal solution. As the number of preferred allocation units increases, the algorithm has the opportunity to discover more successfully allocated tasks and lower average worker overhead. However, when the number is set to 125, the situation worsens, possibly due to increased competition, which makes it difficult to match more workers to the remaining tasks, thus affecting the overall task allocation efficiency.

The same experiment is conducted with different number of workers and tasks, and it is found that the allocation effect is better when *pri_num* is set to 10%–20% of the number of tasks.

5.5 Comparison Experiment

Based on the determination of the above parameters, *pri_num* is taken as 15% of the number of tasks, $P_{reorg} = 0.2$, $M = 100$. Compare this paper's algorithm with the comparison algorithm when the number of tasks is 500 or the number of workers is 1500, change the number of workers and the number of tasks, respectively, and compare the evaluation indexes, and the results are shown in Fig. 3.

In Fig. 3(a), when the number of tasks is 100, all algorithms can easily achieve near-optimal task allocation due to the low problem complexity. As the number of tasks increases, the TPAC algorithm exhibits higher task allocation efficiency, thanks to its consideration of the importance of the order of task allocation and the use of an optimization function that combines the coverage rate and the number of workers, thus adapting to the growth of the number of tasks more efficiently. In Fig. 3(b), the TPAC algorithm is overall at a lower level in terms of average worker overhead. The MA algorithm focuses on the satisfaction of both parties, leading to a generally greater

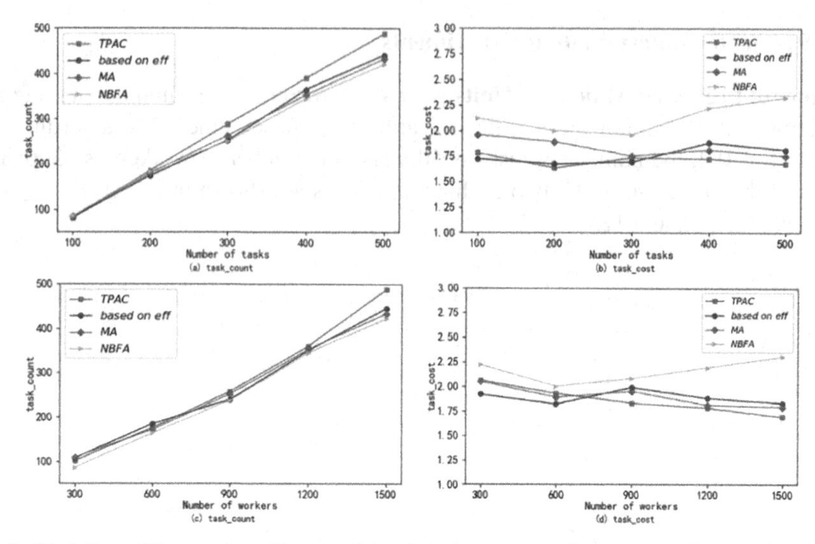

Fig. 3. Variation of the number of successful task assignments and average worker overhead with increasing number of tasks or workers

number of workers selected than the other three algorithms. In Fig. 3(c)(d), the number of tasks is certain. When the number of workers is small, the TPAC algorithm's learning optimization scope in allocating tasks is limited, and it cannot find the optimal allocation scheme better, resulting in a lower number of successful allocations than the other algorithms. But the comparison algorithms are based on the allocation of the worker-related attributes, which is less affected by the size of the workers. However, the final allocation shows a better trend as the number of workers increases.

Therefore, the comprehensive performance of TPAC algorithm is better than the three compared algorithms. In a certain scale of many-to-many task allocation scenarios, it improves the number of tasks to be allocated and provides an effective solution for realizing efficient crowdsourcing task allocation.

6 Conclusion

This paper proposes a software crowdsourcing allocation algorithm based on task priority, distinct from existing methods. The proposed algorithm calculates task priority based on skill rarity and uses this priority along with the ICA to maximize the number of assignments while minimizing worker usage. This approach improves platform allocation efficiency and is validated using a real dataset. Future work will focus on task allocation under task prioritization when workers can participate in multiple tasks simultaneously, expanding the applicability of the allocation mechanism.

References

1. Howe, J.: The rise of crowdsourcing. Wired **14**(6) (2006)

2. Peng, X., Babar, M.A., Ebert, C.: Collaborative software development platforms for crowdsourcing. IEEE Softw. **31**(2), 30–36 (2014)
3. Lazaretti, A.Z., Santos, L.M., Basilio, G.P., et al.: Software crowdsourcing platforms. IEEE Softw. **33**(6), 112–116 (2016)
4. Chen, B., Wang, L., Jiang, X., et al.: Survey of task assignment for crowd-based cooperative computing. Comput. Eng. Appl. **57**(20), 1–12 (2021)
5. Xiao, L.: Research on Task Allocation Technology in Crowdsourcing. Hangzhou University of Electronic Science and Technology (2022)
6. Yu, X., Li, G., Zheng, Y., et al.: CrowdOTA: an online task assignment system in crowdsourcing. In: Proceedings of the 34th IEEE International Conference on Data Engineering (2018)
7. Ma, H., Chen, Y., Tang, W., et al.: Survey of research progress on crowdsourcing task assignment for evaluation of workers' ability. Comput. Appl. **41**(8) (2021)
8. Wu, Y., Zeng, J.R., Peng, H., et al.: Survey on incentive mechanisms for crowd sensing. J. Softw. **27**(8), 2025–2047 (2016)
9. Yu, D., Wang, Y., Zhou, Z.: Software crowdsourcing task allocation algorithm based on dynamic utility. IEEE Access **7**, 33094–33106 (2019)
10. Ma, H., Chen, Y., Huang, Z., et al.: Software crowdsourcing tasks assignment supporting fuzzy measurement of workers' qualification and role collaboration. J. Natl. Univ. Def. Technol. **44**(5), 124–133 (2022)
11. Qiao, L., Tang, F.L., Liu, J.C.: Feedback-based high-quality task assignment in collaborative crowdsourcing. In: 2018 IEEE 32nd International Conference on Advanced Information Networking and Applications (AINA), pp. 1139–1146. IEEE (2018)
12. Wan Y.: Research on Task Allocation Algorithm for Crowdsourcing Platform Based on Game Theory. Huaibei Normal University (2023)
13. Zhang, P., Zhang, Y.: Truthful incentive mechanism for multi-task assignment in crowdsourcing. In: Sixth International Conference on Advanced Electronic Materials, Computers, and Software Engineering (AEMCSE 2023), vol. 12787, pp. 552–558 (2023)
14. Zheng, Y., Wang, J., Li, G., et al.: QASCA: a quality-aware task assignment system for crowdsourcing applications. In: Proceedings of the ACM SIGMOD International Conference on Management of Data (2015)
15. Zhang, P., Fu, X.: Incentive mechanism of crowdsourcing multi-task assignment against malicious bidding. Comput. Appl. **44**(01), 261–268 (2024)
16. Wang Y.: Research on Utility-based Multitask Assignment Method in Software Crowdsourcing Environment. Hubei University (2019)
17. Zhang, J.: Research on collective allocation method of crowdsourcing complex tasks and worker strategy synchronization mechanism. Nanjing Univ. Finance Econ. (2023)
18. Zhang, J.: Collective assignment of complex crowdsourcing tasks based on the KM algorithm. Acad. J. Comput. Inf. Sci. **6**(5), 14–20 (2023)
19. Bo, Y., Liu, Y., Xin, B., et al.: TTAF: a two-tier task assignment framework for cooperative unit-based crowdsourcing systems. J. Netw. Comput. Appl. **218**, 103719 (2023)
20. Yadav, A., Sairam, A.S., Kumar, A.: Concurrent team formation for multiple tasks in crowdsourcing platform. In: Proceedings of the GLOBECOM 2017 - 2017 IEEE Global Communications Conference, pp. 1–7. IEEE (2017)
21. Zhu, H.: Maximizing group performance while minimizing budget. IEEE Trans. Syst. Man Cybern. Syst. **50**(2), 633–645 (2020)
22. Nabila, S.A., Kabir, A.S., Palash, R., et al.: Reputation-aware optimal team formation for collaborative software crowdsourcing in industry 5.0. J. King Saud Univ. Comput. Inf. Sci. **35**(8), 101710 (2023)

23. Atashpaz-Gargari, E., Lucas, C.: Imperialist competitive algorithm: an algorithm for optimization inspired by imperialistic competition. In: IEEE Congress on Evolutionary Computation, pp. 4661–4667. IEEE (2007)
24. Ravi, J., Rajkumar, N., Viji, C., et al.: Investigation of task scheduling in cloud computing by using imperialist competitive and crow search algorithms. Procedia Comput. Sci. **230**, 879–889 (2023)
25. Kashikolaei, G., Hosseinabadi, R., Saemi, B., et al.: An enhancement of task scheduling in cloud computing based on imperialist competitive algorithm and firefly algorithm. J. Supercomput. **76**(8), 6302–6329 (2020)
26. Zhao, K., Yu, D., Zhang, W.: Priority calculation method of software crowdsourcing task release. Comput. Appl. **38**(7), 2032–2036 (2018)
27. Li, X., Fang, W., Zhu, S., et al.: An adaptive binary quantum-behaved particle swarm optimization algorithm for the multidimensional knapsack problem. Swarm Evol. Comput. **86**, 101494 (2024)
28. Li, B., Tang, Z.B.: Multiple level binary imperialist competitive algorithm for solving heterogeneous multiple knapsack problem. Comput. Appl. **43**(9), 2855–2867 (2023)
29. Guo, J.Y., Fu, X.D., Yue, K., et al.: Task allocation of crowdsourcing for maximizing satisfaction with preference matching. Comput. Eng. Sci. **44**(1), 16–26 (2022)
30. Ma, H., Tang, W., Zhu, H., et al.: Resource utilization aware collaborative optimization of IaaS cloud service composition for data-intensive applications. IEEE Trans. Syst. Man Cybern. Syst. **51**(2), 1322–1333 (2021)

Resource Management for GPT-Based Model Deployed on Clouds: Challenges, Solutions, and Future Directions

Yongkang Dang[1], Yiyuan He[1,2], Minxian Xu[1(✉)], and Kejiang Ye[1]

[1] Shenzhen Institute of Advanced Technology, Chinese Academy of Sciences, Shenzhen, Guangdong, China
{yk.dang,yy.he2,mx.xu,kj.ye}@siat.ac.cn

[2] Southern University of Science and Technology, Shenzhen, Guangdong, China

Abstract. The widespread adoption of large language models (LLMs), such as the Generative Pre-trained Transformer (GPT), on cloud computing platforms (e.g., Azure) has resulted in a substantial increase in resource demand. This increase poses significant challenges for resource management within cloud environments. This paper aims to highlight these challenges by initially delineating the unique characteristics of resource management for GPT-based models. Subsequently, we analyze the specific challenges faced by resource management when applied to GPT-based models deployed on cloud platforms, and we propose algorithms for resource profiling and prediction concerning inference requests. To facilitate effective resource management, we present a comprehensive resource management framework that includes resource profiling and forecasting methodologies specifically designed for GPT-based models. Additionally, we discuss the future directions for resource management in the context of GPT-based models, emphasizing potential areas for further exploration and improvement. Through this analysis, we aim to provide valuable insights into resource management for GPT-based models deployed in cloud environments.

Keywords: GPT-based Models · Cloud Computing · Task Profiling · Resource Management Framework

1 Introduction

The GPT-based model refers to an artificial intelligence model characterized by a substantial number of parameters and intricate structures. Through extensive training on large datasets, this model exhibits a high level of proficiency in executing sophisticated natural language processing tasks. Owing to its large scale and resource-demanding attributes, the GPT-based model is typically deployed on public cloud platforms, such as Azure and Google Cloud, rather than on edge devices or smaller hardware setups. This deployment on cloud platforms introduces distinct challenges in resource allocation. This section aims to delineate the unique characteristics of resource management (RM) for the GPT-based model and establish relevant evaluation metrics.

T. Zhu et al. (Eds.): ICA3PP 2024, LNCS 15252, pp. 95–105, 2025.
https://doi.org/10.1007/978-981-96-1528-5_7

1.1 Unique Characteristics of RM for GPT-Based Models

Upon conducting thorough research and comparing it with traditional applications, we have identified the following unique characteristics of resource management for GPT-based models:

Large-Scale Computational Demands: The GPT-based model, which consists of billions of parameters, demands substantial computational resources for both training and inference procedures. Consequently, this complexity adds a layer of intricacy to resource management.

High Storage Demands: The GPT-based model's extensive parameter size necessitates a substantial amount of storage space to store model parameters and intermediate computation outcomes. Executing these models can rapidly deplete all accessible memory on standard hardware.

High-speed Network Demands: During model training, the GPT-based model handles vast datasets and performs complex computations and parameter optimization. In the inference phase, it needs to generate outputs based on inputs and provide results in real-time or near real-time conditions. Therefore, efficient training and inference necessitate high-speed network bandwidth.

Dynamic Resource Demands: The resource requirements of GPT-based models can fluctuate over time and across various tasks. The complexity and volume of tasks contribute to these dynamic changes in resource demands.

Long Training and Inference Processes: Due to the complexity and large-scale of the GPT-based model, its training and inference processes are typically longer than general AI models. Resource management must consider how to maintain system stability and performance over an extended duration.

1.2 Evaluation Metrics for RM for GPT-Based Models

To effectively evaluate the resource management of GPT-based models, the following metrics can be considered:

Resource Utilization: It refers to the degree to which the model effectively utilizes available resources during the training or inference process. Evaluating resource utilization involves ensuring that the model maximizes the use of available resources to enhance efficiency and reduce resource wastage.

Time Efficiency: It refers to the time taken by the model to complete a specified set of tasks (e.g. makespan metric used in traditional task scheduling). For GPT-based models, time efficiency encompasses both training and inference durations. Higher time efficiency indicates that the model is capable of completing training and inference tasks in a more expedited manner, consequently enhancing overall production efficiency.

Cost Efficiency: The cost of a GPT-based model primarily includes computational, storage, and network transmission expenses [8]. Reducing computational costs can lower resource investment. Storage costs are influenced by the model's size and storage capacity requirements. Lower network costs reflect more efficient resource usage.

Our work positions as a review paper, compared with existing review, survey and taxonomy papers related to surveying LLM model mechanisms [7,9,10], we focus on resource management perspective for LLMs. To the best of our knowledge, our work is the first review work to discuss the resource management issues for GPT-based model in cloud environments. The main **contributions** of this paper are as follows:

- We summarize the specific challenges in resource management for GPT-based model and provide a detailed description of these challenges.
- We present a comprehensive and general resource management framework for the GPT-based model that comprises seven different functional components.
- We propose and implement a machine learning (ML)-based method for resource profiling and prediction of LLM inference tasks, and validate its feasibility through experiments.

2 Challenges in RM for GPT-Based Model

By identifying the unique characteristics of resource management for GPT-based models, we have summarized the specific challenges associated with deploying these models in cloud environments [3,5,6]. Key challenges include performance prediction and control, global manageability, resource heterogeneity, scalable resource management systems, and the complexities of model and data parallelism. We will explore these challenges in detail in the following sections.

2.1 Performance Prediction and Control

Different tasks typically exhibit varying complexities, resource demands, and performance expectations for the model. Even under identical workloads and resource configurations, the model's performance can be influenced by the specific characteristics of the tasks and the nature of the data. Moreover, changes in workloads and resource configurations can lead to differences in resource allocation, data parallelism, and other factors that impact performance. These variables contribute to the challenge of predicting and managing the model's behavior and performance across different workloads and resource environments.

2.2 Global Manageability

Global manageability involves the effective management and coordination of resources in large and complex cloud environments. For GPT-based model applications, the main challenges in achieving global manageability include:
Resource Scheduling and Allocation: Due to high resource demands, there is a necessity for efficient dynamic scheduling algorithms to fulfill both user and service-level requirements.
Resource Monitoring and Optimization: Real-time monitoring and automated optimization are crucial for maintaining efficient resource utilization and ensuring effective load balancing.

2.3 Resource Heterogeneity

Resource heterogeneity refers to the variations in performance, scale, power consumption, and cost across different types of resources. For the GPT-based model, resource heterogeneity poses the following challenges:
Resource Dependencies: In GPT-based models, the interdependencies between different types of resources necessitate the use of suitable allocation algorithms to ensure effective synergy and optimal performance.
Resource Interoperability: Resource interoperability is unavoidable, so it is necessary to establish standards and protocols to ensure seamless integration and interaction among resources, and to address data and model transfer and sharing across different resources.

2.4 Scalable Resource Management System

The rapid advancement of GPT-based models presents significant challenges for data centers, particularly in managing concurrent requests and ensuring high throughput. To effectively support model execution and handle extensive datasets, it is essential to develop a highly scalable computing and storage infrastructure. Furthermore, the system must be capable of dynamically allocating and flexibly expanding resources in response to demand, thereby accommodating the diverse scales and complexities associated with various application scenarios.

2.5 Model Parallelism and Data Parallelism

Parallelism mainly consist of model parallelism and data parallelism. In model parallelism, the primary challenges include:
Model Partitioning: GPT-based models are typically large, containing billions of parameters. Partitioning them into sub-models for parallel processing requires careful consideration of internal dependencies and communication needs to ensure both correctness and efficiency in parallel computation.
Synchronization and Communication Overhead: Sub-models on different devices need to synchronize and communicate to transfer gradient information and update parameters. This process can introduce additional computational and communication overhead, potentially impacting the efficiency of parallel computation.

In data parallelism, the main challenges include:
Data Partitioning and Distribution: Dividing large-scale training data and distributing it across different devices is a complex task. It requires careful attention to data balance and distribution efficiency to maintain optimal performance during parallel training.
Data Synchronization and Consistency: Effective data synchronization mechanisms are crucial for ensuring accurate parameter updates and maintaining model consistency across all devices involved in the training process.

3 RM Framework for GPT-Based Model

In response to the challenges faced by the GPT-based model, we propose a comprehensive resource management framework to efficiently manage key resources. This framework comprises several key components: *Resource Monitor*, *GPT Task Scheduler*, *Resource Allocator*, *GPT Task Profiler*, *Synchronizer*, *QoS Manager*, and *Resource Adaptor*. Figure 1 demonstrates our resource management framework for the GPT-based models. The following sections will introduce each component in detail.

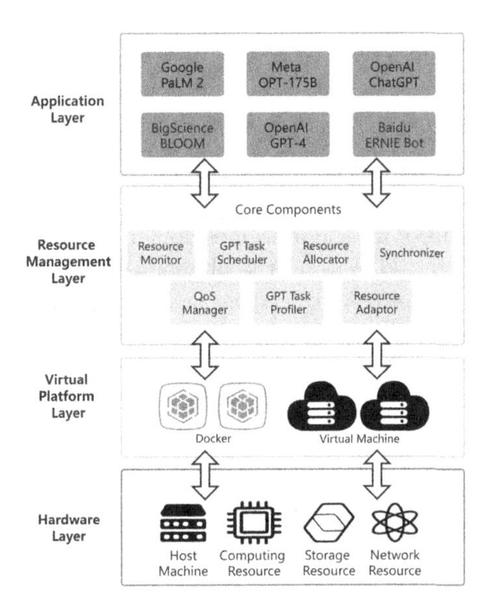

Fig. 1. Resource Management Framework for GPT-based Models

Resource Monitor: This component tracks system resources in real-time, collecting and analyzing usage and performance data. It provides feedback and reports to inform task scheduling and resource allocation decisions, and offers visual displays of resource usage and performance trends.
GPT Task Scheduler: This component manages task scheduling based on requests from the GPT task queue. It identifies suitable instances of GPT-based models for various tasks by taking into account factors such as task priority and resource requirements. Additionally, it utilizes appropriate scheduling algorithms to establish the execution order of the tasks.
Resource Allocator: This component dynamically manages resources based on task needs, system availability, and load conditions. It uses efficient allocation strategies to fulfill task needs and utilize prediction models to estimate resource requirements and system load for more precise allocation and adjustments.

GPT Task Profiler: This component extracts and analyzes the attributes, resource requirements, and QoS needs of GPT tasks, sharing this information with other components to optimize resource management.
Synchronizer: This component ensures efficient resource allocation and smooth execution of GPT tasks by utilizing distributed consistency protocols.
QoS Manager: This component develops optimization strategies based on the QoS requirements of GPT tasks and system constraints, guiding task scheduling and resource allocation. It also monitors real-time performance, making adjustments as needed to maintain the desired QoS levels.
Resource Adaptor: This component automatically adjusts resource allocation for GPT tasks based on system load, GPT task demands, and resource usage using adaptive algorithms and prediction models, thereby achieving dynamic resource scaling.

4 Task Profiling and Resource Prediction

As it is difficult to address all the aforementioned challenges in a single paper, in this work, we focus on addressing the challenges for GPT tasks profiling by precisely profiling the resources utilized by each task. Since the ChatGLM2-6B model is open source and requires less memory for deployment, we chose it as the target model of our study. Our approach leverages K-Prototypes [4] clustering to categorize GPT tasks based on their resource consumption patterns. Following clustering, we employ multiple machine learning algorithms for training and prediction. This method facilitates a detailed and accurate profiling of GPT tasks, thereby optimizing resource allocation and improving scheduling efficiency. This approach consists of 6 steps as shown in Fig. 2 and details are as follows:

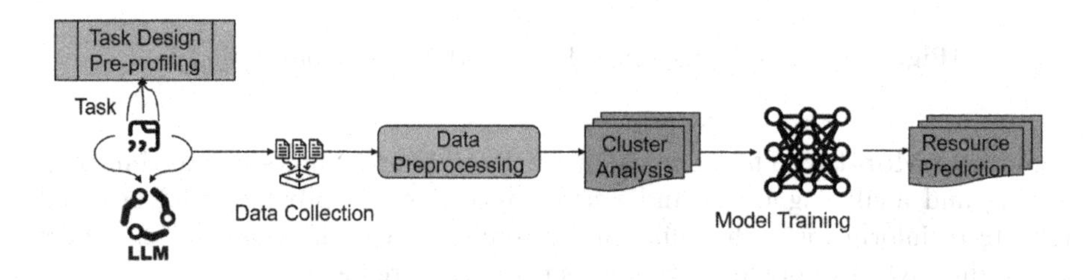

Fig. 2. Task profiling and resource prediction details

4.1 Task Design and Pre-profiling

To comprehensively evaluate the performance of the ChatGLM2-6B model in answering various questions, we designed a diverse set of questions based on their length, type, and language. A total of 182 questions were created, ranging from 5 to 431 tokens in length[1]. The question types include text generation, summarization, translation, common sense Q&A, mathematics and logic, coding skills, and specialized fields, with questions in both Chinese and English. Figure 3 and Fig. 4 illustrate the distribution of question types and languages, respectively.

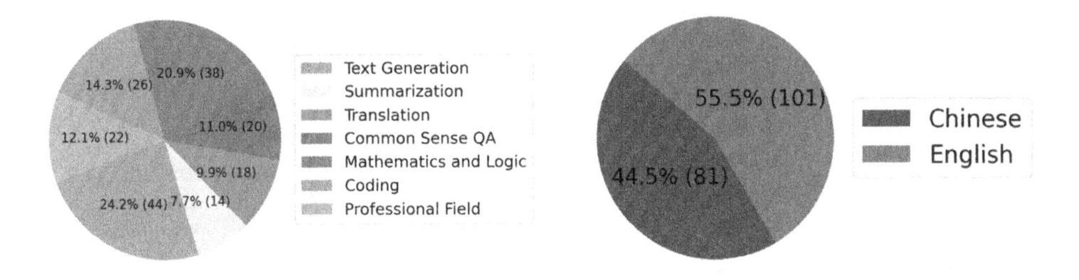

Fig. 3. Type distribution of questions

Fig. 4. Language distribution of questions

4.2 Data Collection

During data collection, pre-prepared questions are sent to the model, and responses are recorded. Meanwhile, GPU utilization and memory usage are monitored and recorded. Each question is repeated 3 times to eliminate randomness.

4.3 Data Preprocessing

To facilitate subsequent operations, we extract key metrics (average and maximum GPU utilization, RAM usage, and total inference time) from the original telemetry data. These metrics were then combined with the attributes of the corresponding questions (length, type, and language) to form a dataset comprising 182 data.

4.4 Cluster Analysis

The K-Prototypes clustering algorithm is specifically developed for datasets that include both numeric and categorical features. It effectively handles these diverse data types by employing distinct distance functions tailored for numeric and categorical attributes. The algorithm partitions the data points into K clusters based on a defined similarity measure, with the objective of maximizing the

[1] Detailed information of our designed questions can be found at: https://github.com/HYIUYOU/Resource-Management-for-GPT-based-Model-Deployed-on-Clouds.

similarity between each data point and its corresponding cluster center, while simultaneously minimizing the similarity among data points in different clusters.

Using K-Prototypes, we have divided 182 data into 6 clusters. The number of samples in each cluster is shown in Fig. 5, and the proportion of samples from each cluster within the total dataset is demonstrated in Fig. 6.

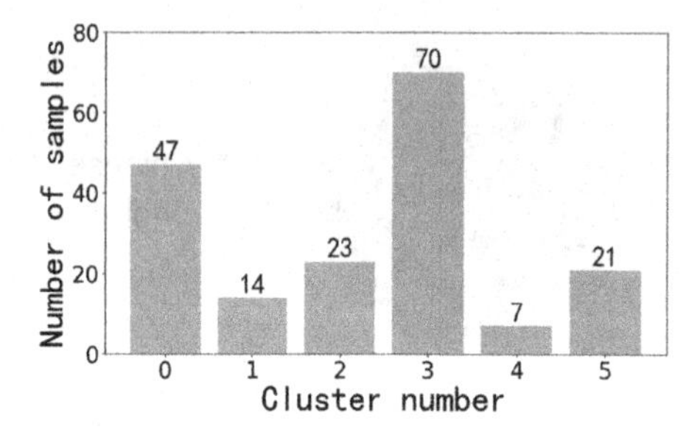

Fig. 5. Clustering results

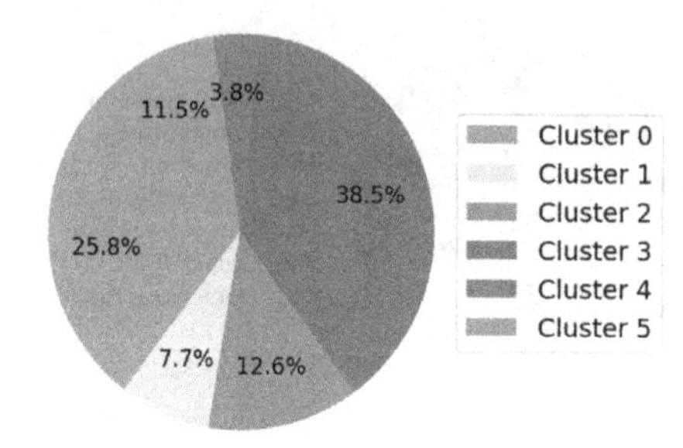

Fig. 6. Cluster sample number distribution

4.5 Model Training and Resource Prediction

After clustering was completed, we use three ML-based algorithms for training and predicting within each cluster: Random Forest (RF), Gradient Boosting (GB) and LightGBM (LGBM) [1,2]. Each sample within a cluster is split into a training set and a test set, with 70% of the data used for training and 30% for testing. We then perform individual predictions for four resource usage metrics: maximum GPU utilization, average GPU utilization, maximum Memory usage, and total inference time. For each test sample in the cluster, we calculate and record two metrics widely used: Relative Error (RE) and Mean Squared Error (MSE).

RE is defined as the proportion of the difference between the predicted ($\hat{y}$) and actual (y) values relative to the actual value, given by:

$$\mathrm{RE} = \frac{|y - \hat{y}|}{|y|}. \tag{1}$$

MSE is the average of the squared differences between the predicted and actual values, formulated as:

$$\mathrm{MSE} = \frac{1}{n}\sum_{i=1}^{n}(y_i - \hat{y}_i)^2. \tag{2}$$

4.6 Evaluations

Given the diversity and data sufficiency of samples in Clusters 0 and 3, we record two evaluation metrics (RE and MSE) for these clusters with three ML-based approaches as introduced in Sect. 4.5. To visually analyze the prediction results, we also plot the cumulative distribution function (CDF) of RE for test sample predictions in each cluster. In addition, we summarize the MSE of the predicted test samples in each cluster in Table 1. The results are as follows:

Table 1. MSE for different ML-based approaches and clusters

Cluster	Max GPU utilization			Avg GPU utilization			Max memory used			Total inference time		
	GB	LGBM	RF	GB	LGBM	RF	GB	LGBM	RF	GB	LGBM	RF
0	32.94	**21.17**	40.07	**56.3**	73.51	73.56	**21.62**	68.52	38.58	4.97	**4.86**	5.25
3	**58.6**	261.09	78.47	**110.62**	297.51	114.18	64.92	88.33	**59.1**	12.63	19.82	**11.12**

Cluster 0: Figure 7 illustrates the performance of various models. For maximum GPU utilization predictions, over 80% of the samples achieved a RE below 10%, with LGBM excelling at less than 5%. When predicting average GPU utilization, more than 80% of the predictions had a RE under 20%, with GB performing best at under 15%. In maximum memory usage predictions, all models achieve good performance, as every prediction had a RE below 0.12%, with GB leading at just 0.04%. Lastly, for total inference time predictions, the models performed acceptably, with over 80% of predictions within a 30% RE and similar performance across all models.

Cluster 3: As shown in Fig. 8, over 80% of maximum GPU utilization predictions had a RE within 30%, with LGBM underperforming. For average GPU utilization, more than 80% of the predictions had a RE up to 100%, again with LGBM showing poor performance. However, in maximum Memory usage predictions, all models performed admirably, maintaining a RE under 0.75%. For

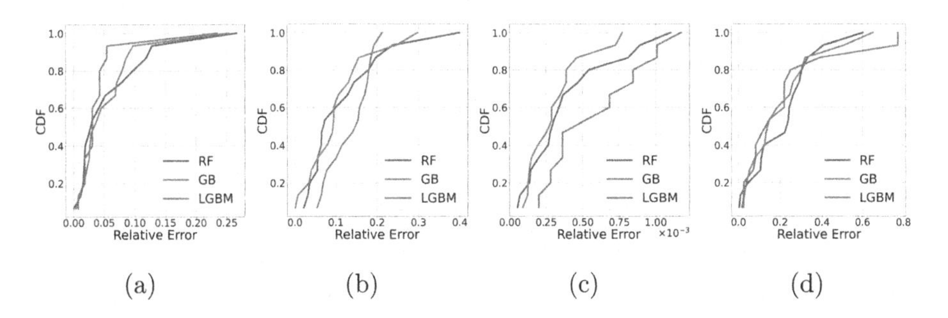

Fig. 7. CDF comparison of RE for Cluster 0. (a) maximum GPU utilization. (b) average GPU utilization. (c) maximum memory used. (d) total inference time.

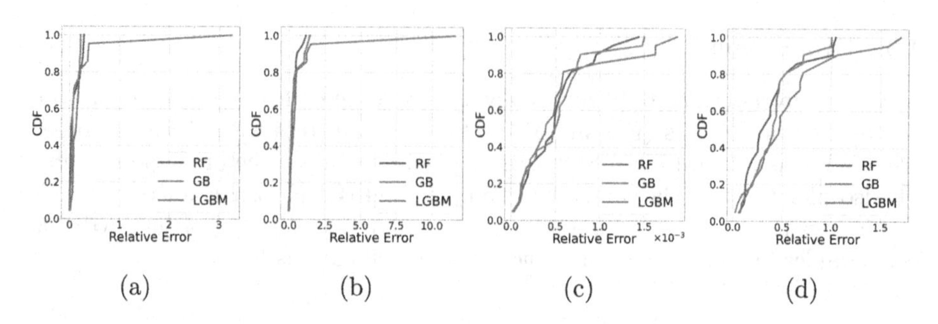

Fig. 8. CDF comparison of RE for Cluster 3. (a) maximum GPU utilization. (b) average GPU utilization. (c) maximum memory used. (d) total inference time.

Total Inference Time, the models' performance was acceptable, with both RF and GB maintaining a RE within 50%, while LGBM lagged at 60%.

To summarize, different models exhibited varying performance across different clusters and prediction metrics. Overall, the GB model demonstrated the best and most stable performance. The RF model followed, showing stable results. The LGBM model, however, showed the least stability, particularly performing poorly in predicting GPU utilization in cluster 3. The results have shown the feasibility to use these approaches for task resource profiling.

5 Conclusions and Future Research Directions

In this paper, we identified unique characteristics of resource management for GPT-based model, and discussed evaluation metrics. We analyzed specific challenges and proposed a comprehensive resource management framework. In addition, we implemented and verified resource characterization and prediction methods for LLM tasks. In this section, we summarize several future research directions for resource management for GPT-based model as follows:

Specialized Hardware for GPT-Based Model: Future research will focus on developing high-performance AI chips (e.g., GPUs, FPGAs) optimized for GPT characteristics to support large-scale parallel computations.

Benchmarks for Performance Evaluation: Future efforts should focus on establishing more comprehensive test suites to evaluate resource management for GPT-based model from multiple dimensions.

Resource Utilization Maximization: Future research needs to investigate more efficient techniques to maximize the utilization of resources by the GPT-based model. This can be achieved through dynamic resource allocation, resource sharing, and parallel computing algorithms designed for GPT-based models.

Scheduling Algorithms and Metrics: Future research should develop specialized scheduling and optimization algorithms for GPT-based model, considering task priorities, resource constraints, and optimization metrics.

Acknowledgments. This work is supported by the National Natural Science Foundation of China (No. 62072451, 62102408), and Guangdong Basic and Applied Basic Research Foundation (No. 2023B1515130002, 2024A1515010251).

References

1. Bentéjac, C., Csörgő, A., Martínez-Muñoz, G.: A comparative analysis of gradient boosting algorithms. Artif. Intell. Rev. **54**, 1937–1967 (2021)
2. Friedman, J.H.: Greedy function approximation: a gradient boosting machine. In: Annals of Statistics, pp. 1189–1232 (2001)
3. Gunasekaran, J.R., Mishra, C.S., Thinakaran, P., Kandemir, M.T., Das, C.R.: Implications of public cloud resource heterogeneity for inference serving. In: Proceedings of the 2020 Sixth International Workshop on Serverless Computing, pp. 7–12 (2020)
4. Ji, J., Pang, W., Zhou, C., Han, X., Wang, Z.: A fuzzy k-prototype clustering algorithm for mixed numeric and categorical data. Knowl.-Based Syst. **30**, 129–135 (2012)
5. Li, Z., et al.: Alpaserve: statistical multiplexing with model parallelism for deep learning serving. arXiv preprint arXiv:2302.11665 (2023)
6. Tuli, S., et al.: Hunter: AI based holistic resource management for sustainable cloud computing. J. Syst. Softw. **184**, 111124 (2022)
7. Xu, M., et al.: A survey of resource-efficient LLM and multimodal foundation models. arXiv preprint arXiv:2401.08092 (2024)
8. Xu, M., Song, C., Wu, H., Gill, S.S., Ye, K., Xu, C.: ESDNN: deep neural network based multivariate workload prediction in cloud computing environments. ACM Trans. Internet Technol. **22**(3), 1–24 (2022)
9. Yin, D., Liu, Y., Zhao, W.X., Wen, J.-R., Xu, J., Du, J.: A survey of resource-efficient LLM and multimodal foundation models. arXiv preprint arXiv:2310.13165 (2023)
10. Zhao, W.X., et al.: A survey of large language models. arXiv preprint arXiv:2303.18223 (2023)

Dynamic Offloading Control for Waste Sorting Based on Deep Q-Network

Jing Wang[2], Xiaoyang Wang[2], Jianxiong Guo[1,2(✉)], Zhiqing Tang[1], Xingjian Ding[3], and Tian Wang[1]

[1] Advanced Institute of Natural Sciences, Beijing Normal University, Zhuhai, China
{jianxiongguo,zhiqingtang,tianwang}@bnu.edu.cn

[2] Guangdong Key Lab of AI and Multi-modal Data Processing, Department of Computer Science, BNU-HKBU United International College, Zhuhai, China
{q030026146,q030026151}@mail.uic.edu.cn

[3] Faculty of Information Technology, Beijing University of Technology, Beijing, China
dxj@bjut.edu.cn

Abstract. With the increasing concern for environmental protection and resource optimization, efficient waste sorting has become a serious challenge today. In this paper, we propose a new offloading control problem that aims to solve waste sorting in wireless bin communication networks. Due to limited computational power, bins belonging to embedded devices rely on simple classification models with varying accuracy. In this scenario, consider a network of intelligent bins, each acting as an independent agent capable of deciding to offload an image to the edge server with a more accurate but resource-intensive model when the local classification is deemed inaccurate. Thus, Our goal is to find a lightweight online offloading policy that can achieve the best possible sorting accuracy while balancing transmission traffic. The method utilizes a Deep Q-network algorithm that enables each intelligent bin to make image-processing decisions autonomously. In the experiment, we validate the effectiveness of improving the performance of waste classification compared with existing flow control methods.

Keywords: Distributed Resource Allocation · Waste Sorting · Wireless Communication Network · Intelligent Waste Management

1 Introduction

In recent years, waste management has grown to be a serious global issue. As stated in [1], more than 2 billion tons of solid waste is produced globally each

This work was supported in part by the National Natural Science Foundation of China (NSFC) under No. 62202055 and No. 62202016, the Start-up Fund from Beijing Normal University under Grant No. 310432104, the Start-up Fund from BNU-HKBU United International College under Grant No. UICR0700018-22, and the Project of Young Innovative Talents of Guangdong Education Department under Grant No. 2022KQNCX102.

T. Zhu et al. (Eds.): ICA3PP 2024, LNCS 15252, pp. 106–117, 2025.
https://doi.org/10.1007/978-981-96-1528-5_8

year, and at least 33% of it is not disposed of environmentally. The data presented in [2] indicates a direct correlation between population growth, economic development, and the corresponding increase in waste generation. Residential waste management has been improved by developing several Internet of Things (IoT) based smart garbage devices and categorization techniques employing Machine Learning (ML) [3–7], including the vision-based intelligent waste bin detection and classification system [8,9]. Effective waste sorting not only contributes to environmental preservation but also facilitates the recovery of valuable resources. However, achieving accurate and efficient waste sorting presents a multifaceted challenge, particularly when dealing with large-scale waste management systems. Conventional waste sorting methods often involve labor-intensive processes and may lack adaptability in dynamically changing scenarios.

Thus, how to obtain an effective offloading policy in edge computing to deal with the energy deficiency of smart bins is an important problem. Deep Q-Network (DQN) algorithm, as a good method to deal with dynamic and high-dimensional input, is often used to optimize offloading policies. Focusing on a few representative examples, a multiuser Mobile Edge Computing (MEC)-aided smart Internet of vehicles (IoV) network is considered in [10], and offloading policies are optimized by combining DQN and Lagrange multiplier. Both [11,12] optimize the joint optimization problem of offloading and resource allocation in MEC through Double DQN, which effectively reduces energy consumption during offloading while achieving similar performance.

In this specific scenario, we consider involving a wireless waste bin communication network in which each bin is an autonomous agent. Each smart bin possesses a relatively weak ML model that trades accuracy for efficiency. An edge server with a stronger model can complement these bins at some point. The images obtained by bins and the bandwidth in this network need to determine which data to offload to the edge server and when. The main difference between these two models is that they have different confidence levels for their outputs. Bins are sometimes insufficient and edge servers are always better than bins. In this case, the smart bin can choose to send its input to the edge server to get a higher confidence answer. Since the bandwidth is limited, there is a need to limit the transmission rate of offloading, namely control the flow of traffic. Qiu *et al.* [13] adopted a Deep Reinforcement Learning (DRL) method combined with a token bucket mechanism to control the offloading rate. However, we find that in waste sorting, we can achieve the same performance by adjusting the reward function in DRL without using a token bucket, which significantly improves the running efficiency. The proposed approach capitalizes on the advancements in DRL and its capacity to empower autonomous agents in decision-making, resulting in efficient and adaptable waste-sorting strategies.

Here, the offloading decision affects current and future rewards (improvements in classification accuracy). Offloaded images are immediately rewarded for higher (expected) accuracy of classification by the edge server. However, because excessive offloading can lead to a significant decrease in rewards, images that should have higher rewards in the future cannot be offloaded. By leveraging

DQN algorithms, agents learn to strike a balance between offloading images for more accurate centralized processing and processing images locally to minimize communication latency. Cooperative interactions between intelligent agents help to form a distributed resource allocation scheme, which improves the overall efficiency of the waste sorting process. Through this research, we can advance intelligent waste management by demonstrating the feasibility of applying DRL in the context of distributed resource allocation for garbage sorting.

Our main contributions in this paper can be summarized as follows: (1) In this paper, we consider a dynamic wireless communication network environment to maximize system performance, the agent has to make a decision between offloading an image for centralized or local processing. (2) We simplify the process of controlling the traffic instead of using a token bucket [13] in the resource allocation process while still maintaining close performance. (3) A comprehensive evaluation using a wide range of image sequences from the ImageNet dataset, illustrating its benefits over competing alternatives.

2 Background and Motivation

The scenario discussed in this paper revolves around a wireless bin communication network, which forms the basis of the waste sorting (WS) problem under our study. In this scenario, a network of smart bins is established where each bin is an autonomous agent. These smart bins are equipped with image processing capabilities and communication functions to make decisions about the processing of waste images. The collaborative location of data and processing provides significant benefits in terms of utilizing distributed computational resources and timeliness of execution. The time for local execution to obtain an answer for image classification can be saved several times compared to execution on an edge server after transmission over a local wireless network.

Accordingly, this timeliness improvement comes at a cost, as the performance of models run in local bins may not be as strong as those that can be run on edge servers. It is interesting to note, however, that the difference in image classification accuracy depends on the used classifiers (weak and strong classifiers). We follow the classification setting in [13], dividing images awaiting recognition into the following three categories: (a) Images that can be accurately classified by both classifiers; (b) Images that are difficult to be accurately classified by both classifiers; and (c) Images that can be handled by the strong classifiers but not by the weak classifiers. The relative proportions of images in each category may vary, but for a typical combination of classifiers [14], more than half of the images are in (a), a small proportion of images are in (b), and the rest are in (c). Here, we want to offload images in (c) instead of (a) and (b) as much as possible. So we need to first identify which images belong to (c).

3 Problem Formulation

Drawing attention to the scenario in the Introduction, images captured by cameras undergo classification by the local (weak) classifier, leading to an offloading decision based on the confidence of that classifier. The formulation of this offloading policy is conceptualized as an online constrained optimization problem.

3.1 Classifier Setup

The symbols W and S respectively represent the weak and strong classifiers deployed on the bin and edge server. For a given image x, the system yields classification outputs $W(x)$ and $S(x)$ in the format of probability distributions encompassing a finite set of potential classes y. For the true class y of the input image x and the classifier output z, an application-specific loss function $L(z; y)$ is defined as a measure of the penalty score for classification error. Thus, the value of the loss depends on whether the image is offloaded or not, which can be expressed as $L(W(x); y)$ is not offloaded for image x and $L(S(x); y)$ otherwise.

At the moment of the offloading decision, we cannot get $S(x)$ and y, thus $L(W(x); y)$ and $L(S(x); y)$ cannot be computed. Therefore, the primary aim of the policy is to maximize the anticipated reward, characterized by the reduction in loss resulting from the offloading decision. It is crucial to note that this reward is influenced not solely by the input arrival process but also significantly by the classifier output process.

3.2 Offload Reward

Suppose that at time t, the true category of image $x[t]$ is $y[t]$, then the classification prediction loss of weak classifier and strong classifier is $L(W(x[t]); y[t])$ and $L(S(x[t]); y[t])$. We define the offloading reward $R[t]$ as the amount of loss that is reduced by offloading the image to the edge:

$$R[t] = L(W(x[t]); y[t]) - L(S(x[t]); y[t]) - N^{log_n N}, \tag{1}$$

where N represents the number of images that have been offloaded at time t. N starts from 0 and is plus 1 each time an image experiences an offload. n is a constant set.

3.3 Decisions

We assume a discrete-time system with an underlying clock that determines when an image can arrive. Image arrivals follow a general arrival interval time process with an arbitrary distribution $F(t)$. This distribution can be chosen to allow for renewal and non-renewal arrival intervals. With such an input process, the policy π for making the offloading decision $a[t]$ at time t may need to take into account the entire input history up to time t, that is $a[t] = \pi(X[t])$, where $X[t]$ is the input history that contains image arrivals and classification results before time t. Our goal is an offloading policy π^* that maximises the sum of benefits over the discount rate $\gamma \in [0, 1)$ for a long time, namely,

$$\pi^* = \arg\max_{\pi} \mathbb{E} \sum\nolimits_{t=0}^{\infty} \gamma^t a[t] R[t]. \tag{2}$$

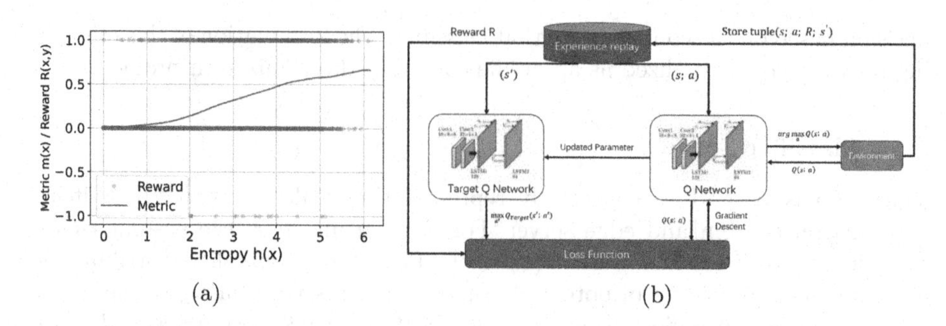

Fig. 1. (a) Depicted is a mapping (illustrated by the red curve) that correlates the entropy of the output from a weak classifier with the offloading metric, accompanied by the actual rewards for images in the training set, represented by purple dots [13]. (b) Structure diagram of DQN Algorithm. (Color figure online)

When no image arrives at time t, we set $x[t]$ to zero and the corresponding classification output to zero. At this point, both the action $a[t]$ and reward $R[t]$ are set to zero. This ensures that the input history $X[t]$ contains information about past image arrival intervals as well as the classification output for each image arrival, so the policy makes decisions only at the time of image arrival.

4 Solution

In this section, we now describe the approach we rely on to derive π^*. The policy assumes a given pair of image classifiers W, S, access to representative training data, and seeks to specify actions that maximize an expected discounted reward as expressed in Eq. (2). There are several challenges in realizing π^*.

First, to improve classification accuracy by utilizing the edge server's strong classifier, we need to identify images with positive load rewards. Based on Eq. (1), the reward associated with the input $x(t)$ depends on the outputs $W(x[t])$ and $S(x[t])$ of the weak and strong classifiers, as well as knowledge of the true class $y(t)$ of the input. But neither $S(x[t])$ nor $y(t)$ is available when an offload decision needs to be made. We learn an estimate of the offloading reward $R[t]$ through a method that relies on the offloading metric $m(x)$ [15].

4.1 Offloading Metric

At each image arrival, only the output $W(x)$ of the weak classifier is available information. Calculate the offloading metric $m(x)$ according to the method in [15]. The offloading metric $m(x)$ represents then an estimate for the corresponding offloading reward R. The entropy $h(z)$ of a classification output z is:

$$h(z) = -\sum_{y \in \mathbb{Y}} z_y \log z_y, \tag{3}$$

which captures the classifier's confidence in its result. This entropy is then mapped to an expected offloading reward using a standard radial basis function kernel:

$$f(\bar{h}) = \frac{\sum_{k=1}^{K} \sigma(\bar{h}, h_k) \times R_k}{\sum_{k=1}^{K} \sigma(\bar{h}, h_k)}, \quad (4)$$

where $\bar{h} = h(z)$ for classification output z, $\sigma(\bar{h}, h_k) = \exp(-\lambda(\bar{h} - h_k)^2)$, and R_k is the reward from the kth sample in the training set with h_k its entropy. Figure 1(a) shows the entropy of the weak classifier output to the offloading rewards [13], which indicates the weak classifier is uncertain about its decision,

By setting $m(x) = f(h(W(x)))$, we choose an expected reward that is essentially a weighted average over the entire training set of K images of reward values for training set inputs with similar entropy values, where images with entropy values closer to that of image x are assigned higher weights.

4.2 DQN Algorithm

With the metric $m(x)$ of image x in hand, the policy's goal is to decide whether to offload it given also the system state as captured in $X(t)$, the history of image arrivals and classification outputs. In the remainder of this section, we provide a brief overview of DQN as shown in Fig. 1(b). Q-learning is a standard RL approach for devising policies that maximize a discounted expected reward summed over an infinite horizon as expressed in Eq. (2). It relies on estimating a Q-value, $Q(s;a)$ as a measure of this reward, assuming that the current state is s, and the policy takes action a. As mentioned above, in our setting, s consists of the arrival and classification history X, while a is the offloading decision.

Estimating Q-values relies on a Q-value function, which in DQN is in the form of a deep neural network. Denoting this network as Q, it learns Q-values during a training phase through a standard Q-value update. Specifically, denoting the current DQN as Q^- let

$$Q^+(s;a) = R(s;a;s') + \gamma \max_{a'} Q^-(s',a'), \quad (5)$$

where s' is the state following action a at state s, $R(s;a;s')$ is the reward from this transition (available during the training phase) with γ the discount factor of Eq. (2), and both a and a' are selected from the set of feasible actions in the corresponding states s and s'. The value $Q^+(s;a)$ is used as the "ground truth", with the difference between $Q^+(s;a)$ and $Q^-(s;a)$ representing a loss function to minimize, which can be realized by updating the weights of the DQN through standard gradient descent. The approach ultimately computes Q-values for all combinations of inputs (state s) and possible actions a, and the resulting policy greedily takes the action with maximum Q-value in each state:

$$\pi(x) = \arg\max_{a} Q(s,a). \quad (6)$$

The challenges in learning the policy of Eq. (6) are the size of the state space and the possibility of correlation and nonstationary input distributions, which can all affect convergence to varying degrees.

5 Evaluation

In this section, we demonstrate that the DQN-based policy can achieve similar results as using token buckets in [13] by adding a weighting function for the number of offloadings to the reward function as shown in Eq. (1). For this policy, we can estimate Q-values efficiently with negligible overhead in embedded devices. To achieve this objective, we establish a testbed that mimics an authentic edge computing environment. In our approach, we engage in simulations and conduct comprehensive experiments. These experiments are designed to assess the policy's operational efficiency on embedded devices and to evaluate its performance across various configurations.

5.1 Experimental Setup

Classification Task. Our classification metric is the top-5 loss (or error). It assigns a penalty of 0 if the image is in the five most likely classes returned by the classifier and 1 otherwise. The strong classifier in our edge server is that of [14] with a computational footprint of 595 MFlops. Our weak classifier is a "homegrown" 16-layer model acting on low-resolution 64×64 images with 13 convolutional layers (8 with 1×1 kernels and 5 with 3×3kernels) and 3 fully connected layers. In this study, the dataset pivotal to our analysis is sourced from [16]. This seminal work provides a comprehensive dataset tailored for the classification of waste materials, making it particularly relevant for our exploration of trash categorization. This dataset has six classes consisting of glass, paper, metal, plastic, cardboard, and trash. Each image is presented on a clean white background version of a single object, which facilitates our use of a single image to identify the waste category of an object.

DQN Configuration. We use the dataset with 2500 waste images (6 categories). We evenly split the validation set into three subsets; two are used as training sets and one is used as the testing set. We generate a training sequence of 108 images from the training sets along with corresponding inter-arrival times and metrics. This sequence is stored in the replay buffer from which we randomly sample (with replacement) input history segments with a fixed length history window of $T = 97$ to train DQN.

DQN Evaluation. When evaluating DQN, we vary image arrival patterns, classification output correlation, and parameters and compare DQN to several benchmarks. Besides, we also contrasted the effect of token buckets on the overall model by conducting experiments and comparing DQN to two practical policies. The first is the MDP policy introduced in [15]. It is oblivious to any structure in either the image arrival process or the classifier output (it assumes that they are i.i.d.), but is cognizant of the token bucket state and attempts to adapt its decisions based on the number of available tokens and its estimate of the long-term image arrival rate. The second, denoted as Baseline, is a fixed threshold policy commonly adopted by many works in the model cascade framework. The baseline uses the same threshold as a lower bound, i.e., attempting to offload images with offloading metrics above the $(1-r)^{th}$ percentile.

5.2 Runtime Efficiency

To evaluate the practicality of our policy based on DQN, we deploy it within a testbed comprising an embedded device and an edge server, interconnected via Wi-Fi. The evaluation of its overhead involved measuring and comparing the runtime execution time of the policy on the embedded device against the time consumed by various components in a comprehensive end-to-end classification task. Subsequently, we provide a concise description of our testbed configuration and the methodology employed for these measurements. To quantify the overhead that DQN imposes, we measure where time is spent across different components of the classification pipeline. The embedded device first classifies every image using its weak classifier. Then it executes the DQN model to estimate the Q-values before making an offloading decision that accounts for the current state. Offloaded images are transmitted to the edge server over the network and finally classified by the strong classifier. Hence, a full classification task includes four main stages: (i) weak classifier inference, (ii) DQN inference, (iii) network transmission, and (iv) strong classifier inference, which all contribute to how long it takes to complete. After the experiment, DQN only takes 0.25 ms on average, the Weak Classifier takes 17.15 ms, Transmission takes 43.37 ms, and the Strong Classifier takes 11.73 ms. This is just over 1% of the time spent in the weak classifier, and for offloaded images, it is less than a third of a percent of the total classification pipeline time. This demonstrates that the benefits DQN affords impose a minimal overhead.

5.3 Policy Performance

In this section, we evaluate DQN's performance across a range of scenarios, which illustrate its ability to learn complex input structures and highlight how this affects its offloading decisions. In the end, we proceed in two stages. First, we introduce complexity in only one dimension of the input structure, i.e., correlation is present in either classification outputs or image arrivals. This facilitates developing insight into how such structure affects DQN's decisions. In the second phase, we created a new model with token buckets on top of the original model and compared the differences between the two models.

Deterministic Image Arrivals and Correlated Classification Outputs. To explore DQN's ability to learn about the presence of correlation in classification outputs, we first fix the irrelevant parameters and vary the two hyperparameters of our sequence generator to realize different levels of classification output correlation: The sampling spread *sp* is varied from 0 (single image) to 1 (full dataset and, therefore, no correlation), while the reset probability *rprob* is

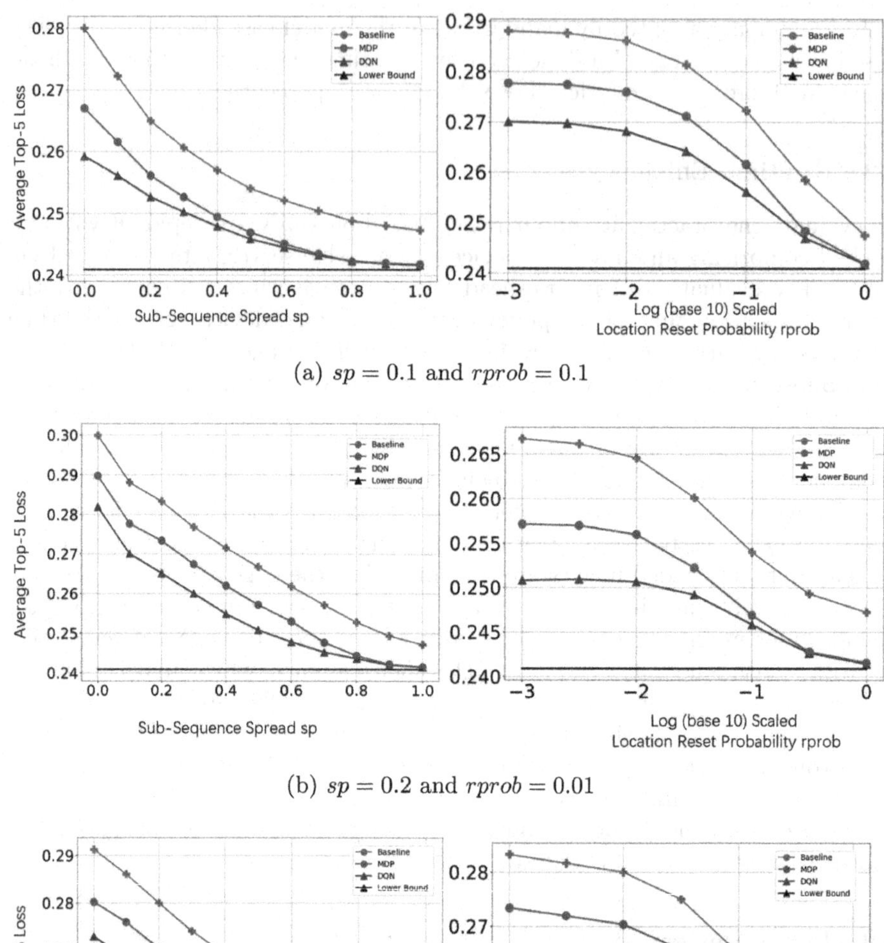

(a) $sp = 0.1$ and $rprob = 0.1$

(b) $sp = 0.2$ and $rprob = 0.01$

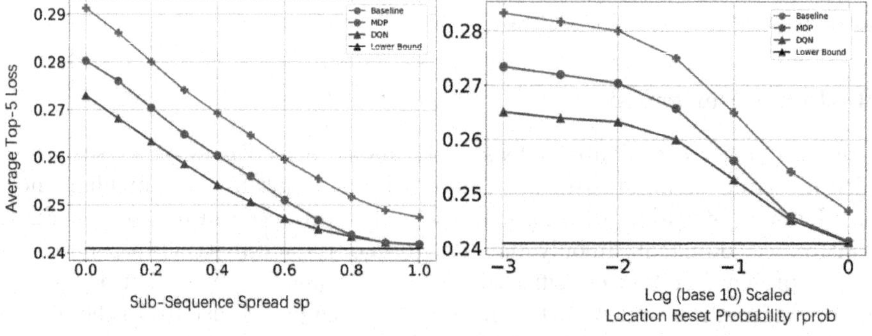

(c) $sp = 0.5$ and $rprob = 0.001$

Fig. 2. Offloading policies performance as a function of classifier output correlation with different parameters.

varied from 10^{-3} to 1 (no correlation). Firstly, we obtain the results of Fig. 2(a) by setting $sp = 0.1$ and $rprob = 0.1$. As expected, when either sp or $rprob$ are large so that classification output correlation is minimal, both DQN and MDP perform similarly and approach the performance of the lower bound. However, when classification output correlation is present, DQN consistently outperforms MDP (and the Baseline). Secondly, as shown in Fig. 2(b) and Fig. 2(c), as correlation increases, performance degrades when compared to the lower bound, but this is not surprising given the offload constraints. Because the model will gradually tend to be processed locally due to the rise of the number of offloading, the whole model will tend to offload the images that can be processed locally when the number of offloading is low and rarely choose to offload when the local processing results are poor when the number of offloading is high.

Effect of Token Buckets. Token buckets are commonly used for network flow control and rate limiting [13,17,18]. In the model, it is used as a decision metric to help the DQN determine when to perform data processing or offloading tasks. This is accomplished through a two-parameter token bucket (r, b) in each device, which controls both short and long-term offloading rates. Specifically, tokens are replenished at a rate of $r \leq 1$, (fractional) tokens per unit of time, and can be accumulated up to a maximum value of b. Every offloading decision requires the availability of and consumes a full token. Consequently, the token rate r, upper-bounds the long-term rate at which images can be offloaded, while the bucket depth b, limits the number of successive such decisions that can be made.

The behavior of the token bucket system can be captured by tracking the evolution of the token count $n[t]$ in the bucket over time, as follows:

$$n[t+1] = \min(b, n[t] - a[t] + r), \tag{7}$$

where $a[t]$ the offloading action at t, which is 1 if an image arrives and is offloaded (these need $n[t] \geq 1$), and 0 otherwise. Again as in [13], we assume that both r and b are rational so that $r = N/P$ and $b = M/P$ for some integers $N \leq P \leq M$. We can then scale up the token count by a factor of P and express it as $\bar{n}$:

$$\bar{n}[t+1] = \min(M, n[t] - P \times a[t] + N). \tag{8}$$

To evaluate the impact after adding token buckets, we contrast the offloading policy changes after adding token buckets. After adding the token bucket, we set the token bucket parameter to: $r = 0.1$, $b = 4$ (r is the recovery rate of the token and b is the depth of the token bucket).

As shown in Fig. 3, the DQN with token budgets to control the flow [13] achieves similar performance with the DQN with our modified reward function under the same parameter setting, where the DQN with our modified reward function is shown as Fig. 3(a) and the DQN with token budgets is shown as Fig. 3(b). This is because we have incorporated a control mechanism into the reward function to prevent excessive offloading. Through this experiment, it shows the effectiveness of our method. Also, our method improves the running efficiency because of its simpler structure. Additionally, in the waste sorting problem, there are fewer categories to be classified, and the general classifier is

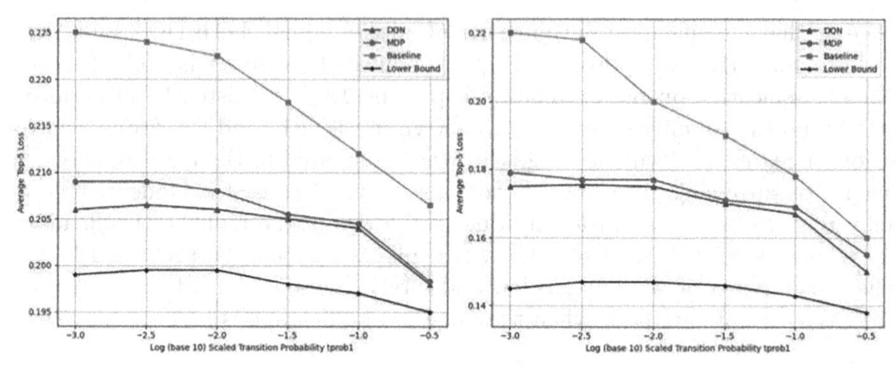

(a) Without token buckets

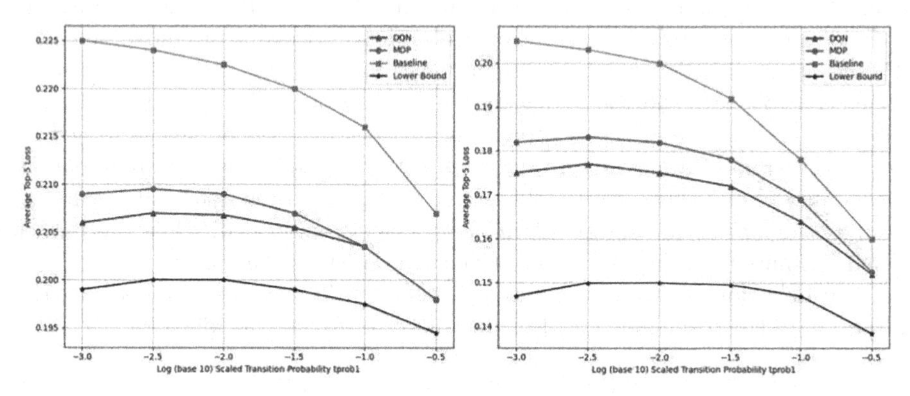

(b) With token buckets

Fig. 3. The model, after adding token buckets, Offloading policies performance as a function of image arrivals correlation.

capable of handling most requirements without resorting to excessive offloading. Thus, DQN with our proposed reward function can be used as an effective approach to control flow traffic in edge computing scenarios.

6 Conclusion

In this paper, we investigate a distributed image classification problem through edge assistance, which helps waste sorting to improve classification accuracy by offloading images to edge servers. By limiting the reward function by the number of offloading, the offloading policy is controlled to achieve a similar effect as using token buckets. We design a DQN-based policy that can optimize the classification accuracy of waste sorting and constrain the offloading decision. Experiments demonstrate that the policy can achieve similar results to the token bucket in the task of waste sorting and image processing.

References

1. Likotiko, E., Matsuda, Y., Yasumoto, K.: Garbage content estimation using internet of things and machine learning. IEEE Access **11**, 13000–13012 (2023)
2. Nowakowski, P., Pamuła, T.: Application of deep learning object classifier to improve e-waste collection planning. Waste Manag. **109**, 1–9 (2020)
3. Malik, M., et al.: Waste classification for sustainable development using image recognition with deep learning neural network models. Sustainability **14**(12), 7222 (2022)
4. Alsubaei, F.S., Al-Wesabi, F.N., Hilal, A.M.: Deep learning-based small object detection and classification model for garbage waste management in smart cities and iot environment. Appl. Sci. **12**(5), 2281 (2022)
5. Wahyutama, A.B., Hwang, M.: Yolo-based object detection for separate collection of recyclables and capacity monitoring of trash bins. Electronics **11**(9), 1323 (2022)
6. Kang, Z., Yang, J., Li, G., Zhang, Z.: An automatic garbage classification system based on deep learning. IEEE Access **8**, 140019–140029 (2020)
7. Nnamoko, N., Barrowclough, J., Procter, J.: Solid waste image classification using deep convolutional neural network. Infrastructures **7**(4), 47 (2022)
8. Salimi, I., Dewantara, B.S.B., Wibowo, I.K.: Visual-based trash detection and classification system for smart trash bin robot. In: 2018 International Electronics Symposium on Knowledge Creation and Intelligent Computing (IES-KCIC), pp. 378–383. IEEE (2018)
9. Chu, Y., et al.: Multilayer hybrid deep-learning method for waste classification and recycling. Comput. Intell. Neurosci. **2018** (2018)
10. Zhang, L., et al.: DQN-based mobile edge computing for smart internet of vehicle. EURASIP J. Adv. Signal Process. **2022**(1), 1–16 (2022)
11. Zhou, H., Jiang, K., Liu, X., Li, X., Leung, V.C.: Deep reinforcement learning for energy-efficient computation offloading in mobile-edge computing. IEEE Internet Things J. **9**(2), 1517–1530 (2021)
12. Huang, L., Feng, X., Zhang, C., Qian, L., Wu, Y.: Deep reinforcement learning-based joint task offloading and bandwidth allocation for multi-user mobile edge computing. Digital Commun. Netw. **5**(1), 10–17 (2019)
13. Qiu, J., Wang, R., Chakrabarti, A., Guérin, R., Lu, C.: Adaptive edge offloading for image classification under rate limit. IEEE Trans. Comput. Aided Des. Integr. Circuits Syst. **41**(11), 3886–3897 (2022)
14. Cai, H., Gan, C., Wang, T., Zhang, Z., Han, S.: Once-for-all: train one network and specialize it for efficient deployment. arXiv preprint arXiv:1908.09791 (2019)
15. Chakrabarti, A., Guérin, R., Lu, C., Liu, J.: Real-time edge classification: optimal offloading under token bucket constraints. In: 2021 IEEE/ACM Symposium on Edge Computing (SEC), pp. 41–54. IEEE (2021)
16. Yang, M., Thung, G.: Classification of trash for recyclability status. CS229 Project Report **2016**(1), 3 (2016)
17. Tang, P.P., Tai, T.Y.: Network traffic characterization using token bucket model. In: IEEE INFOCOM 1999, vol. 1, pp. 51–62. IEEE (1999)
18. Liao, J., Wang, J., Li, T., Wang, J., Zhu, X.: A token-bucket based notification traffic control mechanism for ims presence service. Comput. Commun. **34**(10), 1243–1257 (2011)

Hybridization of One- and Two-Point Bandits Convex Optimization in Non-stationary Environments

Gailun Zeng[2,3] and Jianxiong Guo[1,3](✉)

[1] Advanced Institute of Natural Sciences, Beijing Normal University, Zhuhai, China
jianxiongguo@bnu.edu.cn

[2] Hong Kong Baptist University, Hong Kong, China

[3] Guangdong Key Lab of AI and Multi-modal Data Processing, Department of Computer Science, BNU-HKBU United International College, Zhuhai, China
zenggailun@uic.edu.cn

Abstract. Bandit Convex Optimization (BCO) is an imperative analysis framework when dealing with sequential decision-making problems. Considering to balance the computational cost and bounds of regrets, in this paper, we propose a hybridized algorithm of one- and two-point bandit convex models in non-stationary environments and use a more general performance measure *dynamic regret*, which records the cumulative difference between function loss and a feasible comparator sequence during the time horizon T. The path length of a comparator sequence P_T reveals the non-stationarity of environments. Our proposed algorithm builds an upper bound of dynamic regret $\mathcal{O}((1+P_T)^{1/2}[\beta(\lambda T)^{1/2}+((1-\lambda)T)^{3/4}])$, where the parameter λ can dynamically adjust the bound guarantee to balance the computational cost in real applications.

Keywords: Bandit Convex Optimization · Ensemble Online Learning · Dynamic Regret Analysis · Non-stationary Environments

1 Introduction

Bandit Convex Optimization (BCO) is a classic online convex optimization problem [11]. It can be viewed as a sequential decision-making model with partial information (e.g. lack of gradient information). Recently, BCO has inspired a research trend because the characteristic of partial feedback information of environments is in line with many real-life applications, such as recommendation system [14], IOT management [7], and interactive web pages [12]. But lack of

This work was supported in part by the National Natural Science Foundation of China (NSFC) under No. 62202055, the Start-up Fund from Beijing Normal University under Grant No. 310432104, the Start-up Fund from BNU-HKBU United International College under Grant No. UICR0700018-22, and the Project of Young Innovative Talents of Guangdong Education Department under Grant No. 2022KQNCX102.

T. Zhu et al. (Eds.): ICA3PP 2024, LNCS 15252, pp. 118–135, 2025.
https://doi.org/10.1007/978-981-96-1528-5_9

full information brings main challenges, for instance, the lack of the function gradients makes it infeasible to utilize the classic online gradient descent (OGD) algorithm. Based on this, Flaxman *et al.* [9] proposed an unbiased gradient estimator to approximate function gradients and then utilized OGD to obtain an $\mathcal{O}(T^{3/4})$ expected static regret guarantee. The main two types of BCO problems are one-point BCO and two-point BCO where get one or two function values as feedback, there are already a substantial amount of studies [4,5,8,10,13,16] to analyze the theoretical bounds of regrets and algorithms.

Existing studies mainly consider *static regret* as the performance measure. The static regret is compared to a fixed benchmark, and thus, using this performance measure implicitly assumes that there are fixed offline decisions that perform well at all times. However, data distribution often constantly changes in the real environment, so the above assumption is difficult to hold. Thus, another performance measure *(universal) dynamic regret* [23] is more suitable for real life. Unfortunately, studies about BCO with dynamic regret are still a cutting-edge area, almost all existing work only focuses on worse-case dynamic regret which is a specific instance of the universal dynamic regret.

Since dynamic regret is more general, studies based on this measure are precious. Zhao *et al.* [21] established dynamic regret-bound analysis in non-stationary environments for the first time. Their novel algorithm, named PBGD, obtains $\mathcal{O}(T^{3/4}(1+P_T)^{1/2})$ and $\mathcal{O}(T^{1/2}(1+P_T)^{1/2})$ expected dynamic regret for one-point and two-point feedback models, respectively, where T is the time horizon and P_T is the path length of comparator sequence. Although a two-point feedback model can get a tighter upper bound, it suffers a larger computational cost compared with a one-point feedback model. Thus, it is a natural idea whether we can design a hybridized algorithm of one- and two-point feedback models based on the PBGD. The hybridized algorithm can adjust the proportion of the two feedback models in a whole time horizon T, to balance the computational cost and bounds of dynamic regrets from the perspective of practical applications.

However, the hybridization of these two models is not like a simple linear combination since there are several troublesome problems. For instance, the expert weight inheritance between the two-type models needs to be reasonable, meanwhile, we need to still have a feasible method to finish the theoretical analysis of bounds. How to determine the number of step-size pool candidates and specific step sizes for different models are other specific challenges. Based on these, we propose a Hybridized-PBGD algorithm, which can be regarded as a deep extension of study in [21]. Thus, the main contributions of the paper are summarized as the following points:

- We propose a Hybridized-PBGD algorithm, which can completely combine one- and two-point models sequentially. Besides, we build an upper bound of expected dynamic regret $\mathcal{O}((1+P_T)^{1/2}[\beta(\lambda T)^{1/2}+((1-\lambda)T)^{3/4}])$ where λ is a parameter to adjust the bound guarantee.
- We innovatively propose a *intermediate weight initialization strategy* to address the expert weight updating problem at the transition round. The

strategy not only can inherit Phase 1's experienced weights partially but also pave the way for the bound analysis of Phase 2.
- We elaborately determine the same number of candidates and different step sizes for the step-size candidate pool settings, due to the differences between one- and two-point models. This operation maintains the intrinsic characteristics of the original algorithm, which is indispensable for bound analysis.

2 Preliminaries of Subsequent Analysis

2.1 Assumptions

First, the following assumptions are utilized when we analyze the BCO [1,9,21].

Assumption 1 (Bounded Domain). *The feasible domain $\mathcal{W}$ is closed, containing a sphere with radius r centered at the origin, and is contained by a sphere with radius R centered at the origin: $r\mathbb{B} \subseteq \mathcal{W} \subseteq R\mathbb{B}$, where $\mathbb{B} = \{\boldsymbol{w} \in \mathbb{R}^d \mid \|\boldsymbol{w}\|_2 \leq 1\}$.*

Assumption 2 (Bounded Function Value). *The absolute values of functions will not exceed a constant C: $\forall t \in [T]$, $\max_{\boldsymbol{w} \in \mathcal{W}} |f_t(\boldsymbol{w})| \leq C$.*

Assumption 3 (Lipschitz Continuity). *The function is L-Lipschitz continuous over the feasible domain $\mathcal{W}$, i.e., $\forall \boldsymbol{x}_1, \boldsymbol{x}_2 \in \mathcal{W}$, we have $\forall t \in [T]$, $|f_t(\boldsymbol{x}_1) - f_t(\boldsymbol{x}_2)| \leq L\|\boldsymbol{x}_1 - \boldsymbol{x}_2\|_2$.*

2.2 Dynamic Regret

A classic online convex problem can be viewed as an iterative game between the player and the environment. Formally, at iteration round t, the player chooses the decision or model $\boldsymbol{w}_t \in \mathcal{W}$ from the convex set $\mathcal{W} \subseteq \mathbb{R}^d$ as the current round prediction, while the environment chooses the online convex function $f_t : \mathcal{W} \to \mathbb{R}$. At each round t, the player will suffer an instantaneous loss $f_t(\boldsymbol{w}_t)$, obviously, and the player wants to minimize the total cumulative loss $\sum_{t=1}^{T} f_t(\boldsymbol{w}_t)$. A widely used performance measure is called *regret*. In our paper, the *dynamic regret* [23] is used as the performance measure. That is

$$\text{D-Regret}_T(\boldsymbol{u}_1, \ldots, \boldsymbol{u}_T) = \sum_{t=1}^{T} f_t(\boldsymbol{w}_t) - \sum_{t=1}^{T} f_t(\boldsymbol{u}_t), \tag{1}$$

where $\boldsymbol{u}_1, \ldots, \boldsymbol{u}_T \in \mathcal{W}$ is a comparator sequence, and Eq. (1) expresses the difference between the cumulative loss of the player and that of the comparator sequence. In [23], Zinkevich proved a $\mathcal{O}(\sqrt{T}(1 + P_T))$ dynamic regret with a full-information setting, where P_T is the path length of the comparator sequence, which is defined as $P_T(\boldsymbol{u}_1, \ldots, \boldsymbol{u}_T) = \sum_{t=2}^{T} \|\boldsymbol{u}_{t-1} - \boldsymbol{u}_t\|_2$.

2.3 Bandit Convex Optimization

For online gradient descent (OGD) with full information (e.g. the gradients), the classic algorithm [23] updates from $\boldsymbol{w}_1 \in \mathcal{W}$ as follows:

$$\boldsymbol{w}_{t+1} = \Pi_{\mathcal{W}}[\boldsymbol{w}_t - \eta \nabla f_t(\boldsymbol{w}_t)], \tag{2}$$

where $\Pi_W[\cdot]$ represents Euclidean projection to $\mathcal{W}$ and the step size $\eta > 0$. However, in BCO problems, we can not get the function gradients and only get the feedback of function values. It means that Eq. (2) relying on the function gradient is unsuitable. To cope with the challenge, researchers proposed one-point and two-point feedback models in [1,9] respectively, which utilize one or two function values near the point $\boldsymbol{w}_t$ to build gradient estimator $\tilde{g}_t \approx \nabla f_t(\boldsymbol{w}_t)$.

- For a single-point feedback model, to effectively estimate the gradient based on function values, Flaxman *et al.* [9] proposed the following gradient estimation $\tilde{\boldsymbol{g}}_t \in \mathbb{R}^d$ as

$$\tilde{\boldsymbol{g}}_t = \frac{d}{\delta} f_t(\tilde{\boldsymbol{w}}_t + \delta \boldsymbol{s}_t) \cdot \boldsymbol{s}_t, \tag{3}$$

  where $\boldsymbol{s}_t \in \mathbb{R}^d$ is a uniformly randomly selected unit vector and $\delta > 0$ is a perturbation parameter. Note that, Eq. (3) is an unbiased gradient estimator of the smoothness of the loss function $f_t(\cdot)$ (Lemma 3.4 in [20]). Since the smoothed loss function can be expressed as: $\hat{f}(\boldsymbol{w}) = \mathbb{E}_{\boldsymbol{v} \in \mathbb{B}}[f(\boldsymbol{w} + \delta \boldsymbol{v})]$, for $\forall \delta > 0$, we can get $\mathbb{E}_{\boldsymbol{s} \in \mathbb{S}}\left[\frac{d}{\delta} f(\boldsymbol{w} + \delta \boldsymbol{s}) \cdot \boldsymbol{s}\right] = \nabla \hat{f}(\boldsymbol{w})$ ($\mathbb{B} = \{\boldsymbol{w} \in \mathbb{R}^d \mid \|\boldsymbol{w}\|_2 \leq 1\}$ and $\mathbb{S} = \{\boldsymbol{w} \in \mathbb{R}^d \mid \|\boldsymbol{w}\|_2 = 1\}$), which means that the gradient estimator meets $\mathbb{E}_{\boldsymbol{s} \in \mathbb{S}}[\tilde{\boldsymbol{g}}_t] = \nabla \hat{f}_t(\tilde{\boldsymbol{w}}_t)$.
- For the two-point feedback Model, allow to receive two function values as feedback information during iteration. Thus, two symmetric points, $\boldsymbol{w}_t^{(1)} = \tilde{\boldsymbol{w}}_t + \delta \boldsymbol{s}_t$ and $\boldsymbol{w}_t^{(2)} = \tilde{\boldsymbol{w}}_t - \delta \boldsymbol{s}_t$, are used to get $f_t(\boldsymbol{w}_t^{(1)})$ and $f_t(\boldsymbol{w}_t^{(2)})$ as the feedback. The gradient estimator is constructed as

$$\tilde{\boldsymbol{g}}_t = \frac{d}{2\delta}(f_t(\tilde{\boldsymbol{w}}_t + \delta \boldsymbol{s}_t) - f_t(\tilde{\boldsymbol{w}}_t - \delta \boldsymbol{s}_t)) \cdot \boldsymbol{s}_t. \tag{4}$$

  Since the distribution of the $\boldsymbol{s}_t$ is symmetric, Eq. (4) is also unbiased.

As shown in Algorithm 1, **Case 1.** and **Case 2.** illustrate the one-point and two-point BGD algorithm, respectively. Note that $\tilde{\boldsymbol{w}}_{t+1}$ ($\tilde{\boldsymbol{w}}_{t+1} = \Pi_{(1-\alpha)\mathcal{W}}[\tilde{\boldsymbol{w}}_t - \eta \tilde{\boldsymbol{g}}_t]$) is projected on a slightly smaller set $(1 - \alpha)\mathcal{W}$ to ensure that the final decision $\boldsymbol{w}_{t+1}$ is in the feasible region $\mathcal{W}$, instead of projecting on the original feasible region $\mathcal{W}$ (see **3.4.2** the proof of *Thoerem 3.5* in [20]).

3 Algorithm Design

3.1 Ensemble Learning for Bandit Convex Optimization

In previous research, the BGD algorithm needs to predict the path length P_T as the algorithm input, and P_T reflects the dynamic change degree of the environment, which is unknown in practice. Based on this, Zhao *et al.* [20] proposed

Algorithm 1. Bandit Gradient Descent (BGD) [20]

Input: time horizon T, perturbation parameter δ, shrinkage parameter α and step size η ;

1: Let $\tilde{\boldsymbol{w}}_1 = \mathbf{0}$;
2: **for** $t = 1, 2, \cdots, T$ **do**
3: Select a unit vector $\boldsymbol{s}_t$ uniformly at random;
4: {**Case 1. One-Point Feedback Model**};
5: Calculate $\boldsymbol{w}_t = \tilde{\boldsymbol{w}}_t + \delta \boldsymbol{s}_t$;
6: Receive $f_t(\boldsymbol{w}_t)$ as the feedback;
7: Construct the gradient estimator: $\tilde{\boldsymbol{g}}_t = \frac{d}{\delta} f_t(\tilde{\boldsymbol{w}}_t + \delta \boldsymbol{s}_t) \cdot \boldsymbol{s}_t$;
8: Update $\tilde{\boldsymbol{w}}_{t+1}$: $\tilde{\boldsymbol{w}}_{t+1} = \Pi_{(1-\alpha)\mathcal{W}}[\tilde{\boldsymbol{w}}_t - \eta \tilde{\boldsymbol{g}}_t]$;
9: {**Case 2. Two-Point Feedback Model**};
10: Calculate $\boldsymbol{w}_t^{(1)} = \tilde{\boldsymbol{w}}_t + \delta \boldsymbol{s}_t$ and $\boldsymbol{w}_t^{(2)} = \tilde{\boldsymbol{w}}_t - \delta \boldsymbol{s}_t$;
11: Receive $f_t(\boldsymbol{w}_t^{(1)})$ and $f_t(\boldsymbol{w}_t^{(2)})$ as the feedback;
12: Construct the gradient estimator: $\tilde{\boldsymbol{g}}_t = \frac{d}{2\delta}(f_t(\tilde{\boldsymbol{w}}_t + \delta \boldsymbol{s}_t) - f_t(\tilde{\boldsymbol{w}}_t - \delta \boldsymbol{s}_t)) \cdot \boldsymbol{s}_t$;
13: Update $\tilde{\boldsymbol{w}}_{t+1}$: $\tilde{\boldsymbol{w}}_{t+1} = \Pi_{(1-\alpha)\mathcal{W}}[\tilde{\boldsymbol{w}}_t - \eta \tilde{\boldsymbol{g}}_t]$;
14: **end for**

a groundbreaking online ensemble learning framework (named Parameter−free Bandit Gradient Descent, PBGD), which avoids the dependency of P_T.

To deeply analyze, Zhao *et al.* [20] divided the expected dynamic regret of the one-point feedback model into three terms:

$$\mathbb{E}\left[\sum_{t=1}^{T} f_t(\boldsymbol{w}_t)\right] - \sum_{t=1}^{T} f_t(\boldsymbol{u}_t) = \underbrace{\mathbb{E}\left[\sum_{t=1}^{T} \left(\hat{f}_t(\tilde{\boldsymbol{w}}_t) - \hat{f}_t(\boldsymbol{v}_t)\right)\right]}_{\text{term (a)}}$$
$$+ \underbrace{\mathbb{E}\left[\sum_{t=1}^{T} \left(f_t(\boldsymbol{w}_t) - \hat{f}_t(\tilde{\boldsymbol{w}}_t)\right)\right]}_{\text{term (b)}} + \underbrace{\mathbb{E}\left[\sum_{t=1}^{T} \left(\hat{f}_t(\boldsymbol{v}_t) - f_t(\boldsymbol{u}_t)\right)\right]}_{\text{term (c)}}, \quad (5)$$

where $\boldsymbol{v}_1, \ldots, \boldsymbol{v}_T$ is the scaled comparator sequence and $\boldsymbol{v}_t$ is defined as $\boldsymbol{v}_t = (1-\alpha)\boldsymbol{u}_t$. Due to Lemma 1, it can be proved that the term (b) and term (c) will not exceed $2L\delta T$ and $(L\delta + L\alpha R)T$, respectively. As for term (a), it essentially describes the dynamic regret of the smoothed function $\hat{f}_1, \ldots, \hat{f}_T$ about the scaled comparator sequence $\boldsymbol{v}_1, \ldots, \boldsymbol{v}_T$. It relies on step size setting, which is related to path length P_T. Thus, we only need to design algorithms for term (a) to remove the dependency of P_T.

However, online ensemble learning can not directly be utilized with existing BCO algorithms. Assume there are N experts in an ensemble learning framework, then need to get N original function values to get $(\nabla \hat{f}_t(\tilde{\boldsymbol{w}}_t^1), \nabla \hat{f}_t(\tilde{\boldsymbol{w}}_t^2), \ldots, \nabla \hat{f}_t(\tilde{\boldsymbol{w}}_t^N))$ for updating at round t. It is unavailable since the environment can only provide one or two points as feedback at each round. To address this problem, Zhao *et al.* [20] find that a linear function can bound the expected dynamic regret of $\hat{f}_t$.

Proposition 1. $\mathbb{E}[\hat{f}_t(\tilde{\boldsymbol{w}}_t) - \hat{f}_t(\boldsymbol{v}_t)] \leq \mathbb{E}[\langle \tilde{\boldsymbol{g}}_t, \tilde{\boldsymbol{w}}_t - \boldsymbol{v}_t \rangle]$.

Making use of this proposition, a surrogate loss function [20] $\ell_t : (1-\alpha)\mathcal{W} \mapsto \mathbb{R}$, is designed as:

$$\ell_t(\boldsymbol{x}) = \langle \tilde{\boldsymbol{g}}_t, \boldsymbol{x} - \tilde{\boldsymbol{w}}_t \rangle, \tag{6}$$

which can be seen as a linearization of $\hat{f}_t$ at the point $\tilde{\boldsymbol{w}}_t$. Moreover, the surrogate loss function constructed above naturally satisfies the following two properties.

Property 1. $\forall \boldsymbol{v} \in (1-\alpha)\mathcal{W}$, $\mathbb{E}[\hat{f}_t(\tilde{\boldsymbol{w}}_t) - \hat{f}_t(\boldsymbol{v})] \leq \mathbb{E}[\ell_t(\tilde{\boldsymbol{w}}_t) - \ell_t(\boldsymbol{v})]$.

Property 2. $\forall \boldsymbol{w} \in (1-\alpha)\mathcal{W}$, $\nabla \ell_t(\boldsymbol{w}) = \tilde{\boldsymbol{g}}_t$.

Due to Property 1, we only need to design an online algorithm to optimize the surrogate loss function to approximate the dynamic regret. Property 2 means that it is available to initialize N experts since it directly holds that all experts' gradients, namely $\nabla\hat{\ell}_t(\tilde{\boldsymbol{w}}_t^1), \nabla\hat{\ell}_t(\tilde{\boldsymbol{w}}_t^2), \ldots, \nabla\hat{\ell}_t(\tilde{\boldsymbol{w}}_t^N)$, equal to $\tilde{g}_t$.

For the one-point feedback model, Zhao *et al.* [20] proposed a PBGD algorithm, which uses the surrogate loss function (Eq. (6)) to replace the original function f_t (its smoothed version $\hat{f}_t$). For one-point feedback model, refer to *Thoerem 3.5* in [20], the upper bound includes two parameters, η and δ, the optimal parameter setting can be $\eta^* = (dC\tilde{L})^{-1/2}((7R^2 + RP_T)/T)^{3/4}$ and $\delta^* = (dC/\tilde{L})^{1/2}((7R^2+RP_T)/T)^{1/4}$ (to simplify notation, let $\tilde{L} = 3L+LR/r$ represent the effective Lipschitz constant). It is easy to find that step size and perturbation parameter both need to use an unknown path length $P_T = \sum_{t=2}^{T} \|\boldsymbol{u}_t - \boldsymbol{u}_{t-2}\|_2$ as input for the algorithm. Thus, in the PBGD algorithm, the optimal parameter setting is as

$$\delta^{(1)^\dagger} = \left(\frac{dCR}{\tilde{L}}\right)^{\frac{1}{2}} T^{-\frac{1}{4}}, \quad \eta^{(1)^\dagger} = \sqrt{\frac{R(7R^2 + RP_T)}{2dC\tilde{L}}} T^{-\frac{3}{4}}. \tag{7}$$

Here, the $\delta^{(1)^\dagger}$ is not associated with P_T but $\eta^{(1)^\dagger}$ still need the prior knowledge P_T. Next, We will introduce how to remove the dependency of P_T through maintaining multiple experts in the PBGD framework [20].

Step Size Pool. In fact, the path length P_T is nonnegative and bounded, namely $0 \leq P_T \leq 2RT$ (the upper bound $2RT$ gets from Assumption 1). We also know $\eta^{(1)^\dagger} = \sqrt{\frac{R(7R^2+RP_T)}{2dC\tilde{L}}} T^{-\frac{3}{4}}$ (Eq. (7)), thus, the range of $\eta^{(1)^\dagger}$ is

$$\sqrt{\frac{7R^3}{2dC\tilde{L}}} \cdot T^{-\frac{3}{4}} \leq \eta^{(1)^\dagger} \leq \sqrt{\frac{(7+2T)R^3}{2dC\tilde{L}}} \cdot T^{-\frac{3}{4}}.$$

According to the range of values, the step-size candidate pool $\mathcal{P}^{(1)}$ is constructed to discretize the above range as

$$\mathcal{P}^{(1)} = \left\{ \eta_i^{(1)} = 2^{i-1} \sqrt{\frac{7R^3}{2dC\tilde{L}}} \cdot T^{-\frac{3}{4}} \mid i = 1, \ldots, N^{(1)} \right\}, \tag{8}$$

Algorithm 2. PBGD: Expert-algorithm

Input: time horizon T, step sizes $\eta_i^{(1)} \in \mathcal{P}_{(1)}$ and $\eta_i^{(2)} \in \mathcal{P}_{(2)}$, perturbation parameter δ and shrinkage parameter α ;

1: Initialize $\tilde{\boldsymbol{w}}_0^i \in (1-\alpha)\mathcal{W}$ and send $\tilde{\boldsymbol{w}}_0^i$ to Meta ;
2: **for** $t = 1, 2, \cdots, T$ **do**
3:    **if** Use one-point feedback model **then**
4:       $\tilde{\boldsymbol{w}}_{t+1}^i = \Pi_{(1-\alpha)\mathcal{W}}[\tilde{\boldsymbol{w}}_t^i - \eta_i^{(1)}\tilde{\boldsymbol{g}}_t]$;
5:    **else if** Use two-point feedback model **then**
6:       $\tilde{\boldsymbol{w}}_{t+1}^i = \Pi_{(1-\alpha)\mathcal{W}}[\tilde{\boldsymbol{w}}_t^i - \eta_i^{(2)}\tilde{\boldsymbol{g}}_t]$;
7:    **end if**
8:    Send $\tilde{\boldsymbol{w}}_{t+1}^i$ to meta expert;
9: **end for**

where $N^{(1)}$ represents the number of step-size candidates and is set as [20]

$$N^{(1)} = \left\lceil \frac{1}{2} \log_2 \left(1 + \frac{2T}{7}\right) \right\rceil + 1. \tag{9}$$

Since $N^{(1)} = O(\log T)$, the computation cost of maintaining a PBGD framework is acceptable. Besides, if we set a step-size candidate pool due to Eq. (8), it satisfies the following relationship

$$\eta_k^{(1)} \le \eta^{(1)^\dagger} \le \eta_{k+1}^{(1)} = 2\eta_k^{(1)}, \quad k \in \{1, \ldots, N^{(1)} - 1\}. \tag{10}$$

The PBGD framework includes two parts: an expert-algorithm and a meta-algorithm. The following gives an example of the one-point feedback model.

Expert-algorithm. Suppose there are $N^{(1)}$ experts in our PBGD framework. Let $\mathcal{B} = \{B_1, \ldots, B_{N^{(1)}}\}$ denote the expert set, where B_i denotes the i-th expert who use step size $\eta_i^{(1)} \in \mathcal{P}^{(1)}$. Due to the construction of the surrogate function (Eq. (6)), each expert updates their own $\tilde{\boldsymbol{w}}_t^i$ at t round as follows:

$$\tilde{\boldsymbol{w}}_{t+1}^i = \Pi_{(1-\alpha)\mathcal{W}}\left[\tilde{\boldsymbol{w}}_t^i - \eta_i^{(z)}\nabla\ell_t(\tilde{\boldsymbol{w}}_t^i)\right] = \Pi_{(1-\alpha)\mathcal{W}}\left[\tilde{\boldsymbol{w}}_t^i - \eta_i^{(z)}\tilde{\boldsymbol{g}}_t\right], \tag{11}$$

where the superscript $z \in \{1, 2\}$ denotes the step size for one-point or two-point feedback models, respectively. The process of the expert-algorithm is illustrated in Algorithm 2.

Meta-algorithm. In order to combine each expert's outputs (line 8 in Algorithm 2), a meta-algorithm is designed (Algorithm 3.2 *PBDG: Meta-expert* in [20]). They utilized an exponentially weighted average forecaster algorithm [6] to update the weight of each expert's output at t round. The process of the original *PBDG: Meta-expert* algorithm for one-point BCO is equivalent to the algorithm 3 which only includes Phase 2 (namely $\lambda = 0$).

Algorithm 3. Hybridized-PBGD: Meta-algorithm

Input: time horizon T, the pool of candidate step sizes $\mathcal{P}^{(1)}$ and $\mathcal{P}^{(2)}$, adjust parameter λ and learning rate of meta-algorithm ϵ ;
1: Run N expert-algorithms (Alg. 2) simultaneously with different step size $\eta_i^{(z)}$;
2: Initialize the weight for each expert $i \in [N]$: $w_1^i = \frac{N+1}{N} \cdot \frac{1}{i(i+1)}$;
3: **for** $t = 1, 2, \cdots, \lceil \lambda T \rceil$ **do**
4: {**Phase 1. Two-Point Feedback Model**};
5: Receive $\tilde{\boldsymbol{w}}_t^i$ from each expert $i \in [N]$;
6: Get $\tilde{\boldsymbol{w}}_t = \sum_{i \in [N]} w_t^i \tilde{\boldsymbol{w}}_t^i$;
7: Construct the gradient estimator: $\tilde{\boldsymbol{g}}_t = \frac{d}{2\delta}(f_t(\tilde{\boldsymbol{w}}_t + \delta \boldsymbol{s}_t) - f_t(\tilde{\boldsymbol{w}}_t - \delta \boldsymbol{s}_t)) \cdot \boldsymbol{s}_t$;
8: Construct surrogate loss $\ell_t(\cdot)$ due to Eq. (6);
9: Update the weight of each expert $i \in [N]$: $w_{t+1}^i = \frac{w_t^i \exp(-\epsilon \ell_t(\tilde{\boldsymbol{w}}_t^i))}{\sum_{i \in [N]} w_t^i \exp(-\epsilon \ell_t(\tilde{\boldsymbol{w}}_t^i))}$;
10: Update the gradient estimator $\tilde{\boldsymbol{g}}_t$ to all experts;
11: **end for**
12: Initialize the weight for each expert $i \in [N]$: update $w_{T'}^i$ due to Eq. (16);
13: **for** $t = \lceil \lambda T \rceil + 1, \cdots, T$ **do**
14: {**Phase 2. One-Point Feedback Model**};
15: Receive $\tilde{\boldsymbol{w}}_t^i$ from each expert $i \in [N]$;
16: Get $\tilde{\boldsymbol{w}}_t = \sum_{i \in [N]} w_t^i \tilde{\boldsymbol{w}}_t^i$;
17: Construct the gradient estimator: $\tilde{\boldsymbol{g}}_t = \frac{d}{\delta} f_t(\tilde{\boldsymbol{w}}_t + \delta \boldsymbol{s}_t) \cdot \boldsymbol{s}_t$;
18: Construct surrogate loss $\ell_t(\cdot)$ due to Eq. (6);
19: Update the weight of each expert $i \in [N]$: $w_{t+1}^i = \frac{w_t^i \exp(-\epsilon \ell_t(\tilde{\boldsymbol{w}}_t^i))}{\sum_{i \in [N]} w_t^i \exp(-\epsilon \ell_t(\tilde{\boldsymbol{w}}_t^i))}$;
20: Update the gradient estimator $\tilde{\boldsymbol{g}}_t$ to all experts;
21: **end for**

3.2 Hybridized-PBGD

In [20], they used the PBGD algorithm to analyze the upper bounds of dynamic regret for one-point and two-point feedback models, respectively. We also notice that Zhao *et al.* [21] proposed a *MABCO* algorithm in their follow-up work, but this algorithm uses adaptive regret as the metric rather than dynamic regret. Thus, it still lacks theoretical analysis about the upper bounds of dynamic regret in the hybridization of one- and two-point BCO about PBGD. It is natural to explore whether can combine these two feedback models into a whole PBGD algorithm framework. However, it is infeasible to simply concatenate the two models. For instance, at the transfer round t when from one model to another, how to update expert i's weight $w_t^i \rightarrow w_{t+1}^i$ is a challenge. Besides, if not find a feasible solution to update $w_t^i \rightarrow w_{t+1}^i$, the original analysis method of upper bounds may no longer be applicable for phase 2 in Algorithm 3. Based on both the idea and challenges, we propose a Hybridized-PBGD algorithm that combines one- and two-point feedback models into a whole framework with time horizon T, as illustrated in Algorithm 3.

In our algorithm, it is made up of two phases, in order of two-point and one-point feedback models. Note that the Hybridized-PBGD algorithm is symmetric, the sequence will not influence the theoretical analysis. However, it is reasonable to put a two-point feedback model in front since a two-point model with more feedback information benefits the initial process of the meta-algorithm (Algorithm 3). When each expert initializes its weight values of w_1^i at round $t = 1$ (Algorithm 3), more information combination helps the meta-algorithm fit faster. We define an adjusted parameter $\lambda \in [0, 1]$, which is used to adjust the proportion rate of two phases. From $t = 1$ to $t = [\lambda T]$ ($[\cdot]$ represent the integer part of a number), the algorithm executes the two-point feedback model (line $2-11$), while the rest time executes the one-point feedback model (line $12-21$).

For the two-point feedback model, the analysis is similar to the above-mentioned one-point feedback model. Refer to *theorem 3.6* in [20], the optimal step size $\eta^{(2)^\dagger}$ is set as:

$$\eta^{(2)^\dagger} = \sqrt{\frac{7R^2 + RP_T}{2d^2L^2T}}. \tag{12}$$

Step Size Pool. Similarly, use the range of path length P_T, we can get:

$$\sqrt{\frac{7R^2}{2d^2L^2T}} \leq \eta^{(2)^\dagger} \leq \sqrt{\frac{7R^2 + 2R^2T}{2d^2L^2T}}. \tag{13}$$

Thus, we can construct the step-size candidate pool $\mathcal{P}^{(2)}$ as follows:

$$\mathcal{P}^{(2)} = \left\{ \eta_i^{(2)} = 2^{i-1}\sqrt{\frac{7R^2}{2d^2L^2T}} \mid i = 1, \ldots, N^{(2)} \right\}, \tag{14}$$

here $N^{(2)}$ represents the number of step-size candidates and is set as:

$$N^{(2)} = \left\lceil \frac{1}{2}\log_2\left(1 + \frac{2T}{7}\right) \right\rceil + 1. \tag{15}$$

The success of hybridization needs to maintain the same expert number during the two phases. Thus, we use the whole time horizon T to make $N^{(2)} = N^{(1)}$ (for simplicity, use N in the rest of paper). Then, use $T = T_1$ and $T = T_2$ to construct $\mathcal{P}^{(1)}$ and $\mathcal{P}^{(2)}$'s step sizes, respectively. Same as Eq. (10), we can also get the relationship

$$\eta_k^{(2)} \leq \eta^{(2)^\dagger} \leq \eta_{k+1}^{(2)} = 2\eta_k^{(2)}, k \in \{1, \ldots, N-1\}. \tag{16}$$

Expert-Algorithm and Meta-algorithm. Since two models maintain same N experts, it can also use Algorithm 2 and Eq. (11). As for the meta-algorithm, setting $\lambda = 1$ in Algorithm 3 corresponds to the two-point feedback model.

Weight Inheritance. In Algorithm 3, it can get weight parameters (denoted as $w_{[\lambda T]+1}^i$) updated in advance through line 9 after $[\lambda T]$ rounds. From an empirical perspective, those weights learned from Phase 1 are reasonable. However, if we directly use $w_{[\lambda T]+1}^i$ as initial weights for Phase 2, it brings big challenges to the analysis of the upper bound of Phase 2. Since the scaling trick to analyze bounds with the help of Lemma 2 will not be applicable in this case. But, if we directly use $w_{[\lambda T]+1}^i = (N+1)/N \cdot 1/(i(i+1))$ for Phase 2, the weights of Phase 1 seem meaningless since we can not inherit Phase 1's experienced weights. To address this problem to some extent, we propose an intermediate weight initialization strategy.

Definition 1 (Intermediate Weight Initialization Strategy). *Denote by* $T' := \lfloor \lambda T \rfloor + 1$, *we define* $w^i_{T'}$ *as the initial weight for Phase 2 and* $\bar{w}^i_{T'}$ *(i.e.,* $w^i_{\lfloor \lambda T \rfloor + 1}$*) as the weight gotten from the last round in Phase 1. Let* $w^i_{T'}$ *satisfy the following rules:*

$$w^i_{T'} = \begin{cases} \frac{N+1}{N} \cdot \frac{1}{i(i+1)} & \bar{w}^i_{T'} \notin \left[\frac{N+1}{N} \cdot \frac{1}{i(i+1)}, 1\right] \\ \bar{w}^i_{T'} & \bar{w}^i_{T'} \in \left[\frac{N+1}{N} \cdot \frac{1}{i(i+1)}, 1\right]. \end{cases} \tag{17}$$

With this strategy, we can effectively retain part of the empirical weights of Phase 1, meanwhile, can still utilize Lemma 2 for Phase 2's upper bound analysis.

4 Regret Analysis

In this section, we establish the upper bound of the expected dynamic regrets of the Hybridized-PBGD algorithm. We first analyze Phase 1 and Phase 2, respectively, and then summarize the whole upper bound of the algorithm.

Theorem 1. *As shown in Algorithm 3, let* $T_1 = \lfloor \lambda T \rfloor$ *and the expected dynamic regret of Phase 1 satisfies*

$$\begin{aligned} \mathbb{E}[\textit{D-Regret}_{T_1}(\boldsymbol{u}_1, \ldots, \boldsymbol{u}_{T_1})] &= \mathbb{E}\left[\sum_{t=1}^{T_1} \frac{1}{2}(f_t(\boldsymbol{w}_t^{(1)}) + f_t(\boldsymbol{w}_t^{(2)}))\right] - \sum_{t=1}^{T_1} f_t(\boldsymbol{u}_t) \\ &\le dLR\sqrt{2T_1}(2 + 2\ln(1 + \lceil \log_2(1 + P_{T_1}/(7R)) \rceil) + \sqrt{7 + (P_{T_1}/R)}) \\ &= \mathcal{O}({T_1}^{1/2}(1 + P_{T_1})^{1/2}). \end{aligned}$$

Proof. We first introduce two useful lemmas.

Lemma 1 ([20]). *According to the Theorem 3.13 in [20], we have the following conclusion about Eq. (5). Here, we have*

$$\textit{term (b)} = \mathbb{E}\left[\sum_{t=1}^{T} \left(f_t(\boldsymbol{w}_t) - f_t(\tilde{\boldsymbol{w}}_t) + f_t(\tilde{\boldsymbol{w}}_t) - \hat{f}_t(\tilde{\boldsymbol{w}}_t)\right)\right] \le 2L\delta T. \tag{18}$$

$$\begin{aligned} \textit{term (c)} &\le \mathbb{E}\left[\sum_{t=1}^{T} \left(|\hat{f}_t(\boldsymbol{v}_t) - f_t(\boldsymbol{v}_t)| + |f_t(\boldsymbol{v}_t) - f_t(\boldsymbol{u}_t)|\right)\right] \\ &\le \mathbb{E}\left[\sum_{t=1}^{T} (L\delta + L\|\boldsymbol{v}_t - \boldsymbol{u}_t\|_2)\right] \\ &\le \mathbb{E}\left[\sum_{t=1}^{T} (L\delta + L\alpha R)\right] \textit{(set } \alpha = \delta/r) \\ &= \left(L + \frac{LR}{r}\right)\delta T. \end{aligned} \tag{19}$$

Lemma 2 ([20]). *According to the Lemma 3.14 in [20], for each $i \in [N]$, we have the relationship as*

$$\sum_{t=1}^{T} \ell_t(\tilde{\boldsymbol{w}}_t) - \sum_{t=1}^{T} \ell_t(\tilde{\boldsymbol{w}}_t^i) \leq \tilde{G}R\sqrt{2T}\left(1 + \ln\frac{1}{w_1^i}\right), \tag{20}$$

where $\tilde{G} = \sup_{t\in[T]} \|\tilde{\boldsymbol{g}}_t\|_2$.

Now, we can prove the Theorem 1 based on Lemma 1 and Lemma 2.

For the two-point feedback model, the expected dynamic regret can be decomposed as [21]:

$$\begin{aligned}
&\mathbb{E}\left[\sum_{t=1}^{T_1} \frac{1}{2}(f_t(\boldsymbol{w}_t^{(1)}) + f_t(\boldsymbol{w}_t^{(2)}))\right] - \sum_{t=1}^{T_1} f_t(\boldsymbol{u}_t) \\
&= \mathbb{E}\left[\sum_{t=1}^{T_1} \frac{1}{2}(f_t(\tilde{\boldsymbol{w}}_t + \delta\boldsymbol{s}_t) + f_t(\tilde{\boldsymbol{w}}_t - \delta\boldsymbol{s}_t))\right] - \sum_{t=1}^{T_1} f_t(\boldsymbol{u}_t) \\
&\overset{(a)}{\leq} \mathbb{E}\left[\sum_{t=1}^{T_1} f_t(\tilde{\boldsymbol{w}}_t)\right] + L\delta T_1 - \sum_{t=1}^{T_1} f_t(\boldsymbol{u}_t) \\
&= \underbrace{\mathbb{E}\left[\sum_{t=1}^{T_1} \hat{f}_t(\tilde{\boldsymbol{w}}_t) - \sum_{t=1}^{T_1} \hat{f}_t(\boldsymbol{v}_t)\right]}_{\text{term (a)}} + \delta L T_1 \\
&\quad + \underbrace{\mathbb{E}\left[\sum_{t=1}^{T_1} \left(f_t(\tilde{\boldsymbol{w}}_t) - \hat{f}_t(\tilde{\boldsymbol{w}}_t)\right)\right]}_{\text{term (b)}} + \underbrace{\left[\sum_{t=1}^{T_1} \left(\hat{f}_t(\boldsymbol{v}_t) - f_t(\boldsymbol{u}_t)\right)\right]}_{\text{term (c)}}.
\end{aligned} \tag{21}$$

The Ineq. (a) uses the Lipschitz property of online functions:

$$f_t(\tilde{\boldsymbol{w}}_t + \delta\boldsymbol{s}_t) \leq f_t(\tilde{\boldsymbol{w}}_t) + L\|\delta\boldsymbol{s}_t\|_2 = f_t(\tilde{\boldsymbol{w}}_t) + \delta L,$$

the same result holds for $f_t(\tilde{\boldsymbol{w}}_t - \delta\boldsymbol{s}_t)$. Due to Lemma 1, we know that term (b) and term (c) are bounded. Thus, only need to consider term (a).

According to Proposition 1, we have

$$\text{term (a)} \leq \mathbb{E}\left[\sum_{t=1}^{T_1} (\ell_t(\tilde{\boldsymbol{w}}_t) - \ell_t(\boldsymbol{v}_t))\right]. \tag{22}$$

Based on Eq. (22), we can further split the term within the expectation as

$$\begin{aligned}
\sum_{t=1}^{T_1} (\ell_t(\tilde{\boldsymbol{w}}_t) - \ell_t(\boldsymbol{v}_t)) &= \underbrace{\sum_{t=1}^{T_1} \ell_t(\tilde{\boldsymbol{w}}_t) - \sum_{t=1}^{T_1} \ell_t(\tilde{\boldsymbol{w}}_t^k)}_{\text{meta-regret}} \\
&\quad + \underbrace{\sum_{t=1}^{T_1} \ell_t(\tilde{\boldsymbol{w}}_t^k) - \sum_{t=1}^{T_1} \ell_t(\boldsymbol{v}_t)}_{\text{base-regret}},
\end{aligned} \tag{23}$$

which includes two terms and we analyze base-regret in advance.

We choose the optimal expert, denote by $k^{(2)^*} \in \{1, \ldots, N-1\}$, to obtain the tightest possible bound. In Eq. (14), let $T = T_1$ then Eq. (16) holds. Besides, it directly follows:

$$k^{(2)^*} \leq \left\lceil \frac{1}{2} \log_2 \left(1 + \frac{P_{T_1}}{7R}\right) \right\rceil + 1 \overset{(P_{T_1} < P_T)}{\leq} \left\lceil \frac{1}{2} \log_2 \left(1 + \frac{P_T}{7R}\right) \right\rceil + 1. \tag{24}$$

According to the *theorem 3.12* and the process of *theorem 3.10 proof* in [20], the dynamic regret of base-regret term can be shown as

$$\begin{aligned}
\text{base-regret} &= \sum\nolimits_{t=1}^{T_1} \ell_t(\tilde{\boldsymbol{w}}_t^{k^{(2)^*}}) - \sum\nolimits_{t=1}^{T_1} \ell_t(\boldsymbol{v}_t) \\
&\leq \frac{7R^2 + RP_{T_1}}{4\eta_{k^{(2)^*}}} + \frac{\eta_{k^{(2)^*}} \tilde{G}^2 T_1}{2} \\
&\overset{(Eq.16)}{\leq} \frac{7R^2 + RP_{T_1}}{2\eta^{(2)^\dagger}} + \frac{\eta^{(2)^\dagger} d^2 L^2 T_1}{2} \\
&\overset{(Eq.12)}{=} \frac{3\sqrt{2}}{4} dL \sqrt{T_1(7R^2 + RP_{T_1})},
\end{aligned} \tag{25}$$

where $\tilde{G} = \sup_{t \in [T]} \|\tilde{\boldsymbol{g}}_t\|_2 = dL$.

Subsequently, we can utilize Lemma 2 to analyze the dynamic regret of meta-regret:

$$\begin{aligned}
\text{meta-regret} &= \sum\nolimits_{t=1}^{T_1} \ell_t(\tilde{\boldsymbol{w}}_t) - \sum\nolimits_{t=1}^{T_1} \ell_t(\tilde{\boldsymbol{w}}_t^{k^{(2)^*}}) \\
&\overset{\text{Eq. (20)}}{\leq} \tilde{G}R\sqrt{2T_1}\left(1 + \ln(1/w_1^{k^{(2)^*}})\right) \left(\textit{where } w_1^i = \frac{N+1}{N} \frac{1}{i(i+1)}\right) \\
&\leq \tilde{G}R\sqrt{2T_1}\left(1 + \ln k^{(2)^*} + \ln(k^{(2)^*} + 1)\right) \\
&\leq dLR\sqrt{2T_1}(1 + 2\ln(k^{(2)^*} + 1)).
\end{aligned} \tag{26}$$

Thus, the total dynamic regret of term (a) can be shown as

$$\begin{aligned}
\text{term (a)} &\leq \frac{3\sqrt{2}}{4} dL\sqrt{T_1(7R^2 + RP_{T_1})} + dLR\sqrt{2T_1}(1 + 2\ln(k^{(2)^*} + 1)) \\
&\leq dLR\sqrt{2T_1}(1 + 2\ln(k^{(2)^*} + 1) + \sqrt{7 + (P_{T_1}/R)}).
\end{aligned} \tag{27}$$

Finally, the total dynamic regret of Eq. (21) can be shown as

$$\begin{aligned}
\text{dynamic regret} &\leq \text{ term (a)+term (b)+term (c)} + L\delta T_1 \\
&\leq \text{term (a)} + L\delta T_1 + (L\delta + L\alpha R)T_1 + L\delta T_1 \\
&\leq dLR\sqrt{2T_1}(1 + 2\ln(k^{(2)^*} + 1)) + \frac{3\sqrt{2}}{4} dL\sqrt{T_1(7R^2 + RP_{T_1})} \\
&\quad + dLR\sqrt{T_1}
\end{aligned}$$

$$\begin{aligned}&\overset{\text{Eq. (24)}}{\leq} dLR\sqrt{2T_1}(2+2\ln(1+\lceil\log_2(1+P_{T_1}/(7R))\rceil)\\&+\sqrt{7+(P_{T_1}/R)})\\&=\mathcal{O}({T_1}^{1/2}(1+P_{T_1})^{1/2}).\end{aligned} \tag{28}$$

Therefore, this theorem has been proved. ■

Theorem 2. *As shown in Algorithm 3, let $T_2 = T - \lceil\lambda T\rceil$ and the expected dynamic regret of Phase 2 satisfies*

$$\begin{aligned}&\mathbb{E}[\textit{D-Regret}_{T_2}(\boldsymbol{u}_1,\ldots,\boldsymbol{u}_{T_2})]=\mathbb{E}\left[\sum_{t=1}^{T_2} f_t(\boldsymbol{w}_t)\right]-\sum_{t=1}^{T_2} f_t(\boldsymbol{u}_t)\\&\leq\sqrt{2dCR(3L+LR/r)}\cdot {T_2}^{3/4}\\&\quad\cdot(2+2\ln(1+\lceil\log_2(1+P_{T_2}/(7R))\rceil)+\sqrt{7+(P_{T_2}/R)})\\&=\mathcal{O}({T_2}^{3/4}(1+P_{T_2})^{1/2}).\end{aligned} \tag{29}$$

Proof. The analysis of Phase 2 is similar to Phase 1, the expected dynamic regret of the one-point feedback expands as Eq. (5). We also only need to consider term (a), namely analyze the surrogate loss function's dynamic regret:

$$\begin{aligned}\sum_{t=1}^{T_2}(\ell_t(\tilde{\boldsymbol{w}}_t)-\ell_t(\boldsymbol{v}_t))=&\underbrace{\sum_{t=1}^{T_2}\ell_t(\tilde{\boldsymbol{w}}_t)-\sum_{t=1}^{T_2}\ell_t(\tilde{\boldsymbol{w}}_t^k)}_{\text{meta-regret}}\\&+\underbrace{\sum_{t=1}^{T_2}\ell_t(\tilde{\boldsymbol{w}}_t^k)-\sum_{t=1}^{T_2}\ell_t(\boldsymbol{v}_t)}_{\text{base-regret}},\end{aligned} \tag{30}$$

which includes two terms and we analyze base-regret in advance.

Similarly, denote the best expert by $k^{(1)^*} \in \{1,\ldots,N-1\}$. Let $T = T_2$ in Eq. (8), then Eq. (10) holds. Besides, it directly follows:

$$k^{(1)^*}\leq\left\lceil\frac{1}{2}\log_2\left(1+\frac{P_{T_2}}{7R}\right)\right\rceil+1\overset{(P_{T_2}<P_T)}{\leq}\left\lceil\frac{1}{2}\log_2\left(1+\frac{P_T}{7R}\right)\right\rceil+1. \tag{31}$$

Similar with Eq. (25), we can get the dynamic regret of base-regret as:

$$\begin{aligned}\text{base-regret}&=\sum_{t=1}^{T_2}\ell_t(\tilde{\boldsymbol{w}}_t k^{(1)^*})-\sum_{t=1}^{T_2}\ell_t(\boldsymbol{v}_t)\\&\leq\frac{7R^2+RP_{T_2}}{4\eta_{k^{(1)*}}}+\frac{\eta_{k^{(1)*}}\tilde{G}^2T_2}{2}\\&\overset{\text{Eq. (10)}}{\leq}\frac{7R^2+RP_{T_2}}{2\eta^{(1)\dagger}}+\frac{{\eta^{(1)}}^{\dagger}d^2C^2T_2}{2\delta^2}\\&=\sqrt{\frac{(7R^2+RP_{T_2})dC\tilde{L}}{2R}}\cdot {T_2}^{3/4}+\sqrt{\frac{(7R^2+RP_{T_2})dC\tilde{L}}{4R}}\cdot {T_2}^{3/4}\\&=\sqrt{2dC\tilde{L}(7R+P_{T_2})}\cdot {T_2}^{3/4},\end{aligned} \tag{32}$$

where $\tilde{G} = \sup_{t\in[T]}\|\tilde{\boldsymbol{g}}_t\|_2 = dC/\delta$, the last equality can be obtained by substituting the specific setting $\eta^{(1)^\dagger} = \sqrt{R(7R^2 + RP_{T_2})/(2dC\tilde{L})} \cdot {T_2}^{-3/4}$ and $\delta = \delta^{(1)\dagger} = (dCR/\tilde{L})^{1/2}{T_2}^{-1/4}$.

As for meta-regret, we still can use Lemma 2 under the *Intermediate weight initialization strategy* (Eq. (17)). We first consider the case: $w^i_{T'} = \bar{w}^i_{T'}$. We have the dynamic regret as

$$\begin{aligned}
\text{meta-regret} &= \sum\nolimits_{t=1}^{T_2} \ell_t(\tilde{\boldsymbol{w}}_t) - \sum\nolimits_{t=1}^{T_2} \ell_t(\tilde{\boldsymbol{w}}_t^{k^{(1)^*}}) \\
&\le \tilde{G}R\sqrt{2T_2}\left(1 + \ln(1/w_1^{k^{(1)^*}})\right) (\text{here } w_1^{k^{(1)^*}} = w^i_{T'}) \\
&\overset{(a)}{\le} \frac{dCR}{\delta}\sqrt{2T_2}(1 + + \ln[\frac{Nk^{(1)^*}}{N+1}(k^{(1)^*} + 1)]) \\
&\le \frac{dCR}{\delta}\sqrt{2T_2}(1 + \ln k^{(1)^*} + \ln(k^{(1)^*} + 1)) \\
&\le \frac{dCR}{\delta}\sqrt{2T_2}(1 + 2\ln(k^{(1)^*} + 1))(\delta = \delta^{(1)\dagger}) \\
&= \sqrt{2dCR\tilde{L}}{T_2}^{3/4}(1 + 2\ln(k^{(1)^*} + 1)). \qquad (33)
\end{aligned}$$

The Ineq. (a) holds since $w_1^{k^{(1)^*}} > \frac{N+1}{N} \cdot \frac{1}{k^{(1)^*}(k^{(1)^*}+1)}$. Another case is $w^i_{T'} = \frac{N+1}{N} \cdot \frac{1}{k^{(1)^*}(k^{(1)^*}+1)}$, the above results naturally holds.

Finally, the total dynamic regret of Eq. (5) can be shown as

$$\begin{aligned}
&\mathbb{E}\left[\sum\nolimits_{t=1}^{T_2} f_t(\boldsymbol{w}_t)\right] - \sum\nolimits_{t=1}^{T_2} f_t(\boldsymbol{u}_t) \\
&= \text{term (a)} + \text{term (b)} + \text{term (c)} \\
&\le \text{term (a)} + 2L\delta T_2 + (L\delta + L\alpha R)T_2 \\
&\le \sqrt{2dCR\tilde{L}}{T_2}^{3/4}(1 + 2\ln(k^{(1)^*} + 1) + \sqrt{7 + (P_{T_2}/R)}) \\
&\quad + (3L + LR/r)\sqrt{dCR/\tilde{L}}{T_2}^{3/4} \\
&\le \sqrt{2dCR\tilde{L}}{T_2}^{3/4}(2 + 2\ln(k^{(1)^*} + 1) + \sqrt{7 + (P_{T_2}/R)}) \\
&\overset{\text{Eq. (31)}}{\le} \sqrt{2dCR(3L + LR/r)} \cdot {T_2}^{3/4} \\
&\quad \cdot (2 + 2\ln(1 + \lceil\log_2(1 + P_{T_2}/(7R))\rceil) + \sqrt{7 + (P_{T_2}/R)}) \\
&= \mathcal{O}({T_2}^{3/4}(1 + P_{T_2})^{1/2}). \qquad (34)
\end{aligned}$$

Therefore, this theorem has been proven. ■

Theorem 3. *In the Algorithm 3, the total expected dynamic regret of two phases satisfies: Hybridized-PBGD Regret =*

$$= \underbrace{\mathbb{E}\left[\sum\nolimits_{t=1}^{T_1} \frac{1}{2}(f_t(\boldsymbol{w}_t^{(1)}) + f_t(\boldsymbol{w}_t^{(2)}))\right] - \sum\nolimits_{t=1}^{T_1} f_t(\boldsymbol{u}_t)}_{\textit{Phase 1}}$$

$$+\underbrace{\mathbb{E}\left[\sum_{t=1}^{T_2} f_t(\boldsymbol{w}_t)\right] - \sum_{t=1}^{T_2} f_t(\boldsymbol{u}_t)}_{Phase\ 2}$$

$$\le \sqrt{2dLR}\cdot(2+2\ln(1+\lceil\log_2(1+P_T/(7R))\rceil)$$
$$+\sqrt{7+(P_T/R)})\cdot(\sqrt{dLR}T_1^{1/2}+\sqrt{C(3+R/r)}T_2^{3/4})$$
$$=\mathcal{O}((1+P_T)^{1/2}[\beta(\lambda T)^{1/2}+((1-\lambda)T)^{3/4}]), \tag{35}$$

where β is a non-negative constant.

Proof. According to the Theorem 1 and Theorem 2, the whole dynamic regret of Hybridized-PBGD has

$$\text{Hybridized-PBGD Regret}$$
$$=\underbrace{\mathbb{E}\left[\sum_{t=1}^{T_1}\frac{1}{2}(f_t(\boldsymbol{w}_t^{(1)})+f_t(\boldsymbol{w}_t^{(2)}))\right]-\sum_{t=1}^{T_1} f_t(\boldsymbol{u}_t)}_{\text{Phase 1}}$$
$$+\underbrace{\mathbb{E}\left[\sum_{t=1}^{T_2} f_t(\boldsymbol{w}_t)\right] - \sum_{t=1}^{T_2} f_t(\boldsymbol{u}_t)}_{\text{Phase 2}}$$
$$\le \underbrace{dLR\sqrt{2T_1}(2+2\ln(1+\lceil\log_2(1+P_{T_1}/(7R))\rceil)+\sqrt{7+(P_{T_1}/R)})}_{\text{Phase 1}}$$
$$+\underbrace{\sqrt{2dCR\left(3L+\frac{LR}{r}\right)T_2{}^{3/4}}\left(2+2\ln(1+\lceil\log_2(1+\frac{P_{T_2}}{7R})\rceil)+\sqrt{7+\frac{P_{T_2}}{R}}\right)}_{\text{Phase 2}}$$
$$\overset{(a)}{\le} \underbrace{dLR\sqrt{2T_1}(2+2\ln(1+\lceil\log_2(1+P_T/(7R))\rceil)+\sqrt{7+(P_T/R)})}_{\text{Phase 1}}$$
$$+\underbrace{\sqrt{2dCR\left(3L+\frac{L}{r}\right)T_2{}^{3/4}}\left(2+2\ln(1+\lceil\log_2(1+\frac{P_T}{7R})\rceil)+\sqrt{7+\frac{P_T}{R}}\right)}_{\text{Phase 2}}$$
$$=\sqrt{2dLR}\cdot(2+2\ln(1+\lceil\log_2(1+P_T/(7R))\rceil)$$
$$+\sqrt{7+(P_T/R)})\cdot(\sqrt{dLR}T_1^{1/2}+\sqrt{C(3+R/r)}T_2^{3/4})$$
$$=\sqrt{2dLR}\cdot(2+2\ln(1+\lceil\log_2(1+P_T/(7R))\rceil)+\sqrt{7+(P_T/R)})$$
$$\cdot(\sqrt{dLR}(\lambda T)^{1/2}+\sqrt{C(3+R/r)}((1-\lambda)T)^{3/4})$$
$$=\mathcal{O}((1+P_T)^{1/2}[\beta(\lambda T)^{1/2}+((1-\lambda)T)^{3/4}]), \tag{36}$$

where Ineq. (a) uses the property: $P_{T_1}\le P_T$ and $P_{T_2}\le P_T$.
Therefore, this theorem has been proven. ■

Remarks. The theoretical analysis reveals that the phase 1: two-point feedback model brings a smaller upper bound meanwhile it spends more computational cost than phase 2. This shows the meaning of our Hybridized-PBGD, which can flexibly consider the computational cost by setting suitable λ and relaxing the upper bound.

5 Related Work

Bandit Convex Optimization About Static Regret. In early studies, most researchers focus on analyzing static regrets. The fundamental work of Flaxman *et al.* [9] proposed an unbiased gradient estimation method and proved an $\mathcal{O}(T^{3/4})$ expected static regret. With the property of strong convexity [1] or smoothness [16] of functions, researchers get $\mathcal{O}(T^{2/3})$. Recent breakthroughs [5] proved the existence of efficient online algorithms to obtain $\mathcal{O}(d^{9.5}\text{ploy}(\log T)\sqrt{T})$. Another study [13] provided a theoretical guarantee $\mathcal{O}(d^{2.5}\text{ploy}(\log T)\sqrt{T})$ that significantly reduces the dependency on dimensionality. For the two-point feedback model, Agarwal *et al.* [1] build $\mathcal{O}(d^2\sqrt{T})$ and $\mathcal{O}(d^2 \log T)$ for convexity and strong convexity Lipschitz continuous functions, respectively.

Bandit Convex Optimization About Dynamic Regret. Dynamic regret usually includes two types, universal dynamic regret and worse-case dynamic. Surrounding with the worse-case one, there are many studies [2,3,7,15,17,18]. But for the universal dynamic regret, the studies are limited. Thus, Zinkevich [23] proved a $\mathcal{O}(\sqrt{T}(1+P_T))$ dynamic regret with full-information setting. Zhao *et al.* [19,22] proposed an online ensemble learning framework with a full-information setting and used dynamic regret as the performance measure. In BCO problems, Zhao et al. [20,21] first proposed the PBGD algorithm which innovatively avoids the dependency of path length P_T, and proved $O(T^{\frac{3}{4}}(1+P_T)^{\frac{1}{2}})$ and $O(T^{\frac{1}{2}}(1+P_T)^{\frac{1}{2}})$ dynamic regret for one-point and two-point feedback models, respectively.

6 Conclusion

In this paper, we proposed a hybridized PBGD algorithm that utilizes one-point and two-point feedback models together to solve BCO problems. The original PBGD algorithm is an online assembled learning framework for a one-point or two-point feedback bandit convex model, where maintain N experts simultaneously to optimize the dynamic regret of the system effectively. To balance the computational cost and the upper bound of dynamic regrets more flexibly, it's necessary to design a sequentially hybridized algorithm while giving the strict theoretical derivation of the upper bounds. Our proposed algorithm innovatively utilizes the intermediate weight initialization strategy to overcome the challenge of weight updating at the moment of transition between two phases. Besides, we provide detailed theoretical proof that our hybridized PBGD algorithm holds the upper bound of dynamic regret $\mathcal{O}((1+P_T)^{1/2}[\beta(\lambda T)^{1/2}+((1-\lambda)T)^{3/4}])$. The result is also naturally consistent with original situations (one-point or two-point models) when set $\lambda=0$ or 1. Overall, our method effectively combines two models and provides flexible choice by setting λ to help balance computational cost and the upper bound in real-life applications.

In the end, BCO is viewed as an effective method, which can be utilized widely, including but not limited to, recommendation systems dynamically adjusting strategies based on user feedback, dynamic allocation strategies for network traffic, and real-time decision-making in dynamic environments for autonomous driving and robotics. Our hybridized PBGD algorithm provides rigorous theoretical analysis, which is conducive to downstream real-life applications.

References

1. Agarwal, A., Dekel, O., Xiao, L.: Optimal algorithms for online convex optimization with multi-point bandit feedback. In: Colt. pp. 28–40. Citeseer (2010)
2. Auer, P., et al.: Achieving optimal dynamic regret for non-stationary bandits without prior information. In: Conference on Learning Theory, pp. 159–163. PMLR (2019)
3. Besbes, O., Gur, Y., Zeevi, A.: Stochastic multi-armed-bandit problem with non-stationary rewards. Adv. Neural Inf. Process. Syst. **27** (2014)
4. Bubeck, S., Dekel, O., Koren, T., Peres, Y.: Bandit convex optimization: \sqrtt regret in one dimension. In: Conference on Learning Theory, pp. 266–278. PMLR (2015)
5. Bubeck, S., Lee, Y.T., Eldan, R.: Kernel-based methods for bandit convex optimization. In: Proceedings of the 49th Annual ACM SIGACT Symposium on Theory of Computing, pp. 72–85 (2017)
6. Cesa-Bianchi, N., Lugosi, G.: Prediction, Learning, and Games. Cambridge University Press (2006)
7. Chen, T., Giannakis, G.B.: Bandit convex optimization for scalable and dynamic IoT management. IEEE Internet Things J. **6**(1), 1276–1286 (2018)
8. Dekel, O., Eldan, R., Koren, T.: Bandit smooth convex optimization: improving the bias-variance tradeoff. Adv. Neural Inf. Process. Syst. **28** (2015)
9. Flaxman, A.D., Kalai, A.T., McMahan, H.B.: Online convex optimization in the bandit setting: gradient descent without a gradient. In: Proceedings of the Sixteenth Annual ACM-SIAM Symposium on Discrete Algorithms (SODA), pp. 385–394 (2005)
10. Hazan, E., Levy, K.: Bandit convex optimization: towards tight bounds. Adv. Neural Inf. Process. Syst. **27** (2014)
11. Hazan, E., et al.: Introduction to online convex optimization. Found. Trends® Optimiz. **2**(3–4), 157–325 (2016)
12. Hill, D.N., Nassif, H., Liu, Y., Iyer, A., Vishwanathan, S.: An efficient bandit algorithm for realtime multivariate optimization. In: Proceedings of the 23rd ACM SIGKDD International Conference on Knowledge Discovery and Data Mining, pp. 1813–1821 (2017)
13. Lattimore, T.: Improved regret for zeroth-order adversarial bandit convex optimisation. Math. Statist. Learn. **2**(3), 311–334 (2020)

14. Liu, W., Li, S., Zhang, S.: Contextual dependent click bandit algorithm for web recommendation. In: International Computing and Combinatorics Conference, pp. 39–50. Springer (2018)
15. Luo, H., Wei, C.Y., Agarwal, A., Langford, J.: Efficient contextual bandits in non-stationary worlds. In: Conference on Learning Theory, pp. 1739–1776. PMLR (2018)
16. Saha, A., Tewari, A.: Improved regret guarantees for online smooth convex optimization with bandit feedback. In: Proceedings of the Fourteenth International Conference on Artificial Intelligence and Statistics, pp. 636–642. JMLR Workshop and Conference Proceedings (2011)
17. Wei, C.Y., Hong, Y.T., Lu, C.J.: Tracking the best expert in non-stationary stochastic environments. Adv. Neural Inf. Process. Syst. **29** (2016)
18. Yang, T., Zhang, L., Jin, R., Yi, J.: Tracking slowly moving clairvoyant: optimal dynamic regret of online learning with true and noisy gradient. In: International Conference on Machine Learning, pp. 449–457. PMLR (2016)
19. Zhang, Y.J., Zhao, P., Zhou, Z.H.: A simple online algorithm for competing with dynamic comparators. In: Conference on Uncertainty in Artificial Intelligence, pp. 390–399. PMLR (2020)
20. Zhao, P.: Online Ensemble Theories and Methods for Robust Online Learning. Ph.D. thesis, PhD thesis, Nanjing University, Nanjing (2021). Advisor: Zhi-Hua Zhou
21. Zhao, P., Wang, G., Zhang, L., Zhou, Z.H.: Bandit convex optimization in non-stationary environments. J. Mach. Learn. Res. **22**(125), 1–45 (2021)
22. Zhao, P., Zhang, Y.J., Zhang, L., Zhou, Z.H.: Dynamic regret of convex and smooth functions. Adv. Neural. Inf. Process. Syst. **33**, 12510–12520 (2020)
23. Zinkevich, M.: Online convex programming and generalized infinitesimal gradient ascent. In: Proceedings of the 20th International Conference on Machine Learning (ICML), pp. 928–936 (2003)

Style-Specific Music Generation from Image

Chang Xu[1], Xuan Liu[1(✉)], Yu Weng[1], Shan Jiang[1], Xiangyun Tang[1], and Minfeng Qi[2]

[1] key Laboratory of Ethnic Language Intelligent Analysis and Security Governance of MOE, Minzu University of China, Beijing, China
{23302106,liuxuan,wengyu,jshan,xiangyunt}@muc.edu.cn
[2] City University of Macau, Macau, SAR, China
mfqi@cityu.edu.mo

Abstract. Recent researches in deep neural networks have the ability to create different styles of high quality music, which is comparable to music composed by humans. However, few work enables music generation from images. In this paper, we present ImageDJ, a generative model designed to compose music based on specific images. We use an image classification model to categorize images according to various styles and create a style embedding that guides the process of music generation. Our evaluation result shows that ImageDJ can generate high quality musical compositions inspired by various styles captured in images.

Keywords: Generative AI · Deep Learning · Music Generation

1 Introduction

Generating art and music at a human composer level remains challenging in AI. Among various methods for music generation, such as Markov models, genetic algorithms, and generative grammars, neural network [1] models are particularly notable for their flexibility. They can be trained on complex musical patterns in existing datasets, showcasing a wide range of capabilities and potential in music composition.

While many neural network techniques excel at understanding musical patterns from sources like sheet music or audio files, integrating image data into this process is still a challenge. Typically, networks trained on music-related inputs and can produce impressive music, but they seldom incorporate visual information to guide their compositions. In this paper, we introduce ImageDJ, a model based on recurrent neural network that can accept images as extra input. It discerns stylistic cues from images, then generates music that reflects the visual characteristics.

One approach to generating music involves using probabilistic models that treat music as a probability distribution [2]. These models learn from a corpus of training music to predict sequences of notes based on their likelihood, without

T. Zhu et al. (Eds.): ICA3PP 2024, LNCS 15252, pp. 136–145, 2025.
https://doi.org/10.1007/978-981-96-1528-5_10

relying on predefined musical rules. While this method autonomously identifies patterns, it can be complex due to the structure of music, such as the distinction between monodic and polyphonic melodies. For example, Bach's chorales feature polyphony with multiple voices. However, these models generate music based solely on audio data and do not incorporate visual information.

To address the lack of music generation through images, we present our approach to involve image inputs in music generation task. Our main contribution is a generative model capable of producing style-specific symbolic music using images as input. We believe this is useful for filmmakers and music composers who often need customized music to fit specific scenes in their projects. Moreover, our approach could be extended to more complicated inputs. In summary, this paper has the following contributions:

- We propose ImageDJ, a generative model which is capable of generating style-specific music via different images. We employ an image classification model to correlate images into different styles, generating a style embedding that guides the creation of music.
- Based on our proposed approach, we evaluate our model outputs via human participants with musical background, results show that our model can generate high quality music that is as distinguishable as those composed by human artists.

2 Related Works

There have been some comprehensive survey on music generation using deep learning methods [3–5].

Yu's work [6] introduces a music generation method using Transformers [7] with both fine-grained and coarse-grained attention mechanisms. Fine-grained attention captures detailed music structures by focusing on relevant tokens in other bars, while coarse-grained attention reduces computational costs by summarizing other bars, enabling the model to handle music sequences up to three times longer than those managed by models using full attention.

The work in [8] introduces Stochastic Control Guidance (SCG), a novel approach that guides through forward evaluation of rule functions and integrates with pre-trained diffusion models, enabling guidance without training. It also presents a latent diffusion architecture for high-resolution symbolic music generation, which easily combines with SCG. This framework outperforms existing strong baselines in music quality and rule-based control, surpassing current state-of-the-art generators in various scenarios.

LSTM (Long Short-Term Memory) [9] is a type of recurrent neural network (RNN) designed to learn patterns in sequential data like texts and music. Some studies have used LSTM to solve the problems of generating music [10,11]. Johnson [12] introduced Biaxial LSTM, which models polyphonic music by evaluating the likelihood of each note at each time step based on previous steps and notes generated. The generation process goes from the initial time step to the final one and from the lowest to the highest note. The model considers time and note dimensions, using LSTMs along these axes. Thus, Biaxial LSTM includes two key components: the time-axis module and the note-axis module.

In this setup, the model's input includes notes from the previous time step. The note octaves layer expands each note into a tensor that includes the note itself and its neighboring notes within the same octave, adding spatial context. These note features are processed by an LSTM unit with shared weights across notes along the time axis. The time-axis module then combines the note octaves and the recurrent states from the prior time step to produce higher-level features for each note.

The time-axis component draws inspiration from convolutional neural networks (CNNs). It's structured with stacked LSTM units that recurrently process data over time, connected to each note octave (12 pitches above and below every note) in a manner akin to a convolutional kernel. The LSTM units share weights across notes, enabling the time-axis segment to learn features that are invariant to specific notes. This design fosters generalization across different transpositions. The time-axis section outputs features for each note.

Using the note features generated by the time-axis processing, the note-axis LSTM module progresses through the features, starting from the lowest note and moving up to the highest note. It predicts each note based on the information from previously predicted lower notes. The notes-axis module comprises two layers of LSTM units, which recurrently process the notes. Before inputting into this module, the features of each note are combined with those of the lower selected note. If the current note being analyzed is the lowest note, zeros are appended instead. Put simply, To determine the selected note y_i, the features x_i of the i-th note are merged with the features of the previously chosen lower note y_{i-1} before being fed into the note-axis module.

In addition to the main data inputs, the biaxial architecture supplies contextual information to each LSTM within the time-axis module. This contextual input encompasses a portrayal of the ongoing beat within the musical bar. The representation employs delineates the position of the current time step relative to a 4/4 measure in binary format (Fig. 1).

Building upon biaxial LSTM framework, Mao [13] introduces DeepJ, a deep learning model that is capable of composing polyphonic music while considering a single or a blend of multiple composers styles. It integrates mechanisms to ensure the faithful representation of musical style in the model's output. DeepJ uses additional style embedding input to guide music generation so that the output correlates with specific style. The model architecture is shown in Fig. 2. More related work can be found in [14,15]. Models such as DeepJ have successfully produced style outcomes with measure-level structure and distinguishable styles. However, there are few models that involve images as a part of input to control the musical style of output music. Expanding upon Biaxial and DeepJ architecture, we introduce an intuitive and efficient approach to enable style-specific music generation via images input. Due to the significant differences between images and music in terms of their modalities, it's hard to extract musical features in images directly. Therefore, we employ style as a bridge: initially, utilizing

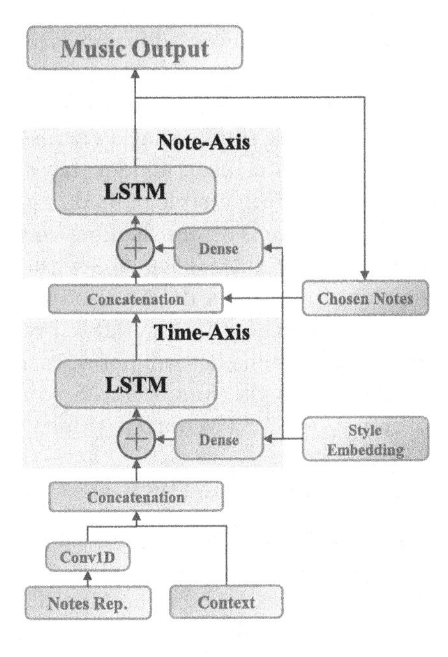

Fig. 1. DeepJ Architecture

images to generate style embedding, which are subsequently employed to guide music generation.

In the subsections, we present our method, data representation and model architecture.

2.1 Image-Style Representation

In the initial DeepJ model, the author assembled a dataset of 23 different music composers, each characterized by their unique artistic style. The music styles are encoded as a one-hot representation across all artists. Musical styles often exhibit shared characteristics; for instance, classical and baroque compositions may display similarities. Hence, DeepJ proposes representing style through a learned distributed representation, rather than a simplistic one-hot encoding.

During the generation process, various styles are mixed by changing the vector representation of style to encompass the desired styles, then normalize the vector to ensure it sums up to one. For example, the representations of composer 1, composer 2, and a blend of both are demonstrated as follows:

$$\begin{bmatrix}1\ 0\ 0\ ...\end{bmatrix}\ \begin{bmatrix}0\ 1\ 0\ ...\end{bmatrix}\ \begin{bmatrix}0.5\ 0.5\ 0\ ...\end{bmatrix}$$

DeepJ also categorizes multiple composers under a specific music genre. For instance, if composers 1 to 4 contribute to the baroque genre, then baroque is denoted as:

$$s_{baroque} = \begin{bmatrix}0.25\ 0.25\ 0.25\ 0.25\ 0\ 0\ ...\end{bmatrix}$$

Expanding on DeepJ, ImageDJ introduces additional framework to accommodate image inputs. In ImageDJ, the image input is firstly processed through Alexnet [16], a convolutional neural network architecture for classification, converting the image into a one-hot style representation across various genres. Next, the one-hot style vector is fed into a linear hidden layer labeled as W, which transforms the one-hot style input into a style embedding.

In each LSTM layer, a connection is established between the style embedding and an additional fully-connected hidden layer $W_l^{'}$ with tanh activation. This connection enables the model to create a detailed, nonlinear representation of the style. In this arrangement, indicated by the index l representing the LSTM layer, a single latent representation influences all notes, exerting its impact before they pass through the LSTM layers via summation. So for the i-th note and its corresponding input features x_i, the LSTM output z_i for that note is determined as follows:

$$z_i = f(x_i + \tanh W_l^{'} W s) \tag{1}$$

In this equation, x_i represents the original input, the symbol "+" denotes element-wise addition, and f denotes the function that executes the LSTM layer on a given input. This ensures consistent application of style conditioning to every note.

2.2 Data Representation

ImageDJ uses the identical note representation as the DeepJ model, besides we add image as extra input. The forthcoming paragraphs introduce the music representation.

The model employs a piano roll format to illustrate MIDI music. In this format, notes at each time step are depicted through a binary vector: a "1" denotes a played note at that index, while "0" indicates silence. Consequently, a musical composition is represented by an N*T binary matrix, where N signifies the playable notes and T represents the time steps. For example, consider a scenario where two notes are sustained for two consecutive time steps, followed by two silent time steps. This would be captured in a matrix encompassing four notes (N=4).

$$t_{play} = \begin{bmatrix} 1\,1\,0\,0 \\ 0\,0\,0\,0 \\ 1\,1\,0\,0 \\ 0\,0\,0\,0 \end{bmatrix}$$

Merely depicting note play doesn't provide a complete picture of MIDI interactions. Holding a note for two time steps and playing it twice are not the same. Note replay occurs when a note is promptly reactivated after it ends, without any gap between two successive actions. This concept is represented by a replay matrix akin to the matrix for note play. Together, note activation and replay define the comprehensive note representation. Additionally, volume of the note(in music theory, dynamics) is also an important aspect to consider. So each note is represented as a 3 length vector: the first element indicates whether the note is played, the second element indicates if this note is replayed, and the third element denotes the dynamics. For instance, the following vector represents a replayed note in dynamics of 0.4:

$$\begin{bmatrix} 1 \; 1 \; 0.4 \end{bmatrix}$$

Polyphonic music may contain more than one note simultaneously, so more than one vector is needed to represent chord. DeepJ addresses this by stacking various pitches of note representation to construct a N*3 matrix to indicate all potential notes at a time step, where N is the scale number of playable notes. The following matrix represent a chord where the first and third notes are being played:

$$\begin{bmatrix} 1 & 0 & 0.5 \\ 0 & 0 & 0 \\ 1 & 0 & 0.5 \\ 0 & 0 & 0 \\ \cdots & \cdots & \cdots \end{bmatrix}$$

2.3 Loss Functions

We use same loss functions Deepj used, which is the additions of three separate loss functions based on if note is played or not: the binary cross-entropy(bce) between t_{play}; the bce between t_{replay}(t_{re} for short) and mean squared error between note dynamics $t_{dynamics}$(t_{dy} for short). Note that when $t_{play} = 0$, which means the current is not played, then the bce(t_{replay}) and mse will not be calculated and losses will be set to zeroes, since it is unnecessary to consider the replay actions and volumes of a unplayed note.

At every time step for each note, our model generates three distinct outputs: play probability, replay probability, and dynamics. We conduct simultaneous training for all three outputs. Play and replay probabilities are addressed as logistic regression tasks, trained using binary cross-entropy loss, following the approach outlined in Biaxial LSTM. Dynamics, on the other hand, undergoes training using mean squared error. Consequently, we introduce the following loss functions:

$$L_{play} = \sum_{t_{play}, y_{play}} t_{play} \log(y_{play}) + (1 - t_{play}) \log(1 - y_{play}) \tag{2}$$

$$L_{replay} = \sum_{t_{play}, t_{re}, y_{re}} t_{play}(t_{re} \log(y_{re}) + (1 - t_{re}) \log(1 - y_{re})) \tag{3}$$

$$L_{dynamics} = \sum_{t_{play}, t_{dy}, y_{dy}} t_{play}(t_{dy} - y_{dy})^2 \tag{4}$$

The final loss function is the combination of the three losses above:

$$Loss = L_{play} + L_{replay} + L_{dynamics} \tag{5}$$

2.4 Architecture

Our model is built upon the design of DeepJ, the key difference is the incorporation of image in the model's input, as shown in Fig. 3. The left side is the original Biaxial architecture, where n_{tn} represents note feature at time step t, a_{tn} represents output prediction, and h_n represents hidden states. where images are fed in and processed through AlexNet for classification, associating them with a particular style. Subsequently, a fully connected hidden layer converts the one-hot style input into a style embedding to directs the generation process.

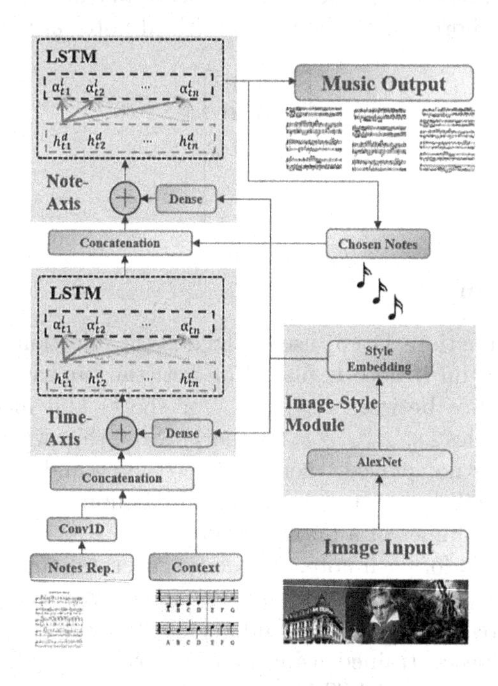

Fig. 2. ImageDJ Architecture

3 Experiments

3.1 Training

We trained ImageDJ on a dataset comprising MIDI music files from 23 composers across the baroque, classical and romantic periods, which is the same dataset the DeepJ was trained on. Additionally, we trained AlexNet on a set of images from three different periods corresponding to three different genres. The MIDI pitch range was truncated to span four octaves, from 36 to 84, and inputs were quantized to have a resolution of 16 time steps per bar. We utilized stochastic gradient descent(SGD) with the Nesterov Adam optimizer [17]. Dropout [18] was employed for regularization. Specifically, our model incorporated a 20% dropout in input layers and a 50% dropout in all non-input layers(except the image-style embedding layer).

3.2 Generation

We generated music samples across three distinct genres: baroque, classical, and romantic. We input images representing each style into the model and generated music samples accordingly. Adopting a similar approach to DeepJ, we employed a method where we randomly sampled from the model's probability distribution, akin to flipping a coin, to determine whether a note should be played. If a decision to play a note was made, we then sampled from another distribution to ascertain if it should be re-attacked. The dynamics of each note were directly dictated by the model. We also devised an adaptive temperature adjustment technique described in DeepJ to avoid model generate long periods of silence. This method involved dynamically adjusting the temperature T of our output sigmoid functions in proportion to the number of consecutive silent time steps the model produced. Specifically, we calculated $T = 0.1t + 1$, where t represents the number of prior consecutive silent outputs. Whenever the model generated a non-silent output at a time step, we reset T to 1. This adaptive adjustment prevented the model from outputting extended periods of silence, enhancing the quality of the generated music.

4 Evaluation

In order to assess the variety of styles produced by ImageDJ, we conducted a survey involving 50 individuals with musical background. Participants were tasked with categorizing music generated by ImageDJ into the genres of baroque, classical, or romantic. The aim of the experiment was to determine the capability of ImageDJ in generating music with distinctive styles based on images, gauging whether humans could accurately identify these genres. During the study, participants received ten music samples, comprising pieces generated by ImageDJ as well as compositions by real composers. Half of the participants were presented with control samples, while the other half received genuine compositions for comparison. Participants accurately identified the styles of ImageDJ outputs 58% of the time, control samples were correctly classified 51% of the time, as shown in Fig. 4. This demonstrates that ImageDJ produces music with styles nearly as distinguishable as those crafted by humans.

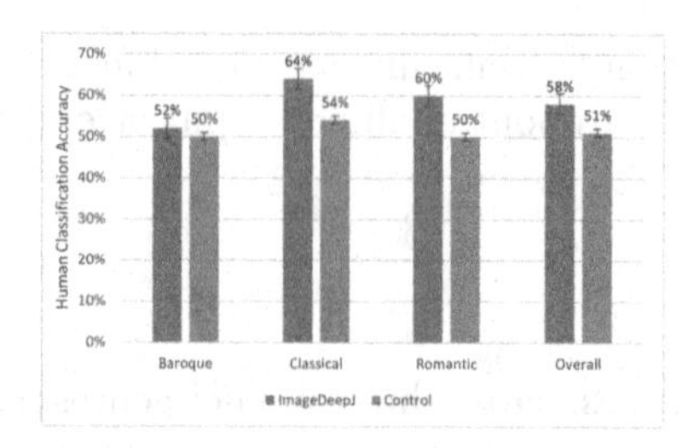

Fig. 3. ImageDJ Architecture

Fig. 4. ImageDJ Architecture

5 Conclusion and Future Work

We introduced a novel model aimed at learning musical styles from images and generating style-specific music using deep neural networks. Our approach utilized a image classification model, converting image to style representation, which are then utilized to direct the music generation process. Experimental results demonstrate that our model effectively generates style-specific music based on the input image styles.

Nevertheless, certain challenges remain, particularly regarding the lack of long-term structure in the generated compositions. Additionally, the style feature extraction process for images lacks complexity, the model can't recognize more nuanced objectives within images. To tackle these issues, we will explore reinforcement learning techniques or adversarial methods to train models capable of incorporating better long-term structure, and investigate more advanced methods to enhance the model's ability to recognize musical features in images.

References

1. Briot, J.-P., Hadjeres, G., Pachet, F.-D.: Deep Learning Techniques for Music Generation, vol. 1. Springer, Heidelberg (2020)
2. Kotecha, N., Young, P.: Generating music using an LSTM network. arXiv preprint arXiv:1804.07300 (2018)

3. Ji, S., Yang, X., Luo, J.: A survey on deep learning for symbolic music generation: representations, algorithms, evaluations, and challenges. ACM Comput. Surv. **56**(1), 1–39 (2023)
4. Ji, S., Luo, J., Yang, X.: A comprehensive survey on deep music generation: multi-level representations, algorithms, evaluations, and future directions. arXiv preprint arXiv:2011.06801 (2020)
5. Kaliakatsos-Papakostas, M., Floros, A., Vrahatis, M.N.: Artificial intelligence methods for music generation: a review and future perspectives. In: Nature-Inspired Computation and Swarm Intelligence, pp. 217–245 (2020)
6. Yu, B., et al.: Museformer: transformer with fine-and coarse-grained attention for music generation. Adv. Neural. Inf. Process. Syst. **35**, 1376–1388 (2022)
7. Vaswani, A., et al.: Attention is all you need. Adv. Neural Inf. Process. Syst. **30** (2017)
8. Huang, Y., et al.: Symbolic music generation with non-differentiable rule guided diffusion. arXiv preprint arXiv:2402.14285 (2024)
9. Graves, A., Graves, A.: Long short-term memory. In: Supervised Sequence Labelling with Recurrent Neural Networks, pp. 37–45 (2020)
10. Shah, F., Naik, T., Vyas, N.: LSTM based music generation. In: 2019 International Conference on Machine Learning and Data Engineering (iCMLDE), pp. 48–53. IEEE (2019)
11. Huang, Y., Huang, X., Cai, Q.: Music generation based on convolution-LSTM. Comput. Inf. Sci. **11**(3), 50–56 (2018)
12. Johnson, D.D.: Generating polyphonic music using tied parallel networks. In: International Conference on Evolutionary and Biologically Inspired Music and Art, pp. 128–143. Springer, Cham (2017)
13. Mao, H.H., Shin, T., Cottrell, G.: DeepJ: style-specific music generation. In: 2018 IEEE 12th International Conference on Semantic Computing (ICSC), pp. 377–382. IEEE (2018)
14. Dong, H.-W., Hsiao, W.-Y., Yang, L.-C., Yang, Y.-H.: Musegan: multi-track sequential generative adversarial networks for symbolic music generation and accompaniment. In: Proceedings of the AAAI Conference on Artificial Intelligence, vol. 32, no. 1 (2018)
15. Copet, J., et al.: Simple and controllable music generation. Adv. Neural Inf. Process. Syst. **36** (2024)
16. Krizhevsky, A., Sutskever, I., Hinton, G.E.: Imagenet classification with deep convolutional neural networks. Adv. Neural Inf. Process. Syst. **25** (2012)
17. Kingma, D.P., Ba, J.: Adam: a method for stochastic optimization. arXiv preprint arXiv:1412.6980 (2014)
18. Srivastava, N., Hinton, G., Krizhevsky, A., Sutskever, I., Salakhutdinov, R.: Dropout: a simple way to prevent neural networks from overfitting. J. Mach. Learn. Res. **15**(1), 1929–1958 (2014)

Multimodal Summarization with Modality-Aware Fusion and Summarization Ranking

Xuming Ye[1], Chaomurilige[1,2](✉), Zheng Liu[1], Haoyu Luo[1], Jun Dong[1], and Yingzhe Luo[3]

[1] Key Laboratory of Ethnic Language Intelligent Analysis and Security Governance of MOE, Minzu University of China, Beijing, China
{xumingye,chaomurilige,liuzheng,22301976,22302097}@muc.edu.cn
[2] Hainan International College of Minzu University of China, Li'an International Education Innovation pilot Zone, Hainan 572499, China
[3] Faculty of Data Science, City University of Macau, Macau, China
yingzheluo@cityu.edu.mo

Abstract. Existing multimodal summarization methods primarily focus on multimodal fusion to efficiently utilize the visual information for summarization. However, they fail to exploit the deep interaction between textual and visual modality. Moreover, optimizing the model by maximum likelihood estimation (MLE) leads to exposure bias, causing the model to generate the next word that is based on the previously generated erroneous words during inference. To address these challenges, we propose a novel modality-aware fusion module (MAF) with a summarization ranking (SumR) training objective. Specifically, the MAF module exploits the interaction in the multimodal input through multiple fusion layers, and SumR aims to align the probability order predicted by the model with actual quality metrics, therefore it is able to reduce the exposure bias problem during inference. Extensive experiments on a large-scale dataset demonstrate that our method outperforms existing models, achieving superior results in both automatic and human evaluation metrics. The generated multimodal summaries provide richer context and enhance user's comprehension by combining the key textual information with the relevant visual content.

Keywords: Multimodal · Summarization · Deep Learning

1 Introduction

Generative artificial intelligence models have demonstrated impressive abilities in understanding and generating human-like text, leading to breakthroughs across various application domains [2,5,14,25]. One particularly promising area is abstractive summarization [3,10,24,26,30,32], where these models generate concise summaries that capture the main content of a document, which thereby

T. Zhu et al. (Eds.): ICA3PP 2024, LNCS 15252, pp. 146–164, 2025.
https://doi.org/10.1007/978-981-96-1528-5_11

enhances the speed and efficiency of information retrieval for the users. Unlike extractive summarization [21,32], which involves selecting and concatenating key sentences from the original text, abstractive summarization [6,13] requires the model to paraphrase and condense the information, often resulting in more coherent and cohesive summaries.

With the increasing diversity of information, there has been more data published in multimodal forms on the internet, such as news articles that typically include both text and images [4,9,17]. Consequently, recent research has begun to focus on multimodal summarization, which aims to leverage both textual and visual information to create more comprehensive and engaging summaries [11,18,31,33,34,36,37]. The output of multimodal summarization typically comprises a brief text summary accompanied by an image that represents the core content, as shown in Fig. 1. Empirical studies have shown that, compared to text-only summaries, multimodal summaries enhance readability and user's experience by providing visual context, which can help clarify the complex information and highlight the key points [11,36,37].

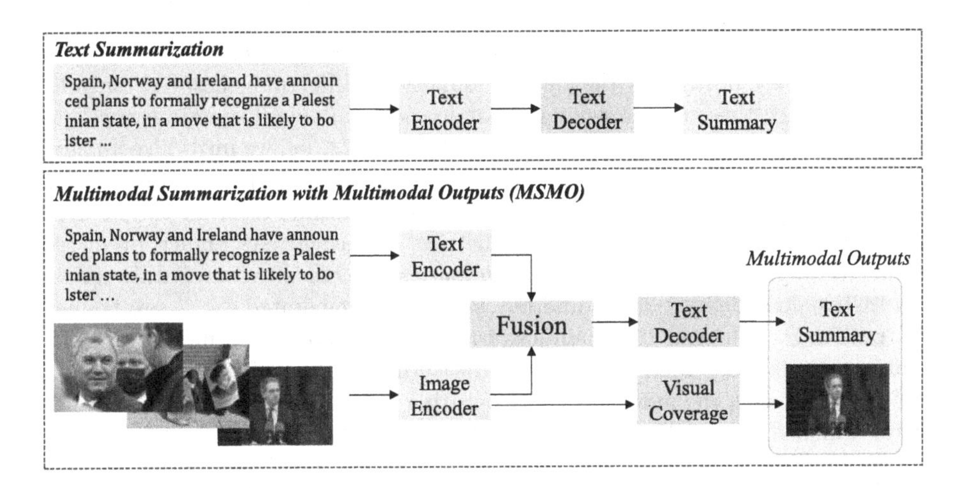

Fig. 1. Overview of text summarization and multimodal summarization with multimodal outputs (MSMO) in general. It is noted that the image in Multimodal Outputs is selected from the document.

Multimodal summarization algorithms generally consist of three main components: feature extraction, modality fusion, and summary decoding. With the advent of pretrained models, feature extraction often employs these models to derive features from different modalities [33]. The key of multimodal summarization is the modality fusion stage. Existing research primarily focuses on effectively integrating image and text information into a unified representation. Techniques such as concatenation, attention-based methods, and gate-based approaches are utilized for fusion [1,23]. However, the training data for multimodal summarization only contains reference text, instead of reference images.

Therefore, due to the absence of an image reference, it is challenging to discern the inner interactions between the textual and visual modality within a single fusion layer. In addition, some works even suffer from the visual modality due to the inefficient fusion [36, 37].

To enhance the quality of multimodal summaries using visual modality, we propose a novel modality-aware fusion module (MAF) combined with a multi-layer fusion strategy. MAF enables a more precise integration of visual and textual information by iteratively refining the contributions of each modality. Employing multiple layers of cross-attention and self-attention progressively refines the interaction between the visual and textual features. Such separation and enhancement lead to more coherent and informative multimodal summaries.

Furthermore, existing multimodal summarization methods typically treat the summarization task as a sequence-to-sequence (Seq2Seq) problem, they learn to generate the summary in an autoregressive manner [3, 22]. These models are trained using maximum likelihood estimation, which maximizes the predictive probability of the next reference word given the preceding sub-sequence. However, during inference, the model must generate the next word based on the already generated sub-sequence, which may contain errors known as exposure bias [22]. In order to maintain performance when these errors exist, the model must accurately determine the relative quality of different generated outputs, especially since there are no reference summaries available during inference.

Inspired by existing Seq2Seq training paradigm [22, 35], we introduce an auxiliary task called Summarization Ranking (SumR) during the training process of our multimodal summarization model. This task aims to align the model's predicted scores with actual quality metrics, ensuring that higher predicted scores correspond to higher quality summaries. Specifically, our SumR uses a contrastive loss defined over different summaries generated by a pre-trained abstractive model. And the training objective of SumR coordinates the order of model probability distribution with the actual quality metric (i.e., ROUGE [19]) by evaluating the model. By incorporating SumR, the model gains the ability to evaluate the quality of its generated summaries, significantly reducing the impact of exposure bias.

In summary, we propose a novel modality-aware fusion module (MAF) and a summarization ranking (SumR) training objective in abstractive multimodal summarization. Extensive experiments on the existing dataset demonstrate that our method can fully exploit visual modality information and generate a more comprehensive multimodal summary. Our contributions are as follows:

- We propose a modality-aware fusion module for abstractive multimodal summarization that enables each modality to complement and enriches the other, which enhances the quality of summary.
- We innovatively use summarization ranking to align the model's probability distribution with the ROUGE evaluation metric, which thereby reduces the exposure bias in the existing multimodal summarization methods.
- Experimental results show that our proposed model outperforms the existing methods in both automatic and manual evaluation metrics.

2 Related Work

Multimodal summarization typically involves three main stages: feature extraction, modality fusion, and summary decoding [31,34,36,37]. In the feature extraction stage, textual and visual features are commonly extracted using separate modules, such as BART [15] for text and VGG [28] for images. However, the core challenge lies in the modality fusion stage, where the integration of these features directly impacts the quality of the generated summary.

Recent works have primarily focused on improving the fusion mechanism. Simple concatenation of multimodal features is a straightforward approach but often struggles with the high-dimensional gaps between different modalities [16]. To address this, multimodal attention mechanisms have been widely adopted, offering a more nuanced integration of information. For instance, Palaskar et al. [23] proposed a multi-source Seq2Seq model with hierarchical attention to create coherent summaries by effectively integrating multimodal data. Similarly, Liu et al. [20] introduced a multi-stage fusion network incorporating a forget gate module to emphasize key visual information. Despite these advances, challenges persist in efficiently utilizing visual data, particularly in scenarios with text-heavy inputs and limited image reference, leading to suboptimal fusion outcomes.

Moreover, the typical training objective in multimodal summarization models is maximum likelihood estimation (MLE) [3,22,33]. MLE focuses on minimizing the difference between the predicted and reference distributions but suffers from exposure bias; during training, the model relies on reference summaries, which are unavailable during inference. While some studies have explored some multi-task frameworks to enhance modality integration [27,29], they fall short of addressing the exposure bias issue.

To overcome these limitations, we propose a novel modality-aware fusion strategy combined with a summary ranking loss. This approach not only refines the fusion process by aligning the textual and visual features but also mitigates the exposure bias, which enhances the overall quality of the generated summaries in multimodal summarization tasks.

3 Methodology

Current multimodal summarization methods mainly focus on improving the visual modality for summarization by introducing multiple subtasks, which is ineffective to improve the quality of summary. Therefore, we propose a modality-aware fusion (MAF) module to learn the cross-modality interaction between two modalities. Moreover, we employ a new training method SumR that uses contrastive loss to reduce exposure bias during inference. Specifically, we align the probability order with the ROUGE score order to ensure that higher-quality summaries are assigned higher probability. The overview of our proposed model is shown in Fig. 2.

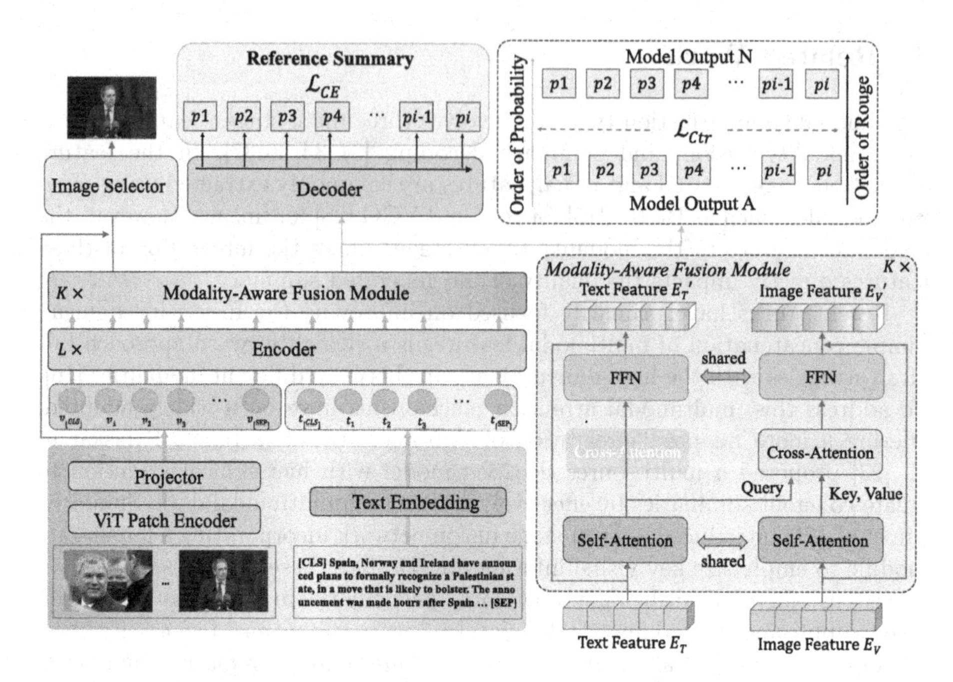

Fig. 2. Overview of our proposed multimodal summarization architecture, which combines text and image features for generating summaries. The model employs BART for text embedding and ViT for image embedding. Text and image features are concatenated and processed through multiple Transformer blocks within BART. Then we use the proposed MAF module to guide the fusion stage of the text and image features. In addition, we add the contrastive loss between the order of probability and the order of rouge, where the model outputs are generated through the beam-search [8].

Given a multimodal document $D = \{T, I\}$, where $T = \{t_1, t_2, \ldots, t_M\}$ is a text sequence with M tokens, and $I = \{I_1, I_2, \ldots, I_N\}$ comprises N images. The proposed architecture summarizes D into a multimodal summary $Y = \{Y_t, Y_i\}$, where $Y_t = \{y_1, y_2, \ldots, y_l\}$ represents a text summary constrained by a length l, and Y_i is the image extracted from I that are most relevant to the generated summary.

3.1 Multimodal Encoders

We use different encoders to process text and images separately, and then concatenate the embeddings into the BART [15] model. Specifically, for text, we convert the input text into tokens and use the embedding layer of BART model to obtain their corresponding embeddings. The embedding layer maps discrete tokens to continuous dense vectors:

$$E_T = [t_{[CLS]}; t_1; ...; t_M; t_{[SEP]}] \tag{1}$$

where $t_{[CLS]}$ and $t_{[SEP]}$ are the special token embeddings introduced to mark the start and the end of each sentence, and M is max length of the text input.

ViT [7] is a vision model based on the Transformer architecture. In our model, we use ViT as our visual backbone. Motivated by Kim [12], we employ a straightforward approach involving linear projection on image patches to derive visual embeddings. Specifically, each input image I_i is segmented into patches and flattened into $I_{patch_{i,k}}$, where the k is the number of patches and i denotes the i-th image. Then, these patches are fedded into ViT model and a linear projection, $I_{patch_{i,k}}$ is embedded into $v_{patch_{i,k}}$:

$$\begin{aligned} v_{patch_{i,k}} &= \text{Projector}(\text{ViT}(I_{patch_{i,k}})) \\ v_i &= [v_{[CLS]}; v_{patch_{i,1}}; ...; v_{patch_{i,k}}] \\ E_I &= [v_1; v_2; ..., v_N] \end{aligned} \tag{2}$$

where $v_{[CLS]}$ are the special token embeddings introduced to mark the start of each image. We concatenate each visual embedding v_i as the global visual feature E_I. Note that the position information of patches is automatically gained in ViT model. Once we obtain the embedding vectors for the text modality and image modality through BART and ViT respectively, we concatenate them and input the result into BART's transformer encoder blocks. We utilize the BART-base model, which consists of 6 layers in its encoder with each layer containing 12 attention heads.

$$h = [h_t; h_v] = f_{enc}([E_T; E_I]) \tag{3}$$

where $[;]$ denotes the concatenation operation, and f_{enc} is the encoder function of BART. We use the hidden state at the last layer of the encoder as the text and image representation.

3.2 Modality-Aware Fusion Module

To improve the usage of multimodal semantics information, different modalities should be fused seamlessly to generate the summary by exploiting the interaction between them. Therefore, we propose a modality-aware fusion module (MAF) to fully leverage the information brought by different modalities, thereby generating a more accurate summary. Specifically, unlike the typical single-layer fusion approach, we utilize multiple fusion layers to explore the relationships between different modalities. Each fusion layer includes a self-attention mechanism, a cross-attention mechanism, and a feed-forward network (FFN). The parameters of the feed-forward network and the self-attention mechanism are shared in each fusion layer.

As shown in Fig. 2, we denote the input of modality-aware fusion module as E_T and E_V to represent the text and visual features, respectively. The self-attention mechanism within each layer first enhances the features within each modality independently. Following the self-attention, cross-attention is applied to capture inter-modality relationships. The text features serve as the query, while the visual features serve as the key and value:

$$\begin{aligned} E'_T &= \text{Atten}(E_T), \quad E'_V = \text{Atten}(E_V) \\ Q &= E'_T W_q, \quad K = E'_V W_k, \quad V = E'_V W_v \\ E'_V &= \text{CA}(Q, K, V) \end{aligned} \tag{4}$$

where the $W_q, W_k, W_v \in \mathbb{R}^{D \times D}$ are learnable parameters, and D is 768 (the dimension of the image and text features). *Atten* is the self-attention mechanism, and CA denotes the cross-attention. This process is repeated across multiple fusion layers to progressively refine the multimodal representation, allowing for deeper interaction between the text and visual modalities. A residual connection is added at each layer to stabilize training and speed up convergence.

3.3 Image Selector

Given the generated summary denoted as R and the set of image captions $\{t_1, t_2, ..., t_N\}$, where each t_i is generated by the CLIP-ViT model, we aim to select the image that best corresponds to the summary R. For each generated caption t_i, we compute the ROUGE-1, ROUGE-2, and ROUGE-L scores relative to the reference summary R. We then calculate the average ROUGE score (AVG-ROUGE) for each caption t_i as the mean of its ROUGE-1, ROUGE-2, and ROUGE-L scores. Finally, we select the caption t_i that achieves the highest average ROUGE score with the summary R. The image corresponding to this caption is selected as the output image. Formally, the best matching caption is determined as follows:

$$t_{\text{best}} = \arg\max_{t_i} \left(\text{AVG-ROUGE}(t_i, R)\right) \tag{5}$$

3.4 Training Objectives

The SumR objective is a critical component of our training methodology, designed to mitigate the exposure bias problem and enhance the quality of generated summaries. The integration of the SumR objective within the training process is accomplished through several key steps which we detail below.

Decoder and Loss Computation. The training process begins with the decoder, which utilizes the encoded hidden states along with the previously generated tokens to predict the next token in the sequence. Initially, the decoder is fed a special start-of-sequence token. It combines the context vector with its current hidden state to generate probabilities for the output token. This sequence prediction continues until an end-of-sequence token is generated or a predefined maximum sequence length is reached.

For each token generated, the model's predictions are compared to the ground truth targets using the cross-entropy loss, which is computed as follows:

$$\mathcal{L}_{CE} = -\frac{1}{N}\sum_{i=1}^{N}\sum_{t=1}^{T}\log P(y_t^{(i)}|y_{<t}^{(i)}, x^{(i)}) \tag{6}$$

where N is the number of samples, T is the length of the reference summary, $y_t^{(i)}$ is the target word at time step t for the ith sample, and $P(y_t^{(i)}|y_{<t}^{(i)}, x^{(i)})$ is the probability predicted by the model for the word $y_t^{(i)}$ given the previous words and the input text $x^{(i)}$.

Integration of SumR Objective. In addition to the standard loss computation, the SumR objective is specifically integrated to enhance the model's ability to rank generated summaries based on their quality, as assessed by real evaluation metrics like ROUGE. To implement this, we use beam search to generate a set of summaries (beam summaries) by a pre-trained model for each input of training data before training. The SumR loss $\mathcal{L}_{SumR}$ is then computed based on the quality ranking of these summaries. For each pair of summaries (T_m and T_n), we compute the SumR loss as follows:

$$\mathcal{L}_{SumR} = \sum_{m}\sum_{n>m}\max(0, g(T_n) - g(T_m) + \delta_{mn}) \tag{7}$$

where $ROUGE(T_m, R) > ROUGE(T_n, R)$, and δ_{mn} is the margin scaled by the rank difference between the beam summaries, defined as $\delta_{mn} = (n - m)\cdot\gamma$, and R is the reference summary, $g(T_m)$ represents the length-normalized estimated log-probability (β is the length penalty):

$$g(T) = \frac{\sum_{t=1}^{L}\log P(y_t^{(i)}|y_{<t}^{(i)}, x^{(i)})}{|T|^{\beta}} \tag{8}$$

To optimize the model, we employ a combination of cross-entropy loss and contrastive loss, which is formulated as:

$$\mathcal{L} = \mathcal{L}_{CE} + \mathcal{L}_{SumR} \tag{9}$$

By integrating the SumR objective in this manner, we ensure that the model not only learns to accurately predict the next token but also improves its ability to generate high-quality summaries that are ranked appropriately based on their actual content quality. This dual-objective approach directly addresses the exposure bias issue while enabling the model to produce more reliable and coherent summaries.

4 Experimental Settings

4.1 Dataset

During the experimental process, the dataset [36] was used, which includes 293,965 training data points, 10,355 validation pairs, and 10,261 test pairs. The author created a similar dataset using the Daily Mail website's online news articles. They summarized the key points of each news article into multiple sentences, maintaining the order in which they appeared in the original text. Each key point was included as part of the summary, and several images were selected to serve as reference images, forming multimodal references.

4.2 Baselines

We report the results of existing multimodal summarization methods. To demonstrate the effectiveness of our proposed method, we compared it with existing multimodal summarization methods and three text summarization methods.

BERTSum [21]: It is a model for extractive and abstractive text summarization based on BERT, with two variants-BertAbs (abstractive) and BertExtAbs (hybrid).

BART [15]: It is a pre-trained model featuring a bidirectional encoder and an autoregressive decoder, allowing it to better understand complex contextual relationships.

ATG/ATL/HAN [36]: They use the global, local and hierarchical image features for multimodal attention, respectively.

MOF [37]: It incorporates a multimodal objective function into the ATG model. We chose the two best-performing variants ($\text{MOF}^{\text{RR}}_{\text{enc}}$ and $\text{MOF}^{\text{RR}}_{\text{dec}}$) of this multimodal objective function for our comparison.

UniMS [33]: It proposes a unified framework for multimodal summarization, introducing a visual-guided decoder to better integrate textual and visual modalities.

SITA [11]: It proposes a novel coarse-to-fine image-text alignment mechanism to identify the most relevant sentence of each image in a document, resembling the role of image captions in capturing visual knowledge and bridging the cross-modal semantic gap.

4.3 Evaluation Metrics

We choose the following evaluation metrics as [36,37]: (1) ROUGE-{1,2,L} is used as the standard evaluation metric for automatic summarization. It calculates the corresponding score by comparing the automatically generated the summary with the reference summary, in order to measure the "similarity" between the two texts; (2) Image Precision (IP) is a commonly used metric for evaluating image selection performance. It defines image accuracy by calculating the ratio between the properly recommended images and the reference ones. Used to measure the ability to accurately select images with high correlation when given a reference image; (3) M_{sim} is a metric for evaluating the correlation between images and texts, is achieved by calculating the maximum similarity between each sentence in the image and the final summary.

5 Experimental Results and Analysis

5.1 Automatic Evaluation

Table 1. Main results of different metrics. R-1, 2, L refers to ROUGE-1, 2, L, IP refers to Image Precision.

Model	R-1	R-2	R-L	IP	M_{sim}
Text Abstractive					
BertAbs [21]	39.02	18.17	33.20	–	–
BertExtAbs [21]	39.88	18.77	38.36	–	–
BART [15]	41.83	19.83	39.74	–	–
Multimodal Abstractive					
ATG [36]*	40.63	18.12	37.53	59.28	25.82
ATL [36]*	40.86	18.27	37.75	62.44	13.26
HAN [36]*	40.82	18.30	37.70	61.83	12.22
MOF^{RR}_{enc} [37]*	41.05	18.29	37.74	62.63	26.23
MOF^{RR}_{dec} [37]*	41.20	18.33	37.80	65.45	26.38
UniMS [33]*	42.94	20.50	40.96	69.38	29.72
SITA [11]*	43.64	20.53	41.03	76.41	33.47
Our Method	**45.61**	**21.74**	**42.07**	**77.19**	**35.15**

Table 1 presents the performance results of various models on abstractive summarization and image selection tasks. The first block of the table lists methods that use text-only inputs, while the second block includes multimodal methods that incorporate both textual and visual inputs. By investigating the results, we have the following observations:

(1) Our proposed method significantly outperforms all other models in both text and image related metrics, achieving an R-L score of 42.07, an IP score of 77.19, and an M_{sim} score of 35.15. The remarkable performance of our method can be attributed to the modality-aware fusion module (MAF) and summarization ranking paradigm. Firstly, our method achieves a 1.02% improvement in the IP score, and an 4.68% increase in M_{sim}, demonstrating the superiority of our multimodal summarization model in leveraging both textual and visual modality. The MAF module allows for the seamless fusion of different modalities by exploiting the interactions between them. Therefore, our model can generate a high quality summary, which is also relevant to the selected image. Secondly, compared to the previous state-of-the-art model SITA, our method improves the R-1 score by 4.5%, the R-2 score by 5.9%, and the R-L score by 2.5%. This results demonstrate the effectiveness of the Summarization Ranking (SumR) training objective, by which can accurately predict the ranking order of generated summaries based on the actual evaluation metrics.

(2) For previous multimodal summarization approaches, ATG and ATL show competitive results, with ATL slightly outperforming ATG by achieving higher scores in the R-1 and IP metrics. The HAN, $\text{MOF}^{\text{RR}}_{\text{enc}}$, and $\text{MOF}^{\text{RR}}_{\text{dec}}$ models show varying degrees of improvement due to their different approaches to handling multimodal inputs. Differ from them, UniMS propose a Unified framework for Multimodal Summarization grounding on BART, that integrates extractive and abstractive objectives, as well as selecting the image output. UniMS improves the ROUGE-L score and IP score by 3.16 and 3.39 over $\text{MOF}^{\text{RR}}_{\text{enc}}$, respectively. Among these, SITA proposed a two-stage alignment strategy to identify the most relevant sentence of each image in a document, resembling the role of image captions in capturing visual knowledge and bridging the cross-modal semantic gap. This alignment strategy allows it to better align and fuse information from both modalities, achieving a significant improvement over previous methods. However, despite these advancements, our method, leveraging the MAF module and SumR training objective, achieves the best overall performance.

(3) In the text abstractive block, BART outperforms both BertAbs and BertExtAbs, indicating strong performance in text-only abstractive summarization. The strong performance of BART can be attributed to its transformer-based architecture, which is pre-trained on a large corpus of text data, allowing it to effectively capture the key information and generate a coherent summary. Thus we chose BART as the backbone of our proposed model.

Overall, the results demonstrate the effectiveness of our advanced multimodal fusion strategy and training objective. By integrating the MAF and SumR, our model not only surpasses existing methods in generating a high quality and relevant textual summary but also excels in image selection, providing a comprehensive enhancement in multimodal summarization tasks.

5.2 Ablation Study

To thoroughly evaluate the individual contributions of different components in our proposed multimodal summarization model, as shown in Table 2, we conduct the ablation study using three key metrics: R-1 (ROUGE-1), R-2 (ROUGE-2), and R-L (ROUGE-L). We design five ablation experiments to assess the impact of different components on the model's performance.

In the MAF(6) and MAF(9) Experiments. To investigate the impact of the number of layers in the modality-aware fusion (MAF) module on model performance, ablation experiments are conducted with the number of MAF layers increased from 3 (as used in our method) to 6 and 9. The results indicate that an increase in the number of layers leads to a slight decrease in performance. This suggests that while adding more fusion layers may enhance interactions between modalities, it may also lead to overfitting or introduce unnecessary complexity, which will even deteriorates the quality of model outputs.

In the w/o MAF Experiment. Without the MAF module significantly reduces the model's performance across all metrics, with R-1 falling to 44.91, R-2 to 21.01, and R-L to 41.62. This substantial decrease underscores the role

Table 2. Ablation study results for our proposed method. The table compares the R-1, R-2, and R-L scores for different configurations. "MAF(6)" and "MAF(9)" indicate configurations where the MAF module has 6 and 9 layers, respectively. "- w/o MAF" represents the model without the MAF module, "- w/o SumR" indicates the model without the SumR training objective, and "- w/o ViT" refers to the model where ViT is replaced by VGG [28] for extracting image features. The results illustrate the impact of each component on the model's performance.

Configuration added	R-1	R-2	R-L
Our Method	**45.61**	**21.74**	**42.29**
MAF(6)	45.55	21.66	41.87
MAF(9)	44.71	20.78	41.32
- w/o MAF	44.91	21.01	41.62
- w/o SumR	43.26	20.52	40.15
- w/o ViT	43.64	20.66	40.37

of the MAF module in effectively integrating textual and visual features, as well as its iterative capability to refine the contributions of each modality, which is crucial for generating a coherent and informative summary.

In the w/o SumR Experiment. Without the training objective of ranking summaries(SumR), the model's performance significantly declines, with R-1, R-2, and R-L scores dropping to 43.26, 20.52, and 40.15, respectively. We attribute this to the role of SumR loss in mitigating exposure bias and enhancing the quality of generated summaries. SumR loss aids the model in learning to evaluate and rank its outputs based on quality metrics, ensuring that higher quality summaries receive higher predicted scores, which is crucial for robust performance during the inference process.

In the w/o ViT Experiment. Evaluating the model with VGG instead of ViT also led to decreased performance, with R-1 at 43.64, R-2 at 20.66, and R-L at 40.37. It confirms the effectiveness of using ViT to incorporate visual information and enrich the textual summaries. Unlike VGG, ViT is designed to capture more complex and nuanced visual features by leveraging self-attention mechanisms, which allows for better integration and understanding of visual context in the summaries. Consequently, replacing ViT with VGG results in less informative and comprehensive summaries.

5.3 Human Evaluation

We further conduct a manual evaluation to assess the quality of the generated multimodal summarization. We randomly sample 300 data points from the test set to rate them between 1 to 3 points on multiple qualitative aspects. For the evaluation, we consider the following metrics:

Table 3. Human evaluation results.

Model	Cov.	Gra.	Con.	IR.
UniMS	2.40	2.54	2.60	2.54
SITA	2.54	2.66	2.66	2.66
Our Method	**2.74**	**2.86**	**2.86**	**2.80**

- **Coverage (Cov.)**: This metric compares the model-generated text summary with the actual text summary to determine if the main points are adequately covered. It ensures that the generated summary encapsulates all the critical information from the source text.
- **Grammar (Gra.)**: This metric examines whether the model-generated text summary is semantically correct. It assesses the grammatical correctness and coherence of the generated summaries, ensuring that they are well-structured and readable.
- **Consistency (Con.)**: This metric measures the factual alignment between the summary and the source document. It checks for any discrepancies or inaccuracies in the generated summary, ensuring that it accurately reflects the content of the source document.
- **Image-Relevance (IR.)**: This metric indicates the text-image relevance of multimodal outputs. It evaluates how well the selected images correspond to the text summary, ensuring that the images are relevant and enhance the overall understanding of the summarized content.

As shown in Table 3, our method achieves the highest scores across all evaluation aspects. Specifically, our method achieves an average coverage score of 2.74, indicating that our summaries more effectively cover the main points of the source documents compared to SITA and UniMS. For grammar, our method achieves an average score of 2.86, demonstrating better semantic correctness in the generated summaries. In terms of consistency, our method scores 2.86, indicating better factual alignment between the summaries and the source documents. Lastly, our method achieves a score of 2.80 in image-relevance, showing a higher text-image relevance than the other models.

5.4 Case Study

The case study illustrated in the Fig. 3 demonstrates the effectiveness of our proposed multimodal summarization model. The figure includes the original article, the reference summary, and the generated summary produced by our model. The reference summary highlights key points in both text and image, while the generated summary similarly integrates text with corresponding images to provide a comprehensive overview. Our model successfully captures essential information and presents it coherently through a combination of text and visuals, thereby enhancing the overall informativeness and contextual relevance of the summaries. This approach not only makes the content more engaging but also helps in better

Multimodal Input

Daphne Oz has entered her ninth month of pregnancy , and the expectant m om is keeping cozy while proudly showing off her large baby bump.

...

The 31-year-old television personality , who is expecting a baby girl around the holidays , took to Instagram over the weekend to share a photo of herself posed with her hands resting on the top of her jutting stomach.

...

The cookbook author and her husband , John Jovanovic , are already parents to a three-year-old daughter , Philomena , and a two-year-old son , Jovan .

Reference Summary

daphne, 31, took to instagram over the weekend to post a snapshot of herself wearing a brown sweater , a pink t-shirt , leggings , and sneakers.
the television personality, who is expecting a baby girl around the holidays, re vealed that she is 36-weeks pregnant.
daphne and her husband, john jovanovic , are already parents to a three-year-o ld daughter, philomena , and a two-year-old son, jovan.

Generated Summary

daphne oz is expecting a baby girl around the holidays.
the 31-year-old television personality took to instagram over the weekend to share a photo of herself showing off her growing stomach.
the cookbook author and her husband are already parents to a three-year-old daughter, philomena, and a two-year old son, jovan.

Fig. 3. A case of multimodal summarization, including the original article with corresponding images, the reference summary, and the generated summary. The highlighted text in the reference summary indicates key information points.

understanding complex information by providing visual context. The generated summary, when compared with the reference summary, shows that our model can effectively leverage multimodal data to produce high-quality summaries that are informative and easy to comprehend.

6 Conclusion

In this paper, we have proposed a novel modality-aware fusion module for multimodal summarization, which seamlessly integrates textual and visual information to produce high-quality summaries. By leveraging multiple layers of cross-attention and self-attention, our approach effectively enhances the mutual supplementation of visual and textual features, resulting in more coherent and informative summaries. Furthermore, we introduced the Summarization Ranking (SumR) task to reduce the exposure bias during inference stage, ensuring that the model's predicted scores align with actual evaluation metrics. Experimental results on a comprehensive dataset validate the effectiveness of our approach, showing significant improvements over existing methods in both automatic and human evaluations. Our work introduced a novel multimodal summarization method, that enhances the information comprehension and engagement, besides paves the way for further advancements in this field.

Acknowledgements. This study was funded in part by the Beijing Municipal Science and Technology Commission under Grant Z231100001723002, and in part by the National Key Research and Development Program of China under Grant 2020YFB1406702-3, in part by the National Natural Science Foundation of China under Grant 61772575 and 62006257.

Disclosure of Interests. The authors have no competing interests to declare that are relevant to the content of this article.

References

1. Atri, Y.K., Pramanick, S., Goyal, V., Chakraborty, T.: See, hear, read: leveraging multimodality with guided attention for abstractive text summarization. Knowl. Based Syst. **227**, 107152 (2021). https://doi.org/10.1016/j.knosys.2021.107152
2. Bau, D., Liu, S., Wang, T., Zhu, J., Torralba, A.: Rewriting a deep generative model. In: Vedaldi, A., Bischof, H., Brox, T., Frahm, J. (eds.) Computer Vision - ECCV 2020 - 16th European Conference, Glasgow, 23–28 August 2020, Proceedings, Part I. LNCS, vol. 12346, pp. 351–369. Springer (2020). https://doi.org/10.1007/978-3-030-58452-8_21
3. Celikyilmaz, A., Bosselut, A., He, X., Choi, Y.: Deep communicating agents for abstractive summarization. In: Walker, M.A., Ji, H., Stent, A. (eds.) Proceedings of the 2018 Conference of the North American Chapter of the Association for Computational Linguistics: Human Language Technologies, NAACL-HLT 2018, New Orleans, 1–6 June 2018, Volume 1 (Long Papers), pp. 1662–1675. Association for Computational Linguistics (2018). https://doi.org/10.18653/v1/n18-1150
4. Chen, Y., et al.: Revisiting multimodal representation in contrastive learning: from patch and token embeddings to finite discrete tokens. In: IEEE/CVF Conference on Computer Vision and Pattern Recognition, CVPR 2023, Vancouver, 17–24 June 2023, pp. 15095–15104. IEEE (2023). https://doi.org/10.1109/CVPR52729.2023.01449
5. Cheng, D., Xu, Z., Li, J., Liu, L., Liu, J., Le, T.D.: Causal inference with conditional instruments using deep generative models. In: Williams, B., Chen, Y., Neville, J. (eds.) Thirty-Seventh AAAI Conference on Artificial Intelligence, AAAI 2023, Thirty-Fifth Conference on Innovative Applications of Artificial Intelligence, IAAI 2023, Thirteenth Symposium on Educational Advances in Artificial Intelligence, EAAI 2023, Washington, 7–14 February 2023, pp. 7122–7130. AAAI Press (2023). https://doi.org/10.1609/aaai.v37i6.25869
6. Choubey, P.K., Fabbri, A.R., Vig, J., Wu, C., Liu, W., Rajani, N.: Cape: contrastive parameter ensembling for reducing hallucination in abstractive summarization. In: Rogers, A., Boyd-Graber, J.L., Okazaki, N. (eds.) Findings of the Association for Computational Linguistics: ACL 2023, Toronto, 9–14 July 2023, pp. 10755–10773. Association for Computational Linguistics (2023). https://doi.org/10.18653/v1/2023.findings-acl.685
7. Dosovitskiy, A., Beyer, L., Kolesnikov, A., Weissenborn, D., Zhai, X., Unterthiner, T., Dehghani, M., Minderer, M., Heigold, G., Gelly, S., Uszkoreit, J., Houlsby, N.: An image is worth 16x16 words: Transformers for image recognition at scale. In: 9th International Conference on Learning Representations, ICLR 2021, Virtual Event, Austria, May 3-7, 2021. OpenReview.net (2021), https://openreview.net/forum?id=YicbFdNTTy

8. Freitag, M., Al-Onaizan, Y.: Beam search strategies for neural machine translation. In: Luong, T., Birch, A., Neubig, G., Finch, A.M. (eds.) Proceedings of the First Workshop on Neural Machine Translation, NMT@ACL 2017, Vancouver, 4 August 2017, pp. 56–60. Association for Computational Linguistics (2017). https://doi.org/10.18653/v1/w17-3207
9. He, B., Wang, J., Qiu, J., Bui, T., Shrivastava, A., Wang, Z.: Align and attend: multimodal summarization with dual contrastive losses. In: IEEE/CVF Conference on Computer Vision and Pattern Recognition, CVPR 2023, Vancouver, 17–24 June 2023, pp. 14867–14878. IEEE (2023). https://doi.org/10.1109/CVPR52729.2023.01428
10. He, J., Kryscinski, W., McCann, B., Rajani, N., Xiong, C.: Ctrlsum: towards generic controllable text summarization. In: Goldberg, Y., Kozareva, Z., Zhang, Y. (eds.) Proceedings of the 2022 Conference on Empirical Methods in Natural Language Processing, EMNLP 2022, Abu Dhabi, 7–11 December 2022, pp. 5879–5915. Association for Computational Linguistics (2022). https://doi.org/10.18653/v1/2022.emnlp-main.396
11. Jiang, C., Xie, R., Ye, W., Sun, J., Zhang, S.: Exploiting pseudo image captions for multimodal summarization. In: Rogers, A., Boyd-Graber, J.L., Okazaki, N. (eds.) Findings of the Association for Computational Linguistics: ACL 2023, Toronto, 9–14 July 2023, pp. 161–175. Association for Computational Linguistics (2023). https://doi.org/10.18653/v1/2023.findings-acl.12
12. Kim, W., Son, B., Kim, I.: Vilt: vision-and-language transformer without convolution or region supervision. In: Meila, M., Zhang, T. (eds.) Proceedings of the 38th International Conference on Machine Learning, ICML 2021, 18–24 July 2021, Virtual Event. Proceedings of Machine Learning Research, vol. 139, pp. 5583–5594. PMLR (2021). http://proceedings.mlr.press/v139/kim21k.html
13. Lam, K.N., Doan, T.G., Pham, K.T., Kalita, J.: Abstractive text summarization using the BRIO training paradigm. In: Rogers, A., Boyd-Graber, J.L., Okazaki, N. (eds.) Findings of the Association for Computational Linguistics: ACL 2023, Toronto, 9–14 July 2023, pp. 92–99. Association for Computational Linguistics (2023). https://doi.org/10.18653/v1/2023.findings-acl.7
14. Lee, H., Lu, J., Tan, Y.: Convergence for score-based generative modeling with polynomial complexity. In: Koyejo, S., Mohamed, S., Agarwal, A., Belgrave, D., Cho, K., Oh, A. (eds.) Advances in Neural Information Processing Systems 35: Annual Conference on Neural Information Processing Systems 2022, NeurIPS 2022, New Orleans, 28 November–9 December 2022 (2022). http://papers.nips.cc/paper_files/paper/2022/hash/8ff87c96935244b63503f542472462b3-Abstract-Conference.html
15. Lewis, M., et al.: BART: denoising sequence-to-sequence pre-training for natural language generation, translation, and comprehension. In: Jurafsky, D., Chai, J., Schluter, N., Tetreault, J.R. (eds.) Proceedings of the 58th Annual Meeting of the Association for Computational Linguistics, ACL 2020, Online, 5–10 July 2020, pp. 7871–7880. Association for Computational Linguistics (2020). https://doi.org/10.18653/v1/2020.acl-main.703

16. Li, H., Yuan, P., Xu, S., Wu, Y., He, X., Zhou, B.: Aspect-aware multimodal summarization for Chinese e-commerce products. In: The Thirty-Fourth AAAI Conference on Artificial Intelligence, AAAI 2020, The Thirty-Second Innovative Applications of Artificial Intelligence Conference, IAAI 2020, The Tenth AAAI Symposium on Educational Advances in Artificial Intelligence, EAAI 2020, New York, 7–12 February 2020, pp. 8188–8195. AAAI Press (2020). https://doi.org/10.1609/aaai.v34i05.6332
17. Li, H., Zhu, J., Ma, C., Zhang, J., Zong, C.: Read, watch, listen, and summarize: Multi-modal summarization for asynchronous text, image, audio and video. IEEE Trans. Knowl. Data Eng. **31**(5), 996–1009 (2019). https://doi.org/10.1109/TKDE.2018.2848260
18. Liang, Y., Meng, F., Wang, J., Xu, J., Chen, Y., Zhou, J.: D^2tv: Dual knowledge distillation and target-oriented vision modeling for many-to-many multimodal summarization. In: Bouamor, H., Pino, J., Bali, K. (eds.) Findings of the Association for Computational Linguistics: EMNLP 2023, Singapore, December 6-10, 2023. pp. 14910–14922. Association for Computational Linguistics (2023), https://aclanthology.org/2023.findings-emnlp.994
19. Lin, C.Y.: ROUGE: a package for automatic evaluation of summaries. In: Text Summarization Branches Out, pp. 74–81. Association for Computational Linguistics, Barcelona (2004). https://aclanthology.org/W04-1013
20. Liu, N., Sun, X., Yu, H., Zhang, W., Xu, G.: Multistage fusion with forget gate for multimodal summarization in open-domain videos. In: Webber, B., Cohn, T., He, Y., Liu, Y. (eds.) Proceedings of the 2020 Conference on Empirical Methods in Natural Language Processing, EMNLP 2020, Online, 16–20 November 2020, pp. 1834–1845. Association for Computational Linguistics (2020). https://doi.org/10.18653/v1/2020.emnlp-main.144
21. Liu, Y., Lapata, M.: Text summarization with pretrained encoders. In: Inui, K., Jiang, J., Ng, V., Wan, X. (eds.) Proceedings of the 2019 Conference on Empirical Methods in Natural Language Processing and the 9th International Joint Conference on Natural Language Processing, EMNLP-IJCNLP 2019, Hong Kong, 3–7 November 2019, pp. 3728–3738. Association for Computational Linguistics (2019). https://doi.org/10.18653/v1/D19-1387
22. Liu, Y., Liu, P., Radev, D.R., Neubig, G.: BRIO: bringing order to abstractive summarization. In: Muresan, S., Nakov, P., Villavicencio, A. (eds.) Proceedings of the 60th Annual Meeting of the Association for Computational Linguistics (Volume 1: Long Papers), ACL 2022, Dublin, 22–27 May 2022, pp. 2890–2903. Association for Computational Linguistics (2022). https://doi.org/10.18653/v1/2022.acl-long.207
23. Palaskar, S., Libovický, J., Gella, S., Metze, F.: Multimodal abstractive summarization for how2 videos. In: Korhonen, A., Traum, D.R., Màrquez, L. (eds.) Proceedings of the 57th Conference of the Association for Computational Linguistics, ACL 2019, Florence, 28 July–2 August 2019, Volume 1: Long Papers, pp. 6587–6596. Association for Computational Linguistics (2019). https://doi.org/10.18653/v1/p19-1659
24. Paulus, R., Xiong, C., Socher, R.: A deep reinforced model for abstractive summarization. In: 6th International Conference on Learning Representations, ICLR 2018, Vancouver, 30 April–3 May 2018, Conference Track Proceedings. OpenReview.net (2018). https://openreview.net/forum?id=HkAClQgA-

25. Ramasinghe, S., Ranasinghe, K.N., Khan, S.H., Barnes, N., Gould, S.: Conditional generative modeling via learning the latent space. In: 9th International Conference on Learning Representations, ICLR 2021, Virtual Event, 3–7 May 2021. OpenReview.net (2021). https://openreview.net/forum?id=VJnrYcnRc6
26. Rush, A.M., Chopra, S., Weston, J.: A neural attention model for abstractive sentence summarization. In: Màrquez, L., Callison-Burch, C., Su, J., Pighin, D., Marton, Y. (eds.) Proceedings of the 2015 Conference on Empirical Methods in Natural Language Processing, EMNLP 2015, Lisbon, 17–21 September 2015, pp. 379–389. The Association for Computational Linguistics (2015). https://doi.org/10.18653/v1/d15-1044
27. Saini, N., Saha, S., Bhattacharyya, P., Mrinal, S., Mishra, S.K.: On multimodal microblog summarization. IEEE Trans. Comput. Soc. Syst. **9**(5), 1317–1329 (2022). https://doi.org/10.1109/TCSS.2021.3110819
28. Simonyan, K., Zisserman, A.: Very deep convolutional networks for large-scale image recognition. In: Bengio, Y., LeCun, Y. (eds.) 3rd International Conference on Learning Representations, ICLR 2015, San Diego, 7–9 May 2015, Conference Track Proceedings (2015). http://arxiv.org/abs/1409.1556
29. Xiao, M., Zhu, J., Lin, H., Zhou, Y., Zong, C.: CFSum coarse-to-fine contribution network for multimodal summarization. In: Rogers, A., Boyd-Graber, J., Okazaki, N. (eds.) Proceedings of the 61st Annual Meeting of the Association for Computational Linguistics (Volume 1: Long Papers), pp. 8538–8553. Association for Computational Linguistics, Toronto (2023). https://doi.org/10.18653/v1/2023.acl-long.476
30. Zeng, Q., et al.: Improving consistency for text summarization with energy functions. In: Bouamor, H., Pino, J., Bali, K. (eds.) Findings of the Association for Computational Linguistics: EMNLP 2023, Singapore, 6–10 December 2023, pp. 11925–11931. Association for Computational Linguistics (2023). https://doi.org/10.18653/v1/2023.findings-emnlp.798
31. Zhang, C., Zhang, Z., Li, J., Liu, Q., Zhu, H.: Ctnr: compress-then-reconstruct approach for multimodal abstractive summarization. In: International Joint Conference on Neural Networks, IJCNN 2021, Shenzhen, 18–22 July 2021, pp. 1–8. IEEE (2021). https://doi.org/10.1109/IJCNN52387.2021.9534082
32. Zhang, H., Liu, X., Zhang, J.: Extractive summarization via chatGPT for faithful summary generation. In: Bouamor, H., Pino, J., Bali, K. (eds.) Findings of the Association for Computational Linguistics: EMNLP 2023, Singapore, 6–10 December 2023, pp. 3270–3278. Association for Computational Linguistics (2023). https://doi.org/10.18653/v1/2023.findings-emnlp.214
33. Zhang, Z., Meng, X., Wang, Y., Jiang, X., Liu, Q., Yang, Z.: Unims: a unified framework for multimodal summarization with knowledge distillation. In: Thirty-Sixth AAAI Conference on Artificial Intelligence, AAAI 2022, Thirty-Fourth Conference on Innovative Applications of Artificial Intelligence, IAAI 2022, The Twelveth Symposium on Educational Advances in Artificial Intelligence, EAAI 2022 Virtual Event, 22 February–1 March 2022, pp. 11757–11764. AAAI Press (2022). https://ojs.aaai.org/index.php/AAAI/article/view/21431
34. Zhang, Z., Shu, C., Chen, Y., Xiao, J., Zhang, Q., Lu, Z.: ICAF: iterative contrastive alignment framework for multimodal abstractive summarization. In: International Joint Conference on Neural Networks, IJCNN 2022, Padua, 18–23 July 2022, pp. 1–8. IEEE (2022). https://doi.org/10.1109/IJCNN55064.2022.9892884

35. Zhao, Y., Khalman, M., Joshi, R., Narayan, S., Saleh, M., Liu, P.J.: Calibrating sequence likelihood improves conditional language generation. In: The Eleventh International Conference on Learning Representations, ICLR 2023, Kigali, 1–5 May 2023. OpenReview.net (2023). https://openreview.net/pdf?id=0qSOodKmJaN
36. Zhu, J., Li, H., Liu, T., Zhou, Y., Zhang, J., Zong, C.: MSMO: multimodal summarization with multimodal output. In: Riloff, E., Chiang, D., Hockenmaier, J., Tsujii, J. (eds.) Proceedings of the 2018 Conference on Empirical Methods in Natural Language Processing, Brussels, 31 October–4 November 2018, pp. 4154–4164. Association for Computational Linguistics (2018). https://doi.org/10.18653/v1/d18-1448
37. Zhu, J., Zhou, Y., Zhang, J., Li, H., Zong, C., Li, C.: Multimodal summarization with guidance of multimodal reference. In: The Thirty-Fourth AAAI Conference on Artificial Intelligence, AAAI 2020, The Thirty-Second Innovative Applications of Artificial Intelligence Conference, IAAI 2020, The Tenth AAAI Symposium on Educational Advances in Artificial Intelligence, EAAI 2020, New York, 7–12 February 2020, pp. 9749–9756. AAAI Press (2020). https://ojs.aaai.org/index.php/AAAI/article/view/6525

Enhancing Text-Image Person Re-identification via Intra-Class Relevance Learning

Wenbin He[1], Yutong Gao[1,2](✉), Wenjian Liu[3], Chaomurilige[1], Zheng Liu[1], and Ao Guo[1]

[1] Key Laboratory of Ethnic Language Intelligent Analysis and Security Governance, Ministry of Education, Minzu University of China, Beijing 100081, China
{hewenbin,chaomurilige,liuzheng,guoao,ytgao92}@muc.edu.cn
[2] Hainan International College of Minzu University of China, Li'an International Education Innovation pilot Zone, Lingshui Li 572499, Hainan, China
[3] City University of Macau, Macau SAR, China
andylau@cityu.edu.mo

Abstract. Text-Image Person Re-Identification (TIReID) is a computer vision task that involves identifying person in images or videos based on textual descriptions. Current works mainly employ Vision Language Pretrained (VLP) models for cross-modal alignment and to reestablish text-image fine-grained associations for re-identification. However, these methods neglect the variations in intra-class sample features, hindering accurate cross-modal alignment and discrimination of fine-grained features for the same person. In this paper, we propose a Cross-modal Intra-Class Learning (CICL) framework to enhance the model's capability of learning fine-grained cross-modal features and intra-class sample variations. We propose an Image-Text Intra-class Relevance Learning (ITRL) method that considers sample relevance during text-image matching, boosting re-identification capability while maintaining fine-grained text-image alignment. Meanwhile, we propose a Bidirectional Masked Matching (BiMM) method that introduces masks to images and texts, prompting the model to attend to different content regions and establish finer-grained cross-modal associations. We test our approach using the CLIP model on three TIReID benchmarks, achieving results that surpass state-of-the-art performance on multiple metrics.

Keywords: Cross-modal Retrieval · Multimodal · Artificial Intelligence

1 Introduction

Text-Image Person Re-identification (TIReID) is a popular challenge in the computer vision that strives to localize person of interest across extensive image or video repositories, guided by natural language descriptions [18]. This task is a

T. Zhu et al. (Eds.): ICA3PP 2024, LNCS 15252, pp. 165–184, 2025.
https://doi.org/10.1007/978-981-96-1528-5_12

sub-task of image text retrieval [23,31] and image-based person re-identification (Re-ID) [9,20,36], which requires solving the complex challenge of strongly and fine-grained associating text descriptions with corresponding visual data [28,57]. This task is widely applied in scenarios where visual cues alone are insufficient for reliable re-identification, such as complex surveillance systems and social media platforms. TIReID not only requires precise retrieval of individual targets, but also poses deeper challenges in identifying them in different environments and visually similar subjects [5]. TIReID has recently attracted increasing attention from researchers and practitioners in the field of multimodal learning and related industries, and it has broad application potential in multimedia analysis, public safety and other fields [46].

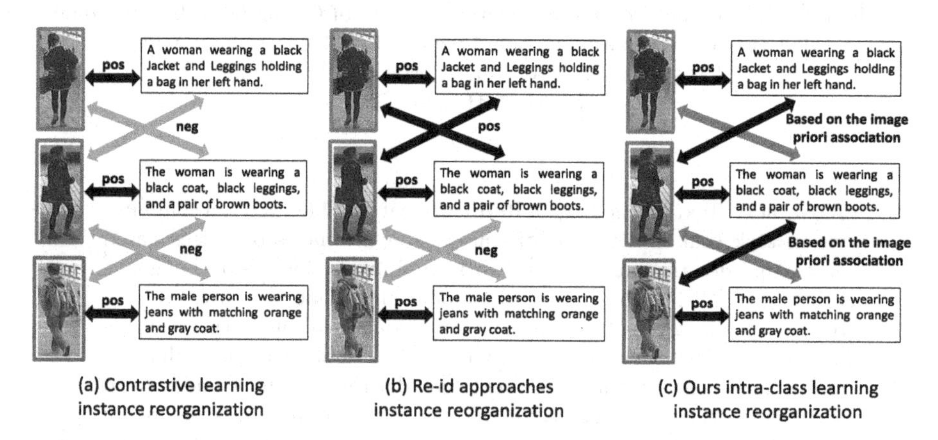

Fig. 1. Comparing the strategies in different methods to construct instance pairs when modal alignment. (a) In the contrastive learning method, only real matching pairs are regarded as positive examples; (b) In the re-id strategy, both images and texts of the same individual are regarded as positive examples; (c) Our method improves the construction of pairs by further considering non-correlated factors.

As a cross-modal retrieval task, TIReID mainly focuses on person individuals as the subjects of retrieval, and can be regarded as a fine-grained intra-class retrieval task [43,57]. The key to enhancing the overall retrieval performance of TIReID lies in the ability to discriminate and learn the subtle similarities and distinctions between different individuals. Cross-modal retrieval frameworks [25,54] usually contain two independent models to process images and text respectively to achieve distributed retrieval and parallel processing. Through training on numerous ground truth image-text pairs, these models can embed the processed image and text outputs into a unified feature space, establishing cross-modal association and alignment.

Benefiting from the latest advancements in multimodal learning, numerous Vision Language Pretrained (VLP) models [2,14,25,37], have demonstrated robust cross-modal alignment capabilities, which can facilitate the text-to-image association for TIReID. However, when applying these VLP models to the TIReID task, their performance is unsatisfactory. This can be attributed to their focus on learning instance-level visual-textual knowledge, which enables capturing features across different environments and categories but fails to establish fine-grained discrimination of intra-class object features.

Recent works [28,34,43] on TIReID have utilized VLP for their strong cross-modal alignment, and focus on the two kind of strategies: forced instance-level alignment for direct mapping between texts and images [25,31], or re-identification to align the same pedestrians across different instances [54,55]. The first strategy mainly adopts contrastive learning methods, focusing on distinguishing different changes between samples and establishing implicit alignment. However, this strategy is more susceptible to changes in out-of-class features such as perspective changes, which hinders the model's learning of the same person features. As shown in Figure 1(a), images of the same person and corresponding text samples have the same semantic features, but are treated as counterexamples during the training process. The re-id approaches, focusing on intra-class consistency, may limit learning of intra-class differences and impair the precise alignment between image and text modalities, as shown in Fig. 1(b). Clustering learning from identical individuals can lead to vague cross-modal associations. Although some new approaches [10,43] have noted these issues, simply combining these strategies in training has not achieved a significant overall improvement.

In this work, we further considered fine-grained text-image alignment and the intra-class sample relationships. We propose a Cross-modal Intra-class Learning (CICL) framework to tackle the challenges of coarse-grained cross-modal feature discrimination and insufficient sample learning in TIReID. It contains a set of sample learning strategies and a set of training learning strategies. To strengthen the model's learning of intra-class person features, we propose an Image-Text Intra-class Relevance Learning (ITRL) method. This method utilizes human priors to guide the model in learning intra-class feature variations and irrelevant changes, thereby enhancing ReID capability while maintaining fine-grained text-image alignment. Moreover, to enhance the model's fine-grained matching capability, we propose a Bidirectional Masked Matching (BiMM) method. It introduces noise or masks to different regions of images and texts, prompting the model to attend to various content regions to establish finer-grained cross-modal associations. We conducted experiments using a CLIP model to evaluate our approach and compare its performance with various training strategies. The results demonstrate that our method not only improves the precise retrieval performance in the Re-ID task but also outperforms most existing methods in overall retrieval performance. Our main contributions can be summarized as follows:

- We introduce a cross-modal intra-class learning (CICL) framework that improves fine-grained feature learning and sample variation handling while maintaining cross-modal alignment ability. This approach utilizes image-based prior and intra-class relevance to boost model retrieval on TIReID.

- In this paper, we propose the Image-Text Intra-class Relevance Learning (ITRL) method to enhance the model's learning of intra-class character features; and propose a newly method Bidirectional Masked Matching (BiMM) to enhance the fine-grained matching ability of the model.
- Extensive experiments on three public benchmark datasets, *i.e.*, CUHK-PEDES [18], ICFG-PEDES [5] and RSTPReid [55] demonstrate the effectiveness of our approach. And the results surpasses the existing state-of-the-art methods in metrics such as mAP and mINP.

2 Related Work

2.1 Person Re-identification Task

Over the past few years, a variety of pedestrian re-identification methods have continuously emerged, as researchers strive to extract distinctive and robust representations from person images, overcoming challenges such as occlusions and background clutter [49,53]. Local representation learning has proven to be an effective and widely used technique in this context, focusing on detailing and unique features. For instance, PCB [32] horizontally segments images to independently analyze local features, while MGN [35] divides images into different scales for learning. Additionally, some methods incorporate spatial attention mechanisms to align body parts [52], and others use additional models to assist in local segmentation and detail extraction [11,29,51]. Extensive efforts have also been made in video-based re-ID [22,48,56], occluded re-ID [33], and unsupervised re-ID [19,30,45], among others. Unfortunately, these advancements are primarily utilized in image-based search scenarios for person re-identification.

2.2 Cross-Modal Retrieval Task

In the cross-modal retrieval task, Vision-Language Pre-training (VLP) models such as CLIP [25] have proven highly effective. They utilize large-scale image-text pairs to achieve semantic alignment between visual and textual data. These models [2,12,14,37] employ robust techniques like Masked Language Modeling (MLM) [12] and Masked Image Modeling (MIM) [1] to enhance feature learning. Techniques such as Image-Text Contrast (ITC) [25] and Image-Text Matching (ITM) [15] facilitate precise alignment and discrimination across modalities. Recent advancements have also explored methods for fine-grained alignment [3,7,16], which focus on the models' ability to discern subtle nuances within classes, a critical aspect for more in-depth retrieval tasks such as TIReID [18]. Moreover, unsupervised learning approaches like SyCoCa [21] are being integrated, combining MLM and MIM for unsupervised fine-grained alignment, making notable progress in image retrieval. These developments underscore the advancements in cross-modal retrieval, emphasizing the strong alignment capabilities of VLP models. However, their direct application to TIReID still faces some limitations.

2.3 Text-Image Person Re-identification Task

Text-Image Person Re-identification (TIReID) is a multimodal task distinct from typical cross-modal retrieval, focusing on fine-grained intra-class distinctions. It was introduced by Li *et al.* [18], TIReID challenges involve matching detailed textual descriptions with corresponding images. Earlier studies [4,38,39,42,55] extensively utilized branches that explicitly exploited human segmentation, body parts, color information, and text phrases. Other works [5,6,27,44] employed attention mechanisms for implicit local feature learning. More recent works utilized ID loss [54] and the Cross-Modal Projection Matching (CMPM) loss [50] to align text and image features within a joint embedding space. Techniques such as Masked Language Modeling (MLM) have also been employed [10] to enhance the learning of textual nuances related to visual features. Despite these efforts, most existing methods overlook the importance of intra-class distinctions, focusing instead on broader alignment strategies. While recent works [8,10,34,43] have begun incorporating advanced vision-language pretraining models and fine-grained alignment methods [28,57], the challenge of achieving precise intra-class differentiation in TIReID remains underexplored. In summary, TIReID has seen considerable methodological advancements but still requires focused efforts on intra-class distinctions to fully address its unique challenges.

3 Method

This chapter provides a detailed introduction to the Cross-modal Intra-class Learning (CICL) framework designed for TIReID, including the definitions of symbols, network architecture design, objective functions, and sample learning strategies. Section 3.1 outlines the overall structure of our training network, while Sect. 3.2 and Sect. 3.3 respectively introduce the innovative training objectives and methods proposed within this framework. Section 3.4 introduces the overall training objectives of our framework.

3.1 Overview of Cross-Modal Intra-Class Learning Framework

Motivated by the partial success of knowledge transfer from CLIP to text-image person retrieval applications, we directly initialize our CICL framework with the complete set of CLIP image and text encoders, as shown in Fig. 2. This initialization aims to strengthen the foundational cross-modal alignment capabilities of our system. To further refine our model during the training phase, we integrate a cross-modal encoder that is specifically engineered to enable robust interaction between image and textual data. The cornerstone of our approach lies in a strong focus on learning fine-grained features and capturing intra-class variations.

Image Encoder. We employ the vision transformer from CLIP as the image encoder. Given an input image $I \in \mathbb{R}^{H \times W \times C}$, it is transformed into a sequence of fixed-size patches. The image is divided into $N = H \times W/P^2$ non-overlapping patches, where P is the patch size. Each patch is linearly projected into a sequence of 1-D tokens, and an additional [cls] token is injected at the beginning of this sequence, forming $\{v_{\text{cls}}, v_1, \ldots, v_N\}$ with positional embedding. These tokens are processed through the image encoder to extract the visual features. The [cls] token is used to capture the global information of the image, resulting in the embedding vector v_{cls}. This vector serves as a global image representation and is projected into the joint image-text embedding space for cross-modal alignment in training and retrieval in inference.

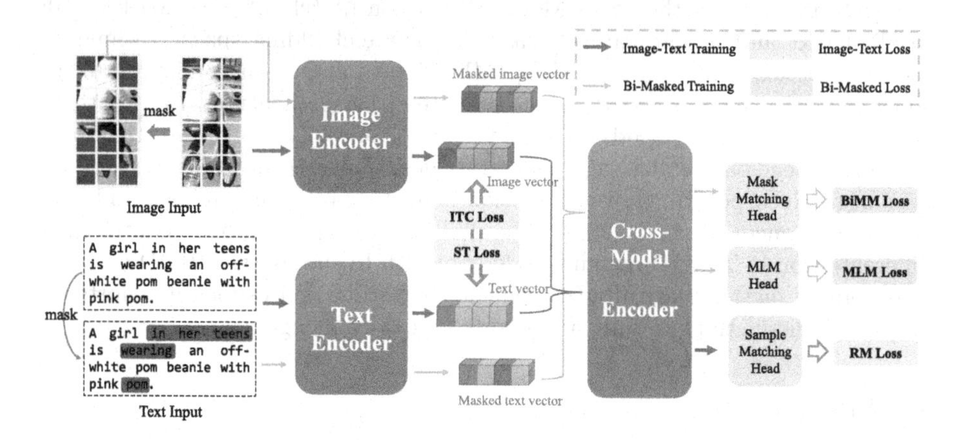

Fig. 2. Overview of the proposed CICL framework. It includes an Image Encoder and a Text Encoder for extracting modality-specific information, utilized in both training and inference phases. Additionally, a Cross-Modal Encoder is integrated to fuse the modal information, which is exclusively used during training. Within this framework, we also implement standard image-text training along with bidirectional masked image-text training, each targeting distinct training objectives.

Text Encoder. We utilize the causal masked transformer from CLIP as the text encoder, and utilize the BPE scheme with a vocabulary size of 49,152 to tokenize the input text descriptions. Each text token is converted into a sequence embedding vector, and then arranged into a sequence with an additional [cls] token appended at the head, forming the sequence $\{w_{\text{cls}}, w_1, \ldots, w_m\}$, where m is the length of text. As same as the image processing, the [cls] token vector is used to represent the overall textual context. The w_{cls} will be linearly projected into the joint image-text embedding space for cross-modal alignment in training and retrieval in inference.

Cross-Modal Encoder. To enhance the interaction between image and text modalities, we adopted a cross-modal interaction encoder similar to IRRA [10]. The encoder structure is similar to the traditional Transformer, but with a multi-head cross-attention layer, using the representation of one modality as the query and the other representation as the keys and values. This design facilitates in-depth integration between modalities. The encoder comprises 4-layer Transformer blocks, taking the output vectors from the image encoder and text encoder as its joint input. This module is only used during the training phase to capture cross-modal interaction information and learn the modal associations between different encoders.

Through this architectural design, we can calculate vectors that capture information in different dimensions by adding various projection heads, as shown in Fig. 2. We utilize these vectors to design different training objectives for enhancing the fine-grained alignment capabilities and overall retrieval performance of the inference model.

3.2 Image-Text Intra-Class Relevance Learning

In cross-modal retrieval with TIReID, it is crucial to establish fine-grained modal alignment and identify distinguishing features between instances. However, traditional approaches, such as image-text contrastive loss and ReID classification loss, often face challenges due to an excessive emphasis on instance-level differences or the blurring of important intra-class variations. To overcome these limitations, we propose the Image-Text Intra-Class Relevance Learning (ITRL) method, which is designed to delicately balance and refine the alignment between text and image pairs within the same individual while clearly distinguishing different individuals.

Different from the strategy of directly constructing positive and negative samples in contrastive learning, in our method, we subdivide text-image pairs into `positive pairs` (ground truth matches from the dataset), `similar pairs` (text and images of the same person), and `negative pairs` (text and images from different individuals), as shown in Fig. 3. This subdivision enables us to more specifically learn the similarities and differences in modal content, akin to inter-class and intra-class learning in image classification task. We designed two new loss functions for training and incorporated human-guided priors to address the interference of irrelevant features (such as environment, camera angles, and lighting) encountered during the modal alignment. Inspiration from contrastive learning, we do not need to modify the dataset input but instead construct subdivided instances during training.

Triplet Similarity Loss. We employ the Triplet loss, commonly used for fine-grained discrimination tasks like face recognition, to model the similarity associations between image-text pairs. Unlike the typical vector distance metric, cross-modal alignment requires considering the distance between vector pairs. Hence, we leverage the cosine similarity angle between image and text vectors as the distance measure. As illustrated in Fig. 3, during training on a mini-batch, we generate triplets with distinctive relationships from the image-text pairs $\{T_1, ..., T_n; I_1, ..., I_n; id_1, ...id_n\}$, and compute their similarity angles, and

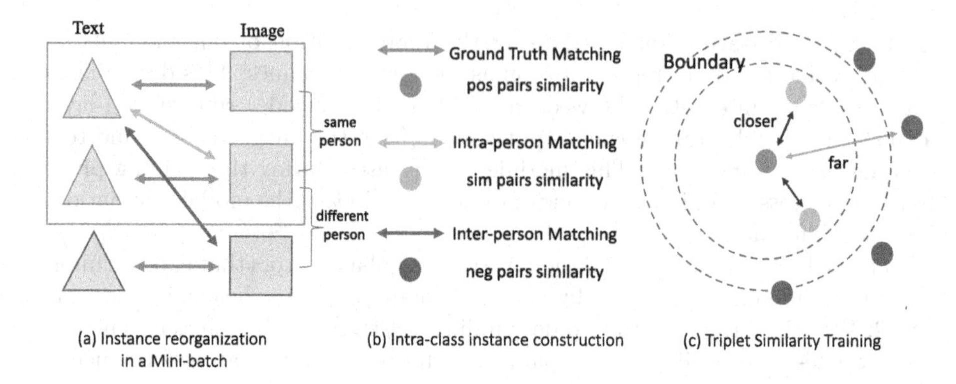

Fig. 3. Strategies and goals for instance recombination in ITRL training. In a batch training, each instance training can be regarded as a Mini-batch. Its text/image can form a new matching pair with other image/text. According to the characteristics of TIReID, we further subdivide them into pos pair, sim pair and neg pair. Our goal is that the pos pair has the highest similarity, followed by the sim pair and the neg pair the smallest. While adding training examples, more detailed goals are learned.

calculate the triplet similarity loss:

$$\mathcal{L}_{ts} = \sum_{i=1}^{n}\sum_{j=1}^{n} \max(d_{min}(O_p, O_s) - d_{max}(O_p, O_n) + \text{margin}, 0) \quad (1)$$

$$O_p = \mathcal{S}(T_i, I_j)_{i=j}, \quad O_s = \mathcal{S}(T_i, I_j)_{id_i=id_j, i\neq j}, \quad O_n = \mathcal{S}(T_i, I_j)_{id_i \neq id_j} \quad (2)$$

where $\mathcal{S}(\cdot)$ calculates the similarity distance between encoded vectors. O_p, O_s, O_n respectively represent positive, similar, and negative match pairs, and id is the unique identifier for each person. $d_{min}(\cdot)$ and $d_{max}(\cdot)$ measure the smallest and largest similarity distances within a batch. The margin is a hyper-parameter used to control the separation between similar and dissimilar instances, and we set it to 0.2, following the practice in Triplet loss as described in [26]. We mainly use this loss to help the model distinguish relevant samples from irrelevant samples, and also use CLIP loss (InfoNCE [24]) to ensure that positive samples achieve the highest similarity.

Relevance Matching Loss. In ReID tasks, person individuals often exhibit consistent features across various observations. With this in mind, we introduce an image-based prior to effectively exploit these features while distinguishing non-relevant factors within class samples. As described in formula 2, we classify `similar pairs` as positive matching instances and utilize a cross-encoder for joint matching learning, employing artificial priors to differentiate the learning process across various samples. We train the model to predict whether images and texts form matching pairs, enhancing its cross-modal reasoning capabilities. This matching process uses cross-entropy loss to compute the relevance matching loss, and it is defined as follows:

$$\mathcal{L}_{rm} = -\sum \alpha_{prior} y \log(\theta_r(I,T)) + (1-\alpha_{prior})(1-y)\log(1-\theta_r(I,T)) \quad (3)$$

where $\theta_r(I,T)$ represents the probability of matching computed by samples matching head. During training, we strategically regroup pairs by selecting the most dissimilar one with $\min(O_s)$ as the similar pair and the most similar one with $\max(O_n)$ as the negative pair, aiming to maximize the information entropy for model learning. The parameter $\alpha_{prior} \in (0,1)$ serves as a prior factor derived from the inherent similarity between image samples. We utilize the Peak Signal-to-Noise Ratio (PSNR) as a metric to assess this similarity prior.

PSNR effectively measures the resemblance between an image and its variant [47], typically producing lower values for images with significant environmental or contextual variations. By incorporating PSNR in the computation of α_{prior}, we guide the learning process to diminish the impact of large intra-class variations that could potentially blur modal alignment during training. We calculate the value of α_{prior} using the PSNR as follows:

$$\alpha_{prior}(I_p, I_s) = norm(10 \cdot log_{10}\frac{\mathrm{MAX}^2}{\mathrm{MSE}(I_p, I_s)}) \quad (4)$$

where MAX represents the maximum possible pixel value of the image, and the MSE$(\cdot)$ calculates the mean squared error between two images. Normalization assigns α_{prior} of 1 to identical images and 0 to unrelated or negative samples.

Taken together, we define the overall loss for ITRL as:

$$\mathcal{L}_{itrl} = \mathcal{L}_{ts} + \mathcal{L}_{rm} \quad (5)$$

3.3 Bidirectional Masked Sample Training

Previous image-text alignment methods typically establish implicit fine-grained associations by MLM or MIM. However, these approaches do not consider bidirectional fine-grained correspondences between images and text, and image reconstruction during training can introduce additional noise. Therefore, we designed an bidirectional masked sample matching training method. By simultaneously masking different regions in both the image and text, forcing model to learn the cross-modal alignment of local features. As illustrated in Fig. 2, we employ the Bidirectional Masked Matching (BiMM) loss to replace the conventional text-guided MIM loss, establishing fine-grained alignments between images and text. Furthermore, we utilize bi-mask MLM loss to enhance the model's ability to align from text to images.

Bidirectional Masked Matching. We adapt the strategy of image-text matching to directly building the alignment between image I and text T. Specifically, we implement a random masking mechanism where portions of the original image and text are masked, generating $\hat{I}$ and $\hat{T}$ respectively. Following [21], we mask 50% of image patches and 15% of text tokens. This selective visibility challenges the model to infer missing information and align the modalities based on the remaining visible data. BiMM can be regarded as a binary classification task. We use cross-entropy loss to calculate matching loss and minimizes it when bi-masked training. The $\mathcal{L}_{bimm}$ is defined as follows:

$$\mathcal{L}_{bimm} = -\sum y \log(\theta_m(\hat{I}, \hat{T})) + (1-y) \log\left(1 - \theta_m(\hat{I}, \hat{T})\right) \tag{6}$$

where $\theta_m(\hat{I}, \hat{T})$ denotes the probability that the masked image $\hat{I}$ and masked text $\hat{T}$ are a match pair. It computed by a mask matching head based on their joint representation processed by the cross-modal encoder. The binary label y indicates whether the original I and T are a true match(1) or a mismatch(0). This loss function evaluates the model's ability to predict correct matches from partially visible regions, enforcing it to match based on partial areas.

Mask Language Modeling. In MLM implementation, we follow the mainstream practice and use visual features to assist in predicting masked text tags to enhance the model's multimodal interaction capabilities. We mask textual tokens with a probability of 15%, replacing them with the special token `[MASK]`. And we change the image masking probability at 0% to align with other methods. The MLM loss function is articulated as follows:

$$\mathcal{L}_{mlm} = \mathbb{E}_{(\hat{I},\hat{T})\sim D} \ell_{mlm}(\theta_t(\hat{I}, \hat{T}), T) \tag{7}$$

where ℓ_{mlm} represents the cross-entropy loss calculated between the predicted tokens from the joint representation and the ground truth values for each masked token. $\theta_t(\hat{I}, \hat{T})$ provides the probability value of predicting each token for the joint input of $\hat{I}$ and $\hat{T}$, which is calculated by the MLM head in the model. We can adjust different image/text mask probabilities to increase the difficulty of model learning and increase the diversity of samples.

3.4 Overall Training Objective

Our training framework integrates multiple objectives to adapt the CLIP model for TIReID task. We employ two training strategies that includes both image-text pairs training and bi-masked image-text pairs training, as shown in Fig. 2.

Following the general practices in re-identification tasks, we incorporate the CMPM loss and ID loss into our training regimen. The ID loss are specifically designed to classify images or texts into distinct groups based on their identities, and the CMPM loss represents the KL divergence from distribution $\mathbf{q}$ to $\mathbf{p}$. The overall training objective of our CICL framework can be denoted as follows:

$$\mathcal{L} = \mathcal{L}_{id} + \mathcal{L}_{cmpm} + \lambda_1 \mathcal{L}_{mlm} + \lambda_2 \mathcal{L}_{bimm} + \lambda_3 \mathcal{L}_{itrl} \tag{8}$$

where λ_1, λ_2 and λ_3 are hyper-parameters used to balance each loss, with all default values set to 1.

During training, our framework utilizes the image encoder, text encoder, and cross-modal encoder. However, during inference, only the image encoder and text encoder are employed, ensuring efficient processing without incurring additional inference time overhead.

4 Experiment

To demonstrate the effectiveness of our proposed CICL framework, we conduct extensive experiments on three challenging TIReID datasets and perform ablation studies to analyze the improvements from each training objective.

4.1 Experimental Setup

Datasets. Our approach is evaluated using three mainstream benchmark datasets, *i.e.*, **CUHK-PEDES** [18], **ICFG-PEDES** [5], and **RSTPReid** [55]. Specifically, CUHK-PEDES [18] contains 40,206 images and 80,440 text descriptions of 13,003 persons. Its training set includes 34,054 images and 68,108 textual descriptions for 11,003 identities. The validation and test sets each cover 1,000 identities, with 3,078 and 3,074 images and 6,158 and 6,156 textual descriptions respectively. The average representation per identity in this dataset is about three images. ICFG-PEDES [5] is derived from MSMT17 [41] and contains 54,522 images of 4,102 identities, each with a descriptive sentence. The dataset divides these into 34,674 image-text pairs for training (3,102 identities) and 19,848 for testing (1,000 identities), averaging 13 images per identity, taken from multiple camera perspectives. RSTPReid [55] is also collected from the MSMT17 [41] dataset, containing 20,505 images of 4,101 identities, each with two sentences. The average number of images per identity is five. The data is organized with 3,701 identities in the training set and 200 identities each in the validation and test sets, ensuring a broad coverage of scenarios for robust evaluation.

Evaluation Metrics. To comprehensively evaluate the retrieval performance, following previous practices [10,57], we adopt the Rank-k metrics (R1, R5, R10), mean Average Precision(mAP), mean Inverse Negative Penalty(mINP) [46], and mean Similarity Distribution(mSD) [57]. The Rank-k metrics measure the probability of retrieving at least one correct match within the top-k ranked results when given a textual query description, providing an immediate measure of retrieval effectiveness. A higher mAP value demonstrates greater overall retrieval accuracy across all queries. Notably, increased mINP and mSD values signify the model's enhanced ability to prioritize correct matches higher in the ranked list and finely discriminate between similar images, respectively.

Implementation Details. The proposed framework follows the standard architecture of CLIP-ViT-B/16 [25], but we additionally incorporate a 4-layer cross-modal encoder for training, similar to the approach used in IRRA [10]. The image and text encoders are initialized with pre-trained CLIP weights, while the cross-modal encoder is randomly initialized. The input images are resized to 384 × 128, the patch size is set to 16 × 16, and the feature dimensions for both visual and textual modalities are set to 512. For training, we set the batch size to 128 and take 7 h to complete on a H800 GPU. The Adam [13] optimizer is used with a weight decay of 1e–4. The initial learning rate is set to 1e-5 with power learning rate decay, and we employ linear warm-up over the first 5 epochs.

Table 1. Performance comparisons with SOTA methods on CUHK-PEDES dataset.

Method	Ref	R1	R5	R10	mAP	mINP	mSD
SAF [17]	ICASSP'22	64.13	82.62	88.40	–	–	–
AXM-Net [6]	AAAI'22	64.44	80.52	86.77	58.73	–	–
IVT [28]	ECCVW'22	65.59	83.11	89.21	60.66	–	–
CFine [43]	TIP'23	69.57	85.93	91.15	–	–	–
TP-TPS [34]	arXiv'23	70.10	86.10	90.98	66.32	–	–
IRRA [10]	CVPR'23	73.15	**89.82**	**93.70**	66.10	50.26	42.29
CFAM [57]	CVPR'24	72.87	88.61	92.87	64.92	–	**50.20**
Baseline	–	68.11	86.92	92.15	60.58	43.72	38.52
CICL(Ours)	–	**74.03**	89.11	93.44	**67.90**	**53.60**	43.56

4.2 Overall Comparsion Results

In this section, we report our experimental results and compare with other SOTA methods on CUHK-PEDES [18], ICFG-PEDES [5] and RSTPReid [55]. Note that, the Baseline in Table 1, Table 2 and Table 3, denotes the CLIP models fine-tuned with the original CLIP loss (InfoNCE [24]). Among recent comparative models, IRRA [10] primarily employs ID loss, MLM loss, and SDM loss to align implicit associations between images and text. CFAM [57] further introduces cross-modal fine-grained aligning learning to align fine-grained content across modalities. Both models are trained based on CLIP-ViT-B/16 [25].

Results on CUHK-PEDES. We first evaluate the proposed method on the most common benchmark, CUHK-PEDES. As shown in Table 1, our CICL outperforms the state-of-the-art methods on several metrics, and achieves 74.03% Rank-1 accuracy, 67.90% mAP and 53.60% mINP, respectively. It already achieves comparable or even better performance compared with many works proposed in recent years, e.g., CFine [43], IRRA [10], CFAM [57]. When compared to the current state-of-the-art IRRA [10], our model demonstrates improvements of +0.88% in R1, +1.80% in mAP, and +3.34% in mINP, indicating enhanced granularity and alignment capabilities, as well as superior overall retrieval performance for the text-image person re-identification task. Furthermore, when evaluated against baseline model, our approach shows substantial enhancements in fine-grained alignment, further evidencing its superior adaptability and effectiveness for the TIReID challenges.

Table 2. Performance comparisons with SOTA methods on ICFG-PEDES dataset.

Method	Ref	R1	R5	R10	mAP	mINP	mSD
LGUR [27]	MM'22	59.02	75.32	81.56	–	–	–
IVT [28]	ECCVW'22	56.04	73.60	80.22	–	–	–
CFine [43]	TIP'23	60.83	76.55	82.42	–	–	–
TP-TPS [34]	arXiv'23	60.64	75.97	81.76	–	–	–
IRRA [10]	CVPR'23	63.46	80.25	85.82	38.06	7.93	23.75
CFAM [57]	CVPR'24	62.17	79.57	85.32	36.34	–	**28.01**
Baseline	-	56.81	75.99	82.60	32.37	5.77	20.21
CICL(Ours)	-	**64.26**	**80.48**	**85.92**	**41.22**	**10.35**	25.77

Results on ICFG-PEDES. As shown in Table 2, we also evaluate our model on the ICFG-PEDES benchmark, which has higher re-identification demands due to multiple images per individual. Our CICL demonstrated great performance improvements on this benchmark, achieving 64.26% in Rank-1 accuracy, 41.22% in mAP, 10.35% in mINP, and 25.77% in mSD. These results outperform many models introduced in recent research. Specifically, compared to the previous state-of-the-art IRRA [10], our model achieved improvements of +0.8% in R1, +3.16% in mAP, and +2.42% in mINP. Further enhancements ranging from 0.2% to 2% were also observed across other metrics. The relatively low mINP [46] metric on ICFG-PEDES underscores the benchmark's rigorous complexity, emphasizing the challenge of identifying the most difficult matching samples.

Table 3. Performance comparisons with SOTA methods on RSTPReid dataset.

Method	Ref	R1	R5	R10	mAP	mINP	mSD
LBUL [40]	MM'22	45.55	68.20	77.85	–	–	–
IVT [28]	ECCVW'22	46.70	70.00	78.80	–	–	–
CFine [43]	TIP'23	50.55	72.50	81.60	–	–	–
TP-TPS [34]	arXiv'23	50.65	72.45	81.20	–	–	–
IRRA [10]	CVPR'23	**60.20**	**81.60**	88.20	47.20	25.34	29.45
CFAM [57]	CVPR'24	59.40	81.35	88.50	46.04	–	**34.27**
Baseline	–	54.75	78.90	86.05	43.52	22.97	27.36
CICL(Ours)	–	59.95	81.25	**89.05**	**48.05**	**27.23**	30.34

Results on RSTPReid. We also present our experimental findings on the RSTPReid dataset, a newly introduced benchmark, as detailed in Table 3. Our model achieves performance comparable to other leading models while exhibiting improvements on certain metrics. Compared to IRRA [10], we observe gains of +0.85% in R10, +0.85% in mAP, and +1.89% in mINP. These modest increases can be attributed to the limited training data available in this dataset, which may constrain the model's generalization ability. Nonetheless, when compared to our baseline, our method still delivers enhancements of +3% to +5%, demonstrating its efficacy even on a newer, less populated dataset.

Overall, based on the test results across three benchmarks, our model demonstrates significant improvements in R@1, mAP, and mINP metrics. Compared to the state-of-the-art, our model achieves maximum +0.88% increase in R@1, a maximum +3.16% improvement in mAP, and a maximum +3.34% gain in mINP. These results clearly indicate substantial enhancements in our model's overall retrieval quality.

4.3 Ablation Study

In this subsection, we evaluate the effectiveness of each component within the CICL framework. We use the CLIP-ViT-B/16 [25] model fine-tuned with InfoNCE loss as the baseline and the IRRA [10] model, fine-tuned with $\mathcal{L}_{mlm} + \mathcal{L}_{cmpm} + \mathcal{L}_{id}$, as a structurally similar model to CICL but without our additional training strategies and objectives. This setup facilitates a detailed ablation study.

To rigorously assess the impact of different components, we conduct an extensive empirical analysis on two public datasets: CUHK-PEDES [18] and ICFG-PEDES [5]. The evaluation metrics include Rank-1, Rank-10, mAP, mINP, and mSD, with results detailed in Table 4.

Table 4. Ablation results of components on CUHK-PEDES and ICFG-PEDES.

No.	Methods	$\mathcal{L}_{bimm}$	$\mathcal{L}_{itrl}$		CUHK-PEDE				ICFG-PEDES			
			$\mathcal{L}_{ts}$	$\mathcal{L}_{rm}$	R1	mAP	mINP	mSD	R1	mAP	mINP	mSD
–	Baseline				68.11	60.58	43.72	38.52	56.81	32.37	5.77	20.21
0	IRRA				73.31	66.69	51.88	42.67	63.37	38.93	8.52	24.35
1	$+\mathcal{L}_{bimm}$	✓			72.92	66.44	51.54	42.74	63.86	39.84	9.02	24.97
2	$+\mathcal{L}_{ts}$		✓		72.56	**67.97**	**54.81**	43.32	62.99	40.20	9.98	25.08
3	$+\mathcal{L}_{rm}$			✓	73.05	67.49	53.43	43.09	62.84	40.10	10.00	25.07
4	$+\mathcal{L}_{ts} + \mathcal{L}_{rm}$		✓	✓	73.46	67.26	52.79	42.97	63.24	40.67	9.93	25.46
5	CICL	✓	✓	✓	**74.03**	67.90	53.60	**43.56**	**64.26**	**41.22**	**10.35**	**25.77**

For intra-class relevance learning (ITRL), we perform an isolated analysis of each loss component. Removing the similarity triplet loss, as seen in the comparison between No.0 and No.2, results in a slight decrease in Rank-k performance, with a reduction of -0.75% in R1. However, this change is accompanied by significant improvements in other metrics such as mean Average Precision (mAP), mean Inverse Negative Penalty (mINP), and mean Similarity Distribution (mSD), with mINP increasing by +2.93%, consistently across both datasets. In the case of relevance matching loss, comparing No.0 to No.3, there is a minor reduction in Rank-k performance of -0.26%, but notable enhancements in mAP, mINP, and mSD, with mAP increasing by +1.05%. Combining both loss components under ITRL, as illustrated by the comparisons between No.2 and No.4, and No.3 and No.4, we observe minimal changes in mAP and mINP, but a clear improvement in Rank-k metrics, with increases of +0.9% and +0.41% in R1 respectively. Furthermore, the comparison between No.0 and No.4 shows little difference in Rank-k values but a substantial increase in mINP of +1.25%. These findings indicate that ITRL significantly enhances the overall retrieval performance by learning inter-sample similarities and distinctions, without compromising cross-modal alignment capabilities.

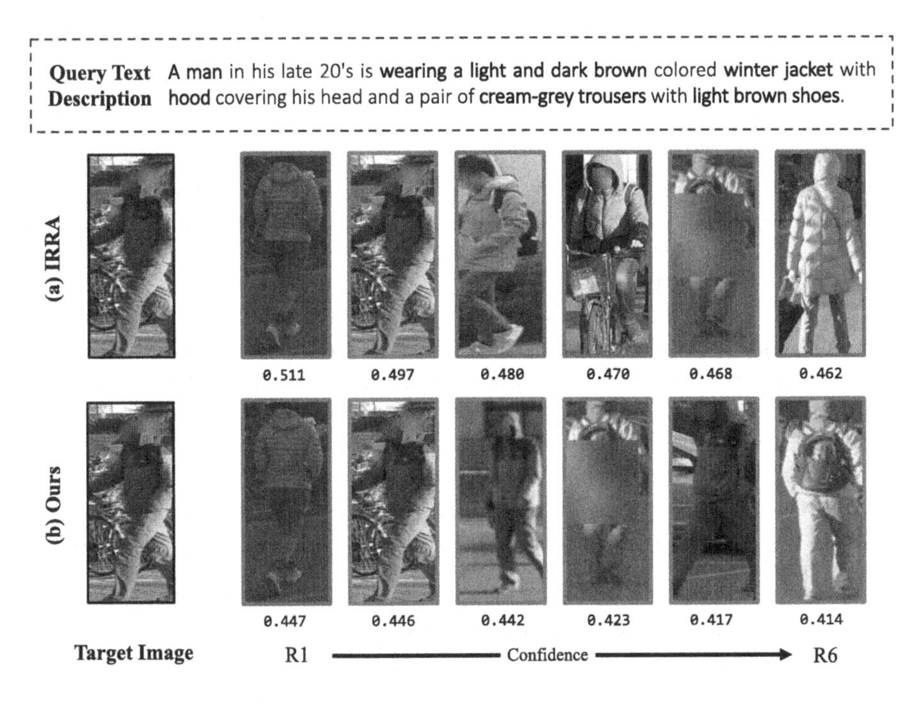

Fig. 4. Comparison of top-6 retrieved results on ICFG-PEDES between IRRA (the first row) and ours CICL (the second row) for same text query. The green/red boxes denote the true/false results. (Color figure online)

Regarding bidirectional masked matching (BiMM), when implemented alone under data-scarce conditions as reflected in the comparison of No.0 and No.1 on CUHK-PEDES, it does not effectively boost overall performance, potentially due to noise introduced by the masking process, with a decrease of 0.39% in R1.

However, with adequate training data, as seen in the comparison of No.0 and No.1 on ICFG-PEDES, there is a marked improvement in model performance, with an increase of +0.49% in R1. In both scenarios, there is an increase in mSD scores of +0.6%. These results suggest that BiMM enhances the model's fine-grained alignment capabilities. Moreover, when BiMM is combined with ITRL, as shown in the comparison between No.4 and No.5, improvements are observed across all metrics, with an increase of +1.02% in R1 and +0.55% in mAP. This affirms that BiMM contributes positively to the TIReID task by enhancing the ability to achieve fine-grained alignment.

4.4 Qualitive Analysis of CICL

As illustrated in Fig. 4, we present several examples to demonstrate the top-6 ranking results achieved by our CICL model. These examples highlight the model's capability in aligning visual and textual data accurately. Even in situations where there are false matches, the top-6 results generally show a high degree of correlation between the visual attributes of the person and the textual descriptions provided.

In this particular test case, we used the same query text to compare the results from the IRRA and our CICL model. The top-6 similarity results, displayed in Fig. 4, indicate that overall, CICL outperforms IRRA even though both models initially failed at the top-1 result. Despite the top results being incorrect, the images did match descriptions of "light and dark brown, cream-grey trousers". Analyzing the confidence levels, as illustrated in Fig. 4(a), IRRA shows higher confidence in its results; however, many images that scored high do not actually meet the description criteria, particularly at ranks 3 and 6. In contrast, as shown in Fig. 4(b), the confidence levels for images retrieved by CICL are more evenly distributed, with even the incorrect top-ranked image not significantly outscoring the correctly identified images. The consistency in confidence levels for similar features across images suggests that CICL is less susceptible to environmental changes and is better at discerning semantic associations between image details and textual queries. This also meets one of our expectations for training objective learning.

Specifically, the CICL model demonstrates its effectiveness by ranking the correct matches higher and assigning them higher similarity scores. This reflects the model's refined ability to discern and emphasize finer details and contextual cues that are mentioned in the textual descriptions. By doing so, CICL significantly improves the chances of retrieving the most relevant and specific matches first, showcasing its robustness in handling complex, real-world scenarios where multiple visual-textual attributes must be aligned accurately.

5 Conclusion

In this paper, we propose a Cross-modal Intra-class Learning (CICL) framework and developed two key methods: Bidirectional Masked Matching (BiMM) and

Image-Text Intra-class Relevance Learning (ITRL). These approaches effectively address the challenge of fine-grained cross-modal feature discrimination and intra-class variation in Text-Image Person Re-Identification (TIReID). Demonstrated by extensive experiments on benchmark datasets, our approach have significantly improved retrieval accuracy and performance. Our exploration in cross-modal intra-class learning has shown promising results in the TIReID task. However, we recognize that there remains untapped potential in the re-id domain, with substantial information of image-text pairs still to be harnessed. We leave the exploration of leveraging such untapped information to future work.

Acknowledgements. This work is supported by Key Laboratory of Ethnic Language Intelligent Analysis and Security Governance of MOE, Minzu University of China, Beijing, China.

Disclosure of Interests. The authors have no competing interests to declare that are relevant to the content of this article.

References

1. Bao, H., Dong, L., Piao, S., Wei, F.: Beit: bert pre-training of image transformers. arXiv preprint arXiv:2106.08254 (2021)
2. Bao, H., et al.: Vlmo: unified vision-language pre-training with mixture-of-modality-experts. Adv. Neural. Inf. Process. Syst. **35**, 32897–32912 (2022)
3. Chen, J., et al.: Eve: efficient vision-language pre-training with masked prediction and modality-aware moe. In: Proceedings of the AAAI Conference on Artificial Intelligence, vol. 38, pp. 1110–1119 (2024)
4. Chen, Y., Zhang, G., Lu, Y., Wang, Z., Zheng, Y.: Tipcb: a simple but effective part-based convolutional baseline for text-based person search. Neurocomputing **494**, 171–181 (2022)
5. Ding, Z., Ding, C., Shao, Z., Tao, D.: Semantically self-aligned network for text-to-image part-aware person re-identification. arXiv preprint arXiv:2107.12666 (2021)
6. Farooq, A., Awais, M., Kittler, J., Khalid, S.S.: Axm-net: implicit cross-modal feature alignment for person re-identification. In: Proceedings of the AAAI Conference on Artificial Intelligence, vol. 36, pp. 4477–4485 (2022)
7. Guo, Q., et al.: M2-encoder: advancing bilingual image-text understanding by large-scale efficient pretraining. arXiv preprint arXiv:2401.15896 (2024)
8. Han, X., He, S., Zhang, L., Xiang, T.: Text-based person search with limited data. arXiv preprint arXiv:2110.10807 (2021)
9. He, S., Luo, H., Wang, P., Wang, F., Li, H., Jiang, W.: Transreid: transformer-based object re-identification. In: Proceedings of the IEEE/CVF International Conference on Computer Vision, pp. 15013–15022 (2021)
10. Jiang, D., Ye, M.: Cross-modal implicit relation reasoning and aligning for text-to-image person retrieval. In: IEEE International Conference on Computer Vision and Pattern Recognition (CVPR) (2023)
11. Kalayeh, M.M., Basaran, E., Gökmen, M., Kamasak, M.E., Shah, M.: Human semantic parsing for person re-identification. In: Proceedings of the IEEE Conference on Computer Vision and Pattern Recognition, pp. 1062–1071 (2018)

12. Kim, W., Son, B., Kim, I.: Vilt: vision-and-language transformer without convolution or region supervision. In: International Conference on Machine Learning, pp. 5583–5594. PMLR (2021)
13. Kingma, D.P., Ba, J.: Adam: a method for stochastic optimization. arXiv preprint arXiv:1412.6980 (2014)
14. Li, J., Selvaraju, R., Gotmare, A., Joty, S., Xiong, C., Hoi, S.C.H.: Align before fuse: vision and language representation learning with momentum distillation. Adv. Neural. Inf. Process. Syst. **34**, 9694–9705 (2021)
15. Li, L.H., Yatskar, M., Yin, D., Hsieh, C.J., Chang, K.W.: Visualbert: a simple and performant baseline for vision and language. arXiv preprint arXiv:1908.03557 (2019)
16. Li, L.H., et al.: Grounded language-image pre-training. In: Proceedings of the IEEE/CVF Conference on Computer Vision and Pattern Recognition, pp. 10965–10975 (2022)
17. Li, S., Cao, M., Zhang, M.: Learning semantic-aligned feature representation for text-based person search. In: ICASSP 2022-2022 IEEE International Conference on Acoustics, Speech and Signal Processing (ICASSP), pp. 2724–2728. IEEE (2022)
18. Li, S., Xiao, T., Li, H., Zhou, B., Yue, D., Wang, X.: Person search with natural language description. In: Proceedings of the IEEE Conference on Computer Vision and Pattern Recognition, pp. 1970–1979 (2017)
19. Lin, Y., Dong, X., Zheng, L., Yan, Y., Yang, Y.: A bottom-up clustering approach to unsupervised person re-identification. In: Proceedings of the AAAI Conference on Artificial Intelligence, vol. 33, pp. 8738–8745 (2019)
20. Luo, H., Gu, Y., Liao, X., Lai, S., Jiang, W.: Bag of tricks and a strong baseline for deep person re-identification. In: Proceedings of the IEEE/CVF Conference on Computer Vision and Pattern Recognition Workshops (2019)
21. Ma, Z., Xu, F., Liu, J., Yang, M., Guo, Q.: Sycoca: symmetrizing contrastive captioners with attentive masking for multimodal alignment. arXiv preprint arXiv:2401.02137 (2024)
22. Meng, J., Wu, A., Zheng, W.S.: Deep asymmetric video-based person re-identification. Pattern Recogn. **93**, 430–441 (2019)
23. Miech, A., Alayrac, J.B., Laptev, I., Sivic, J., Zisserman, A.: Thinking fast and slow: efficient text-to-visual retrieval with transformers. In: Proceedings of the IEEE/CVF Conference on Computer Vision and Pattern Recognition, pp. 9826–9836 (2021)
24. Oord, A.V.D., Li, Y., Vinyals, O.: Representation learning with contrastive predictive coding. arXiv preprint arXiv:1807.03748 (2018)
25. Radford, A., et al.: Learning transferable visual models from natural language supervision. In: International Conference on Machine Learning, pp. 8748–8763. PMLR (2021)
26. Schroff, F., Kalenichenko, D., Philbin, J.: Facenet: a unified embedding for face recognition and clustering. In: Proceedings of the IEEE Conference on Computer Vision and Pattern Recognition, pp. 815–823 (2015)
27. Shao, Z., Zhang, X., Fang, M., Lin, Z., Wang, J., Ding, C.: Learning granularity-unified representations for text-to-image person re-identification. In: Proceedings of the 30th ACM International Conference on Multimedia, pp. 5566–5574 (2022)
28. Shu, X., et al.: See finer, see more: implicit modality alignment for text-based person retrieval. In: European Conference on Computer Vision, pp. 624–641. Springer, Heidelberg (2022). https://doi.org/10.1007/978-3-031-25072-9_42

29. Song, G., Leng, B., Liu, Y., Hetang, C., Cai, S.: Region-based quality estimation network for large-scale person re-identification. In: Proceedings of the AAAI Conference on Artificial Intelligence, vol. 32 (2018)
30. Song, L., et al.: Unsupervised domain adaptive re-identification: theory and practice. Pattern Recogn. **102**, 107173 (2020)
31. Sun, S., Chen, Y.C., Li, L., Wang, S., Fang, Y., Liu, J.: Lightningdot: Pre-training visual-semantic embeddings for real-time image-text retrieval. In: Proceedings of the 2021 Conference of the North American Chapter of the Association for Computational Linguistics: Human Language Technologies, pp. 982–997 (2021)
32. Sun, Y., Zheng, L., Yang, Y., Tian, Q., Wang, S.: Beyond part models: person retrieval with refined part pooling (and a strong convolutional baseline). In: Proceedings of the European Conference on Computer Vision (ECCV), pp. 480–496 (2018)
33. Wang, G., et al.: High-order information matters: learning relation and topology for occluded person re-identification. In: Proceedings of the IEEE/CVF Conference on Computer Vision and Pattern Recognition, pp. 6449–6458 (2020)
34. Wang, G., Yu, F., Li, J., Jia, Q., Ding, S.: Exploiting the textual potential from vision-language pre-training for text-based person search. arXiv preprint arXiv:2303.04497 (2023)
35. Wang, G., Yuan, Y., Chen, X., Li, J., Zhou, X.: Learning discriminative features with multiple granularities for person re-identification. In: Proceedings of the 26th ACM International Conference on Multimedia, pp. 274–282 (2018)
36. Wang, H., Shen, J., Liu, Y., Gao, Y., Gavves, E.: Nformer: robust person re-identification with neighbor transformer. In: Proceedings of the IEEE/CVF Conference on Computer Vision and Pattern Recognition, pp. 7297–7307 (2022)
37. Wang, W., et al.: Image as a foreign language: beit pretraining for vision and vision-language tasks. In: Proceedings of the IEEE/CVF Conference on Computer Vision and Pattern Recognition, pp. 19175–19186 (2023)
38. Wang, Z., Fang, Z., Wang, J., Yang, Y.: *ViTAA*: visual-textual attributes alignment in person search by natural language. In: Vedaldi, A., Bischof, H., Brox, T., Frahm, J.-M. (eds.) ECCV 2020. LNCS, vol. 12357, pp. 402–420. Springer, Cham (2020). https://doi.org/10.1007/978-3-030-58610-2_24
39. Wang, Z., et al.: Caibc: capturing all-round information beyond color for text-based person retrieval. In: Proceedings of the 30th ACM International Conference on Multimedia, pp. 5314–5322 (2022)
40. Wang, Z., et al.: Look before you leap: improving text-based person retrieval by learning a consistent cross-modal common manifold. In: Proceedings of the 30th ACM International Conference on Multimedia, pp. 1984–1992 (2022)
41. Wei, L., Zhang, S., Gao, W., Tian, Q.: Person transfer gan to bridge domain gap for person re-identification. In: Proceedings of the IEEE Conference on Computer Vision and Pattern Recognition, pp. 79–88 (2018)
42. Wu, Y., Yan, Z., Han, X., Li, G., Zou, C., Cui, S.: Lapscore: language-guided person search via color reasoning. In: Proceedings of the IEEE/CVF International Conference on Computer Vision, pp. 1624–1633 (2021)
43. Yan, S., Dong, N., Zhang, L., Tang, J.: Clip-driven fine-grained text-image person re-identification. IEEE Trans. Image Process. (2023)
44. Yan, S., Tang, H., Zhang, L., Tang, J.: Image-specific information suppression and implicit local alignment for text-based person search. IEEE Trans. Neural Netw. Learn. Syst. (2023)

45. Ye, M., Lan, X., Yuen, P.C.: Robust anchor embedding for unsupervised video person re-identification in the wild. In: Proceedings of the European Conference on Computer Vision (ECCV), pp. 170–186 (2018)
46. Ye, M., Shen, J., Lin, G., Xiang, T., Shao, L., Hoi, S.C.: Deep learning for person re-identification: a survey and outlook. IEEE Trans. Pattern Anal. Mach. Intell. **44**(6), 2872–2893 (2021)
47. Zhai, G., Min, X.: Perceptual image quality assessment: a survey. Sci. China Inf. Sci. **63**, 1–52 (2020)
48. Zhang, G., Chen, Y., Dai, Y., Zheng, Y., Wu, Y.: Reference-aided part-aligned feature disentangling for video person re-identification. In: 2021 IEEE International Conference on Multimedia and Expo (ICME), pp. 1–6. IEEE (2021)
49. Zhang, G., Jiang, T., Yang, J., Xu, J., Zheng, Y.: Cross-view kernel collaborative representation classification for person re-identification. Multimedia Tools Appl. **80**(13), 20687–20705 (2021)
50. Zhang, Y., Lu, H.: Deep cross-modal projection learning for image-text matching. In: Proceedings of the European Conference on Computer Vision (ECCV), pp. 686–701 (2018)
51. Zhao, H., et al.: Spindle net: person re-identification with human body region guided feature decomposition and fusion. In: Proceedings of the IEEE Conference on Computer Vision and Pattern Recognition, pp. 1077–1085 (2017)
52. Zhao, L., Li, X., Zhuang, Y., Wang, J.: Deeply-learned part-aligned representations for person re-identification. In: Proceedings of the IEEE International Conference on Computer Vision, pp. 3219–3228 (2017)
53. Zheng, L., Yang, Y., Hauptmann, A.G.: Person re-identification: past, present and future. arXiv preprint arXiv:1610.02984 (2016)
54. Zheng, Z., Zheng, L., Garrett, M., Yang, Y., Xu, M., Shen, Y.D.: Dual-path convolutional image-text embeddings with instance loss. ACM Trans. Multimedia Comput. Commun. Appl. (TOMM) **16**(2), 1–23 (2020)
55. Zhu, A., et al.: DSSL: deep surroundings-person separation learning for text-based person retrieval. In: Proceedings of the 29th ACM International Conference on Multimedia, pp. 209–217 (2021)
56. Zhu, X., Jing, X.Y., You, X., Zhang, X., Zhang, T.: Video-based person re-identification by simultaneously learning intra-video and inter-video distance metrics. IEEE Trans. Image Process. **27**(11), 5683–5695 (2018)
57. Zuo, J., et al.: Ufinebench: towards text-based person retrieval with ultra-fine granularity. In: Proceedings of the IEEE/CVF Conference on Computer Vision and Pattern Recognition, pp. 22010–22019 (2024)

FEDNPAIT: Federated Learning with NADAM and PADAM for Instruction Tuning

Zhipeng Gao(✉), Yichen Li, and Xinlei Yu

The State Key Laboratory of Networking and Switching Technology, Beijing University of Posts and Communications, Beijing, China
{gaozhipeng,liyichen23,yuxinlei518}@bupt.edu.cn

Abstract. Large language models (LLMs) trained on massive publicly accessible data excel in general domain tasks. However, the performance of LLMs in specialized domains such as healthcare and mathematics frequently fails to meet established benchmarks. The primary obstacle to improving this performance is the scarcity of high-quality and domain-specific data due to stringent data privacy regulations. Federated learning, which facilitates the training of LLMs across distributed devices through a decentralized approach that precludes centralized data aggregation presents a potential solution. Nevertheless, federated learning for instruction tuning encounters challenges including data heterogeneity and substantial computational resource requirements. Consequently, this paper proposes FEDNPAIT, an innovative federated learning framework for instruction tuning that incorporates Nesterov-accelerated momentum and a new parameter for controlling the adaptive strength of gradient updates during optimization to improve the performance of LLMs in heterogeneous federated learning environments. Furthermore, FEDNPAIT effectively addresses computational resource limitations through the incorporation of Low-Rank Adaptation. Experimental results indicate that FEDNPAIT effectively mitigates the adverse effects of data heterogeneity in federated environments and outperforms existing federated learning algorithms. Specifically, the results from tuned Llama3-8B demonstrate enhanced generalization and robustness, illustrating how to incorporate complex optimization algorithms into federated training of LLMs, and opening new possibilities for federated instruction tuning in sensitive industries.

Keywords: Federated Learning · Large Language Models · Adaptive Optimization Algorithm

1 Introduction

The development of large language models (LLMs) has revolutionized the field of Natural Language Processing (NLP) in recent years [1]. Examples include

T. Zhu et al. (Eds.): ICA3PP 2024, LNCS 15252, pp. 185–203, 2025.
https://doi.org/10.1007/978-981-96-1528-5_13

OpenAI's ChatGPT [2], Meta's LLaMA [3] and Google's PaLM [4], which have demonstrated outstanding performance in complex tasks such as machine translation and text generation across general domains [5]. However, in specialized domains such as healthcare, finance, or mathematics, the text generation capabilities of LLMs are frequently found to be suboptimal, exhibiting low levels of accuracy. The principal factor underpinning the inadequate performance of LLMs in specialized fields is the requirement for a robust and nuanced understanding of domain-specific knowledge, including the technical terminology that is peculiar to each respective discipline. To address this, the method of 'Instruction Tuning' has been proposed [6]. Instruction Tuning [7] involves guiding LLMs to generate desired outputs through specially designed input instructions, while fine-tuning the pre-trained models using curated, domain-specific datasets. This process enhances the LLMs ability to accurately interpret and generate text relevant to specialized fields. Instruction tuning demands high-quality, comprehensive, and voluminous domain-specific datasets to be effective. The quality of data influences the model's ability to learn and apply the domain-specific knowledge accurately [8]. The breadth of the data ensures that the model can handle a wide range of topics within the domain and the volume of the data contributes to the robustness of the model's performance.

Nevertheless, the acquisition of high-quality, domain-specific data presents significant challenges. In recent years, the promulgation of numerous privacy protection legislations, such as the General Data Protection Regulation (GDPR) [9] enacted by the European Union, has imposed stringent safeguards on personal data through regulatory frameworks. This has rendered the collection of domain-specific data increasingly arduous. To address this issue, the paper employs federated learning to overcome the challenge of utilizing domain-specific data for instruction tuning of LLMs, a challenge caused by privacy concerns [10]. Federated learning enables the collaborative training of machine learning models across numerous dispersed devices without requiring the central aggregation of data [11], helping to maintain privacy while utilizing domain-specific data for instruction tuning of LLMs.

Domain-specific data originates from various devices, where different user habits and institutional practices lead to data heterogeneity. This can significantly degrade the performance of instruction tuning of LLMs. Considering the inherent characteristics of federated learning, the diversity in data sources often leads to the phenomenon known as "client drift" [12] when LLMs from various clients are aggregated. Client drift occurs as a result of discrepancies in data distributions and model updates among the clients, which can lead to suboptimal performance of the aggregated LLMs. Additionally, federated learning is conducted on client devices, and the training of LLMs demands substantial computational resources. Therefore, it is crucial to integrate complex processing algorithms into federated learning to ensure that federated learning on instruction tuning not only maintains the high accuracy and robustness of trained LLMs, but also safeguards the confidentiality and security of individual data sources under the condition of limited computing resources [13].

This integration leverages the benefits of distributed learned data while effectively mitigating the effects of client drift and computational resource limitations.

To enable an exhaustive exploration, we build a comprehensive and practical framework named FEDNPAIT (Federated Learning with NADAM and PADAM for Instruction Tuning)as shown in Fig. 1. This framework encapsulates a robust integration of federated learning algorithms, optimization techniques, and processes involving the training and evaluation of LLMs. It is designed to be a comprehensive framework covering the entire process from instruction fine-tuning of LLMs to their final performance evaluation. To enhance the accuracy and robustness of instruction-tuned LLMs, we drew inspiration from NADAM (Nesterov-accelerated Adaptive Moment Estimation) [14] and PADAM (Proposed Algorithm for Adaptive Momentum) [15], incorporating their principles into the federated learning server-side optimization for instruction tuning. Within the scope of our survey, this represents the first instance where the principles of NADAM and PADAM algorithms have been adapted for the application of federated learning to LLMs. By doing so, we aim to demonstrate the feasibility of deploying advanced machine learning strategies in decentralized environments, thereby enhancing the broader understanding of how to achieve scalable model training without compromising data security or model accuracy. To address the issue of computational resource limitation, we employ a Low-Rank Adaptation known as LoRA [16], which involves the integration of trainable low-rank matrices into the Transformer model. This substantially reduces the number of trainable parameters, thereby achieving a reduction in computational and memory costs while maintaining high model performance. This thorough approach guarantees that the large model not only meets the predefined standards of performance but also adapts efficiently to diverse application scenarios, thereby reinforcing the robustness and applicability of the FEDNPAIT framework in advancing the field of machine learning.

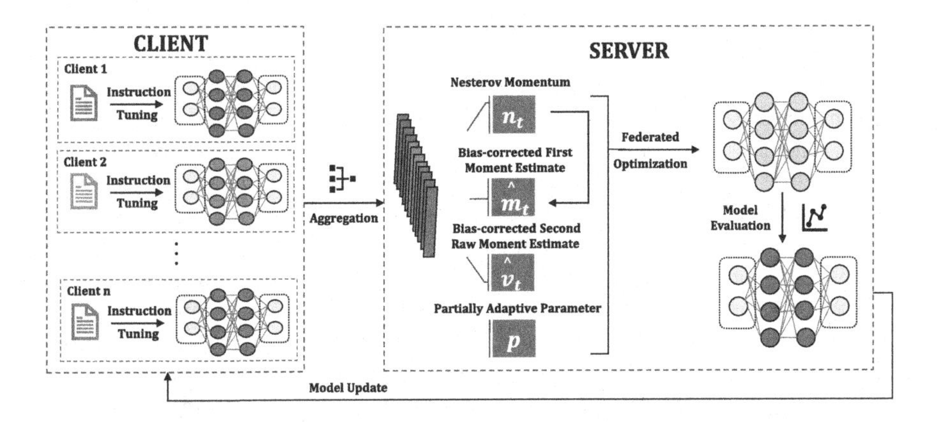

Fig. 1. Framework of FEDNPAIT

Main Contributions. Our contributions are as follows:

- We propose FEDNPAIT, a federated learning framework for instruction tuning of LLMs, which addresses the privacy issues of domain-specific data and improves the accuracy and robustness of LLMs on domain-specific tasks.
- We propose an innovative federated adaptive optimization algorithm that adjusts the bias-corrected moment estimate and optimizes low-rank adaptation. This algorithm enhances the generalization ability of LLMs in heterogeneous federated learning scenarios and effectively addresses computational resource constraints.
- We conduct a comprehensive empirical study based on FEDNPAIT, showing that LLMs tuned by our method on heterogeneous datasets obtain advanced performance.

2 Related Work

Large Language Models. With the development of AI-Generated Content (AIGC), many global companies and institutions have developed their own LLMs [17]. The training process for these LLMs typically involves four primary stages. Initially, a substantial amount of high-quality data from various domains is amassed and preprocessed [18]. Subsequently, foundational models such as the Transformer [19] and BERT [20] are chosen and refined through modifications and optimizations. The third stage entails autoregressive pre-training using the aforementioned data. The culminating step involves fine-tuning the pre-trained model with domain-specific, high-quality datasets to enhance its adaptability to human requirements [21]. The performance of these models is directly proportional to the quantity and quality of the data used [22]. For instance, Llemma, which is trained on multiple high-quality mathematical datasets has shown powerful performance in the field of mathematics [23]. However, Pablo et al. indicate a gradual decline in the availability of high-quality public data [24]. In the process of instruction tuning, there is a phenomenon of high-quality data exhaustion.

Federated Learning. Federated learning is a distributed machine learning approach that allows multiple parties to collaboratively train machine learning models while keeping data on their own devices [11]. This approach has received a lot of attention because of its ability to protect data privacy and address the challenges of data island [25]. However, federated learning demands to aggregate model parameters from diverse clients, which gives rise to the challenge of heterogeneity in the aggregated parameters. Li et al. proposed the convergence of the FedAvg algorithm and demonstrated that data heterogeneity reduces the speed of federated learning convergence [26]. To address this, Kingma et al. proposed an improved optimization method, which combined the smooth characteristics of momentum method and the advantages of adaptive learning rate technology by calculating the adaptive learning rate from the estimation of first-order and second-order gradient moments [27]. This algorithm was especially suitable for dealing with heterogeneous datasets, and it had significant effects in accelerating convergence and maintaining algorithm stability. Chen et al. sought to bridge the generalization gap of adaptive methods like Adam by introducing partial adaptivity in the learning rate, thus combining the strengths of Adam and SGD with momentum for better performance in training deep networks [15]. Reddi et al. propose an enhancement to federated learning by integrating adaptive optimization techniques such as ADAGRAD, ADAM, and YOGI [13]. This approach aims to address the challenges of client heterogeneity and the efficiency of communication, demonstrating significant improvements in convergence rates and overall performance in federated settings. However, the federated learning optimization algorithms do not explore the direction of improving the convergence speed and the generalization ability of the model.

Adaptive Moment Estimation. NADAM is an optimization algorithm that modifies the traditional ADAM optimizer by incorporating Nesterov momentum. This enhancement aims to improve convergence speeds and accelerate the model training process. PADAM introduces a parameter p, which controls the adaptive intensity of gradient updates during optimization. These adjustments are particularly beneficial in the context of federated learning, where reducing the time for model aggregation can markedly increase overall training efficiency. This modification is also intended to enhance the model's generalization ability, especially in the scenario of LLMs.

Federated Learning and Large Language Models. Currently, scholarly investigations are exploring the possibility of combining large models with federated learning. The research direction focuses on how to leverage federated learning frameworks to efficiently train and optimize large language models while preserving data privacy. FATE-LLM [28] investigated federated fine-tuning of LLMs but did not address instruction tuning. Shepherd [29] focused on federated instruction tuning employing the FedAvg algorithm, yet it lacks an advanced optimization strategy. OpenFedLLM [30] introduced numerous federated optimization algorithms for LLMs but did not address enhancing their convergence speed or generalization capabilities. Addressing these gaps, the proposed FEDNPAIT model comprehensively considers the aforementioned issues, aiming to optimize the convergence efficiency and generalization robustness of federated learning in LLMs scenarios.

3 Preliminaries

3.1 Federated Learning

Federated learning is a distributed machine learning approach that protects user privacy and improves the overall learning efficiency of the model by training the model locally on each client and only sending model updates to the server instead of sharing raw data. The basic process of federated learning is as follows:

1. Model Initialization. The server initializes the global model parameters θ_0.

2. Local Updates. In each communication round t, the server randomly selects a subset of clients and sends the current global model θ_t to them. Each selected client k updates the model parameters locally using their data through local training (typically Stochastic Gradient Descent SGD).

$$\theta_{t+1}^k = \theta_t - \eta \nabla F_k(\theta_t) \tag{1}$$

where η is the learning rate and $\nabla F_k(\theta_t)$ is the gradient of the model w_t with respect to the local data of client k.

3. Aggregating Updates. All participating clients send their updates back to the server. The server aggregates these updates to form a new global model. This is typically done by computing a weighted average:

$$\theta_{t+1} = \sum_{k=1}^{K} \frac{n_k}{n} \theta_{t+1}^k \tag{2}$$

where n_k is the number of data points for client k, n is the total number of data points across all clients, and K is the number of clients participating in the current round.

4. Repeat Process. Repeat steps 2 and 3 until the termination criteria are satisfied, which include achieving a specified number of iterations or observing no further enhancement in model performance.

3.2 Adaptive Moment Estimation

Adaptive Moment Estimation (ADAM) is an optimization algorithm designed for first-order gradient-based optimization of stochastic objective functions. It incorporates adaptive learning rates for each parameter based on estimates of first and second moments of the gradients. This approach is particularly effective for handling large-scale and non-stationary problems.

1. Initialization. The Adam algorithm begins with the initialization of parameters. The parameter vector θ is initialized at θ_0. Additionally, the first moment vector m and the second moment vector v are initialized to zero, represented as $m_0 = 0$ and $v_0 = 0$ respectively. The time step t is also set to zero, represented as $t_0 = 0$.

2. Iteration. At each time step t, Adam performs several computations to update the parameters. It first calculates the gradient g_t with respect to the objective function at θ_{t-1}, followed by updating the biased estimates of the first and second moments. The updates are given by:

$$m_t = \beta_1 m_{t-1} + (1 - \beta_1) g_t \tag{3}$$

$$v_t = \beta_2 v_{t-1} + (1 - \beta_2) g_t^2 \tag{4}$$

where β_1 and β_2 are hyperparameters which control the exponential decay rates of these moment estimates, affecting their sensitivity to new gradients versus old gradients.

These moment estimates are then corrected for bias:

$$\hat{m}_t = \frac{m_t}{1 - \beta_1^t} \tag{5}$$

$$\hat{v}_t = \frac{v_t}{1 - \beta_2^t} \tag{6}$$

The parameter update at each step is computed using these bias-corrected estimates:

$$\theta_t = \theta_{t-1} - \frac{\alpha \hat{m}_t}{\sqrt{\hat{v}_t} + \epsilon} \tag{7}$$

where α is the learning rate which controls the size of the update step and ϵ is a minor constant to avoid dividing by zero.

3. Termination. The iteration process is repeated until a stopping criterion is met, which might be based on the number of iterations, a threshold in improvement of the objective function, or other convergence criteria.

3.3 Low-Rank Adaptation

Low-Rank Adaptation (LoRA) is a parameter-efficient model adaptation approach. By injecting a trainable low-rank matrix into the weight matrix of the Transformer model, it can achieve fast adaptation of downstream tasks.

1. Retain Pre-Trained Weights. In traditional fine-tuning, all model weights are updated. However, in LoRA, the pre-trained weight matrix W_0 is kept constant without modification, ensuring stability and performance during adaptation.

2. Insert Low-Rank Matrices. While weight matrices in Transformer models are generally full-rank, the core concept of LoRA is to represent weight changes with two low-rank matrices. These two matrices are A and B, where A is initialized with random Gaussian values and B is initialized with zeros. The weight update is given by:

$$W = W_0 + BA \tag{8}$$

A has dimensions $r \times k$ where k is the output dimension. B has dimensions $d \times r$ where d is the input dimension and r is the rank.

3. Re-parameterization and Scaling. During training, only the low-rank matrices A and B are optimized while the pre-trained weight matrix W_0 remains fixed. The forward pass in the network is calculated as follows:

$$h = W_0x + (BA)x = W_0x + B(Ax) \tag{9}$$

To ensure training stability and control learning speed, LoRA applies scaling when calculating the weight updates. The scaling is given by:

$$\Delta W = \frac{BA}{r}\alpha \tag{10}$$

where α is a constant to control scaling, often linked to the learning rate.

4. Application in Transformer Models and Inference. In Transformer models, LoRA primarily focuses on the self-attention mechanism's weight matrices, including W_q, W_k, W_v and W_o. LoRA can selectively insert low-rank matrices into these weight matrices to achieve parameter-efficient adaptation, with common applications targeting the query and value matrices. During model deployment and inference, LoRA can merge pre-trained weights with updated low-rank matrices to ensure there is no additional inference latency.

4 Method

4.1 Instruction Tuning

Due to the limited computational resources available to federated learning clients, we chose the LoRA approach for fine-tuning to ensure that each client can manage the training costs. This approach addresses the challenge of adapting large language models to downstream tasks without requiring a complete parameter fine-tuning process. By freezing the pre-trained model's weights and adding trainable low-rank matrices to each layer of the model, which significantly reduces the number of trainable parameters, resulting in a more efficient and scalable training process. Denote $W_0 \in \mathbb{R}^{m \times n}$ as one weight matrix of the base model. The weight update formula is as follows:

$$W_0 + \Delta W = W_0 + BA \tag{11}$$

where $A \in \mathbb{R}^{m \times r}$, $B \in \mathbb{R}^{r \times n}$, and $r \ll \min(m, n)$.

Instruction Tuning is a technology designed to enhance the ability of LLMs to perform specific tasks in accordance with human instructions. By fine-tuning these models on datasets that include both instructions and their expected outputs, this technology significantly improves the models' capability to follow instructions, resulting in more controllable behavior in practical applications and the ability to rapidly adapt to new tasks or domains. Furthermore, Instruction Tuning aids in optimizing the performance of models on specific tasks and reducing erroneous responses, thereby bridging the gap between traditional pre-training and actual user requirements, enhancing the practicality and efficiency of the models. The federated instruction tuning process is shown in Fig. 2.

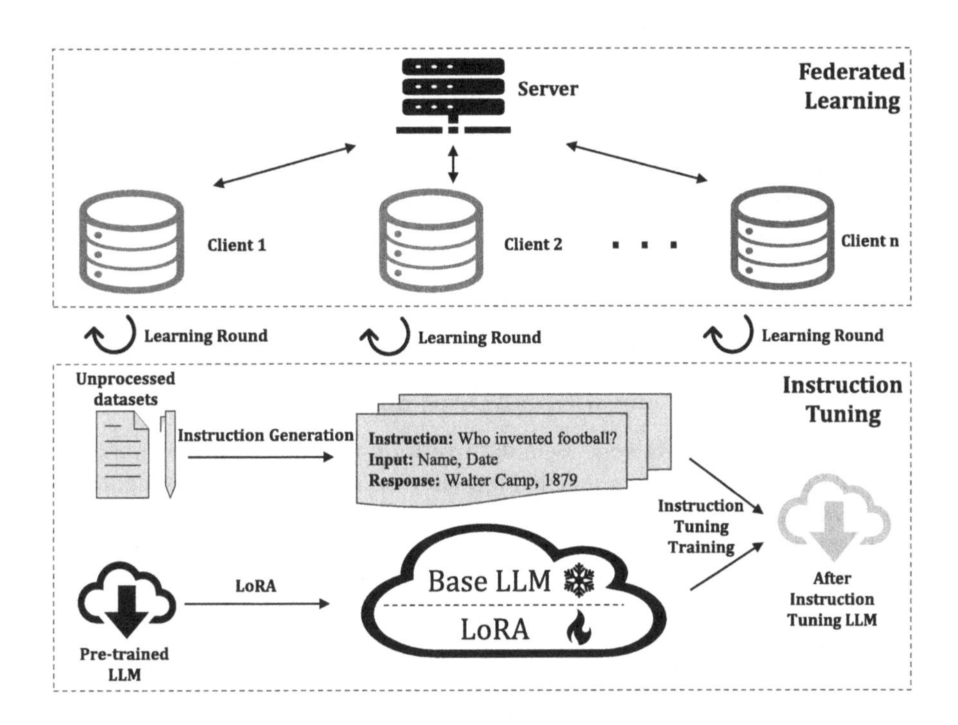

Fig. 2. Federated Instruction Tuning

The core objective of Instruction Tuning is to enhance the ability of LLMs to follow human instructions. Our methodological approach encompasses the following two critical steps:

1. Construction of the Instruction Dataset. Instruction Tuning necessitates a dataset comprising pairs of instructions (Instructions) and their expected outputs (Outputs). Typically, this dataset can be represented as a series of (Instruction, Output) pairs:

$$D = \{(I_1, O_1), (I_2, O_2), \ldots, (I_n, O_n)\} \tag{12}$$

where I_j denotes the jth instruction and O_j is the corresponding expected output.

2. Model with LoRA Fine-Tuning. Upon acquiring the instruction dataset, the LLM is further trained to adapt to these instructions, employing techniques like Low-Rank Adaptation (LoRA) to reduce the number of parameters that need training. The fine-tuning process can be formalized as minimizing the loss function between the predicted outputs and the actual outputs:

$$\text{minimize } L(\theta) = \sum_{(I_j, O_j) \in D} F(f_\theta(I_j), O_j) \tag{13}$$

where $f_\theta(I)$ represents the output of the model f under parameters θ, F is the loss function which typically cross-entropy loss or other loss functions suitable for sequence prediction.

3. Termination. The iterative process is repeated until the preset number of training cycles is reached.

Through this systematic iterative process, the model gradually adapts to follow complex and diverse human instructions, thereby enhancing its efficacy and reliability in practical applications.

4.2 Federated Learning with NADAM and PADAM

In this section, we selectively assimilate certain theoretical underpinnings of NADAM and PADAM into the federated learning algorithm. In this algorithm, each client computes the gradient of its data locally and uses a local optimizer to update the model parameters. These local updates are then aggregated to update the global model parameters via a momentum mechanism and a correction term. This algorithm realizes model training under the premise of protecting user privacy through federated learning. It leverages the advantages of momentum and adaptive learning rate to accelerate convergence and improve performance.

Our algorithmic innovation primarily lies in the aggregation of local updates. Inspired by NADAM and PADAM, we have made enhancements on the server side by optimizing the bias-corrected first moment estimate and adjusting the bias-corrected second raw moment estimate. Similar to the updates of first and second-order moment estimates in the ADAM algorithm, our approach incorporates the concept of NADAM during the correction of the first-order moment. The essence of Nesterov accelerated gradient lies in projecting the current momentum vector towards the expected update direction before updating the parameters, i.e., the update is akin to taking a step forward in the direction of the previous momentum vector and another step forward in the direction of the current gradient. This methodology can be interpreted as a prediction of the forthcoming position, allowing gradient computation at this predicted location. Consequently, compared to conventional momentum methods, Nesterov accelerated gradient can more efficiently utilize gradient information as it already accounts for the influence of momentum prior to parameter updates. The corrections are provided as follows:

$$n_t = \beta_1 m_t + (1 - \beta_1) g_t \tag{14}$$

$$\hat{m}_t = \beta_1 \frac{m_t}{1 - \beta_1^t} + \frac{1 - \beta_1}{1 - \beta_1^t} n_t \tag{15}$$

$$\hat{v}_t = \frac{v_t}{1 - \beta_2^t} \tag{16}$$

where n_t is denoted as the Nesterov accelerated gradient of that training run, $\hat{m}_t$ is the bias-corrected first moment estimate, and $\hat{v}_t$ is the bias-corrected second moment estimate.

The computation of the second moment resembles the ADAM algorithm, but we have incorporated the principles of the PADAM algorithm during parameter updates. In ADAM, the second moment estimate $\hat{v}_t$ is directly used to adjust the learning rate, allowing it to adapt to each dimension of the parameters. However, this highly adaptive approach can sometimes be too aggressive, resulting in poor generalization on test data. Since our algorithm involves training LLMs, the model's generalization ability is a crucial evaluation metric. By introducing p, PADAM aims to find a balance between adaptive optimization and less adaptive methods (such as SGD). The optimal value of p is usually determined through a series of experiments and can be validated across different datasets and model architectures through cross-validation. The process is as follows:

$$\theta_{t+1} = \theta_t - \frac{\alpha \hat{m}_t}{(\hat{v}_t)^p + \epsilon} \tag{17}$$

where p is a hyperparameter that ranges from 0 to 0.5 and controls the strength of adaptability in gradient updates. When p approaches 0, the algorithm tends toward traditional SGD, which may lead to slower convergence but could yield better generalization in some cases. When p approaches 0.5 (the value in ADAM), the algorithm is closer to ADAM, retaining the advantages of fast convergence but potentially showing insufficient generalization in certain tasks. Therefore, during the experimental phase, it is necessary to determine the appropriate value for p.

Algorithm 1 provides pseudo-code for our methods. During the training step, an iterative process similar to federated learning is adopted by our algorithm, which consists of four iterative steps: global model downloading, local model training, local model uploading, and global model aggregating. Crucially, in the final step of global model aggregating, the NADAM algorithm is employed to expedite and optimize the integration process, leveraging its adaptive learning rate capabilities to achieve superior convergence rates. Then the PADAM algorithm is employed to improve the generalization ability of the model while maintaining convergence rates. In the evaluation step, the performance of the large model is scrutinized using a diverse array of evaluation datasets. By employing various metrics and benchmarks, FEDNPAIT ensures a comprehensive evaluation.

Algorithm 1. Federated Learning with NADAM and PADAM

1: **Initialization:** θ_0, parameters $\beta_1, \beta_2 \in [0, 1)$, $p \in (0, 0.5)$
2: **Initialize accumulators:** $m_0 \leftarrow 0$, $v_0 \leftarrow 0$
3: **for** $t = 0, \ldots, T-1$ **do**
4: Sample a subset of clients $C_t \sim (\mathcal{D}, \mathcal{K})$
5: **for** each client $k \in C_t$ **in parallel do**
6: Compute gradient $g_t \leftarrow \nabla F_k(\theta_t)$
7: $\theta_{t+1}^k \leftarrow \theta_t - \eta g_t$
8: **end for**
9: $m_t \leftarrow \beta_1 m_{t-1} + (1-\beta_1) g_t$
10: $v_t \leftarrow \beta_2 v_{t-1} + (1-\beta_2) g_t^2$
11: $n_t \leftarrow \beta_1 m_t + (1-\beta_1) g_t$
12: $\hat{m}_t \leftarrow \beta_1 \frac{m_t}{1-\beta_1^t} + \frac{1-\beta_1}{1-\beta_1^t} n_t$
13: $\hat{v}_t \leftarrow \frac{v_t}{1-\beta_2^t}$
14: $\theta_{t+1} \leftarrow \theta_t - \frac{\alpha \hat{m}_t}{(\hat{v}_t)^p + \epsilon}$
15: **end for**

5 Experiment

5.1 Basic Setups

Heterogeneity of Data. In federated learning, the data is distributed across multiple clients or servers and the model is trained collaboratively without the need to centralize the data. Consequently, we employ a non-independent and identically distributed (non-IID) sampling strategy to approximate the heterogeneous preferences of diverse users for a spectrum of entities. Our experiment uses the number of clients and the dataset as input parameters. Label indices are generated contingent upon the topical categorization intrinsic to the dataset, followed by the systematic updating of the dataset's labels with these corresponding indices. Subsequently, the dataset is divided into multiple shards, each containing an equal number of data points. Next, the dataset is sorted by the index label to evenly distribute the data set among the clients. Each client randomly selects an equal number of shards as its local dataset. Finally, the method returns a list containing a subset of each client's dataset to simulate the imbalance and heterogeneity of real data distribution.

Training Datasets. We utilize datasets from Hugging Face to ensure compatibility with a wide range of datasets in FEDNPAIT. We select Yahoo Datasets (Yahoo Answers Topics[1]) as our experimental dataset due to its extensive and high-quality data covering a comprehensive spectrum of themes encompassing ten distinct categories such as Society & Culture, Science & Mathematics, Health, Education & Reference, etc. These categories have been meticulously selected to satisfy the prerequisites for non-IID data, ensuring a diversified and representative sample that is essential for robust statistical analysis.

[1] https://huggingface.co/datasets/yahoo_answers_topics.

Subsequently, we randomly sample the Yahoo dataset and distribute the resulting data to 20 clients in a non-IID manner. Figure 3 illustrates the distribution result of our allocated datasets.

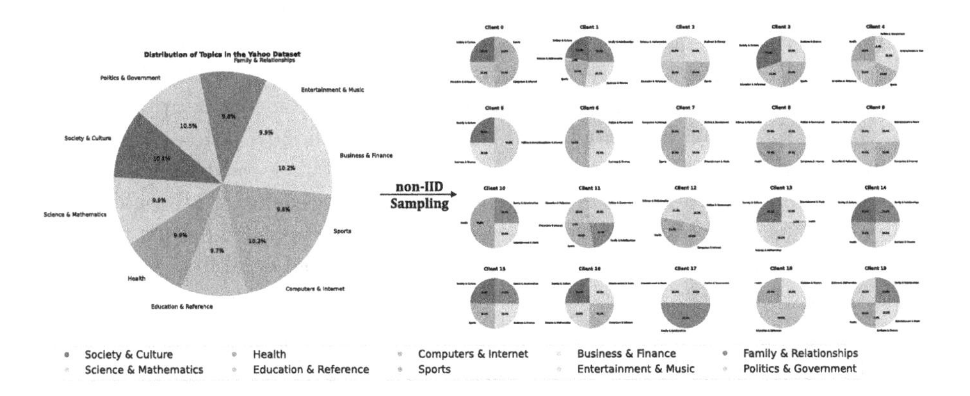

Fig. 3. Yahoo Dataset with non-IID Sampling

Training Details. In this study, we utilized an 8-billion-parameter LLM to serve as a benchmark for our experimental evaluation. To enhance computational efficiency within federated learning, the models underwent a quantization process converting floating-point numbers to eight-bit integers (int8). Concurrently, we implemented the LoRA technique, which adjusts the model's existing layers without adding extra trainable parameters, thus maintaining a balance between model complexity and performance. Our experimental setup consisted of a computing environment equipped with four NVIDIA GeForce RTX 3090 graphics processing units. The base model for our experiments was the pre-trained Llama3-8B[2], renowned for its robustness and efficacy in NLP tasks. Throughout the experimental phase, we conducted 200 communication rounds of FL to ensure thoroughness and reliability of the results. For instruction tuning, we adopted the Alpaca [31] template for formatting as shown in Table 1.

Table 1. Template for Federated Instruction Tuning

Below is an instruction that describes a task. Write a response that appropriately completes the request.
Instruction:
{Instruction}
Response:

[2] https://huggingface.co/meta-llama/Meta-Llama-3-8B.

5.2 P-Value Determination

In the design of the Padam optimization algorithm, a concept known as the "partially adaptive parameter" p was introduced to balance the advantages of both Adam and SGD. Specifically, the value of p determines the extent to which the learning rate is adaptively adjusted during training: lower values of p make the optimization process more similar to SGD, enhancing generalization ability; whereas higher values make it more akin to ADAM, speeding up optimization. Therefore, different models and datasets may require different p values to achieve optimal training results. In our experiments, we used the LLaMA-3 model and Yahoo dataset and conducted a series of tests using traditional training approaches. By observing the loss curves associated with different p values (as shown in Fig. 4), we found that when $p = \frac{1}{4}$, the model exhibited a relatively lower average loss and better generalization ability compared to settings of $p = \frac{1}{8}$ and $p = \frac{1}{16}$. Based on these findings, we chose $p = \frac{1}{4}$ as the parameter setting for the optimization process. This choice was made after considering a combination of the algorithm's convergence speed and the model's generalization ability.

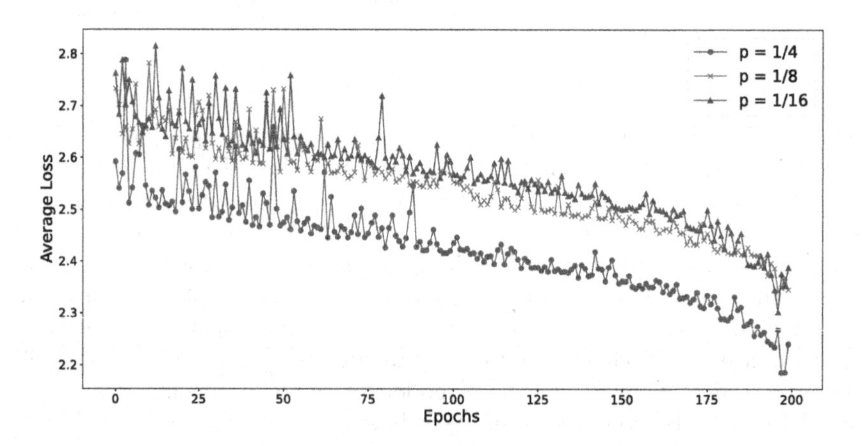

Fig. 4. Average Loss with Different Choices of p for Training Llama-3 on Yahoo Dataset

5.3 Model Evaluation on MMLU

We trained LLaMA-3 on the Yahoo dataset using various methods such as local FedAvg, FedAdam, and FedNPAIT. MMLU (Massive Multitask Language Understanding) [32] evaluation dataset is selected to test the performance of the model. MMLU is a benchmark designed to assess the knowledge acquired by language models during the pretraining phase. It evaluates the model's performance on a wide range of topics in zero-shot and few-shot settings, which closely resembles the way human capabilities are assessed. It offers professional questions at varying levels of difficulty from elementary to advanced. The purpose is to measure world knowledge and problem-solving skills, making it a comprehensive tool to identify strengths and weaknesses of models across different domains.

The results of the evaluation are shown in Table 2. "Base" refers to the pre-trained LLaMA-3 model without any instruction fine-tuning. "Local" refers to training involving only a single client without the use of any federated learning algorithms; to avoid the randomness in experimental results, we tested three clients (named Local-1, Local-2, Local-3 respectively). FedAvg, FedAdam, and FedNPAIT refer to the models obtained after aggregation using the corresponding federated learning algorithms.

Table 2. Federated instruction tuning on diverse domain. M- denotes MMLU benchmark where A: abstract algebra H: history J: jurisprudence N: nutrition P: philosophy S: sociology V: virology. The data in the table are the accuracy rates.

Evaluation	M-A	M-H	M-J	M-N	M-P	M-S	M-V	Avg.
Base	0.320	0.226	0.713	0.729	0.682	0.816	0.542	0.575
Local-1	0.290	0.226	0.621	0.669	0.621	0.766	0.524	0.531
Local-2	0.290	0.172	0.694	0.683	0.666	0.821	0.536	0.552
Local-3	0.247	0.234	0.713	0.693	0.646	0.786	0.536	0.551
FedAvg	0.310	0.226	0.732	0.722	0.711	0.836	0.542	0.583
FedAdam	0.270	0.221	0.741	0.732	**0.723**	0.837	0.542	0.581
FedNPAIT (ours)	**0.350**	**0.241**	**0.769**	**0.745**	0.714	**0.851**	**0.554**	**0.603**

It can be seen from Table 2 that all federated learning algorithms consistently outperformed the Local algorithm. Furthermore, models trained locally displayed accuracy rates on some metrics that were even lower than the pre-trained model (Base). This indicates that overfitting may occur due to local training alone. Within the federated learning algorithms, the models trained under our FedNPAIT framework generally surpassed those trained using the FedAdam and FedAvg algorithms, achieving the best results on 6 out of 7 metrics. This indicates that our algorithm not only maintains the benefits of aggregating model parameters in federated learning but also enhances the generalization ability of the model.

5.4 Model Evaluation on MT-Bench

MT-Bench [33] is a benchmarking framework designed to evaluate multilingual translation systems. This framework provides a comprehensive multi-domain evaluation platform, enabling systems to be tested across various languages and tasks. The dataset employed by the MT-Bench method encompasses problems in eight distinct domains: coding, extraction, humanities, math, roleplay, reasoning, writing, STEM (science, technology, engineering, mathematics). We selected representative problems from MT-Bench and engaged GPT-4 to score different training methods. After completing the scoring, we processed the results. We set the performance of pre-trained LLaMA-3 (Base) to 1 for each domain while the performance of other methods is scored relative to the Base. This approach offers a clearer comparison of how different methods perform across various domains. The specific results are depicted in Fig. 5.

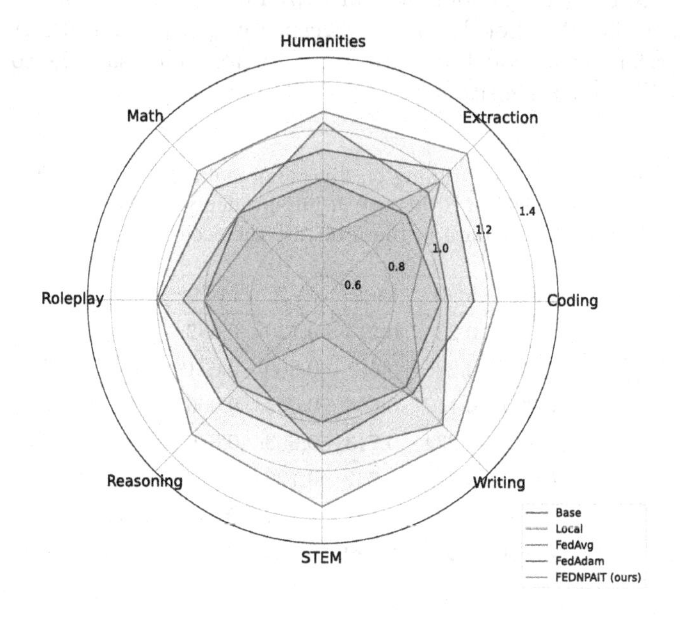

Fig. 5. Federated Instruction Tuning

From Fig. 5, it is evident that the Local method, involving only local client participation in training, leads to a phenomenon known as "overfitting to specific tasks" in the trained model, resulting in instability. In contrast, the three methods employing federated learning demonstrate higher stability. Compared with FedAvg and FedAdam methods, the model trained by the FEDNPAIT method performs better in these domains, and the performance and generalization ability of the model are improved.

5.5 Model Comparison with Other LLMs

In the field of AIGC, ChatGPT-4o [34] is widely regarded as one of the most powerful models currently available. However, this does not imply that our model is always inferior to ChatGPT-4o. As demonstrated in Table 3, while ChatGPT-4o possesses exceptional capabilities, FEDNPAIT exhibits higher accuracy in addressing specific questions. Unlike ChatGPT-4o, which may provide erroneous responses, and ChatGPT-3.5, which only offers search methods without direct answers, FEDNPAIT successfully provides direct responses to instructions.

Table 3. Response demonstration from Yahoo dataset. The fact that the nearest US post office in Se 5th Ave Portland is 204 SW 5th AVE Portland OR 97204-1897. FedNPAIT (ours) correctly identifies the instruments. In contrast, ChatGPT does not provide the correct answer.

Where is the nearest US post office in Se 5th Ave, Portland?	
LLaMA-3 (base)	Portland Main Post Office: 1140 SW 6th Ave Portland OR 97204
Local-1	**204 SW 5th AVE Portland OR 97204-1897**
Local-2	**204 SW 5th AVE Portlanod OR 97204-1897**
Local-3	Portland Main Post Office: 1140 SW 6th Ave Portland OR 97204
ChatGPT-3.5	To find the nearest US Post Office to SE 5th Ave in Portland, you can use the official USPS website or a mapping service like Google Maps
ChatGPT-4o	East Portland Post Office located at 1020 SE 7th Ave Portland OR 97214
FedNPAIT (ours)	**204 SW 5th AVE Portland OR 97204-1897**

This performance disparity underscores the importance of utilizing high-quality and diverse datasets in model training. High-quality data not only enhances the accuracy of models but also improves their ability to process complex instructions. However, due to factors such as data privacy, collecting high-quality datasets has become increasingly challenging. FEDNPAIT demonstrates its potential in data handling and model optimization, particularly when dealing with data that involves high privacy or specialized knowledge.

6 Conclusion

In this paper, we presented the FEDNPAIT framework, which selectively assimilates the NADAM and PADAM optimization algorithms into federated learning specifically for instruction tuning of LLMs. Our methodology directly addresses challenges such as data privacy and client data heterogeneity, which are prevalent in decentralized learning environments. By deploying an adaptive optimization algorithm, FEDNPAIT significantly enhances the performance and generalization capabilities of LLMs without compromising the privacy of the underlying data. The results from our experiments demonstrate that FEDNPAIT outperforms traditional federated learning methods, such as FedAvg and FedAdam, in terms of stability and accuracy across various domains. This improvement highlights the framework's ability to manage non-IID data effectively and adapt to the diverse computational capabilities of different clients. The assimilation of NADAM and PADAM not only speeds up the convergence but also ensures that the learning is robust, which is critical for the practical deployment of federated learning systems.

The experimental results further validate that FEDNPAIT outperforms existing methods, offering a promising solution for decentralized training scenarios and making it a viable option for industries requiring stringent data privacy measures. Future work will focus on refining the optimization techniques and further exploring its scalability and efficiency, particularly in scenarios involving even larger networks and more complex model architectures. Moreover, assessing the impact of extreme data heterogeneity on the performance of framework will provide deeper insights into its operational limits and optimization potential.

Acknowledgements. This work was supported by the China State Grid Corporation Limited Science and Technology Project under Grant 5700-202341275A-1-1-ZN. The corresponding author is Zhipeng Gao.

References

1. Brown, T., et al.: Language models are few-shot learners. Adv. Neural. Inf. Process. Syst. **33**, 1877–1901 (2020)
2. Radford, A., Narasimhan, K., Salimans, T., Sutskever, I., et al.: Improving language understanding by generative pre-training. OpenAI (2018)
3. Touvron, H., et al.: Llama 2: open foundation and fine-tuned chat models. arXiv preprint arXiv:2307.09288 (2023)
4. Touvron, H., Martin, L., Stone, K., Albert, P., Almahairi, A., Babaei, Y.: Palm: scaling language modeling with pathways. J. Mach. Learn. Res. **24**(240), 1–113 (2023)
5. Li, Z., Haroutunian, L., Tumuluri, R., Cohen, P., Haffari, G.: Improving cross-domain low-resource text generation through LLM post-editing: a programmer-interpreter approach. arXiv preprint arXiv:2402.04609 (2024)
6. Liu, H., Li, C., Wu, Q., Lee, Y.J.: Visual instruction tuning. Adv. Neural Inf. Process. Syst. **36** (2024)
7. Zhang, S., et al.: Instruction tuning for large language models: a survey. arXiv preprint arXiv:2308.10792 (2023)
8. Xia, M., Malladi, S., Gururangan, S., Arora, S., Chen, D.: Less: selecting influential data for targeted instruction tuning. arXiv preprint arXiv:2402.04333 (2024)
9. Goldberg, S.G., Johnson, G.A., Shriver, S.K.: Regulating privacy online: an economic evaluation of the GDPR. Am. Econ. J. Econ. Pol. **16**(1), 325–358 (2024)
10. Yao, Y., Duan, J., Xu, K., Cai, Y., Sun, Z., Zhang, Y.: A survey on large language model (LLM) security and privacy: the good, the bad, and the ugly. High-Conf. Comput. **100211** (2024)
11. Konečný, J., McMahan, H.B., Yu, F.X., Richtárik, P., Suresh, A.T., Bacon, D.: Federated learning: strategies for improving communication efficiency. arXiv preprint arXiv:1610.05492 (2016)
12. Karimireddy, S.P., Kale, S., Mohri, M., Reddi, S.J., Stich, S.U., Suresh, A.T.: Scaffold: stochastic controlled averaging for on-device federated learning, **2**(6) (2019). arXiv preprint arXiv:1910.06378
13. Reddi, S., et al.: Adaptive federated optimization. arXiv preprint arXiv:2003.00295 (2020)
14. Dozat, T.: Incorporating nesterov momentum into adam (2016)
15. Chen, J., Zhou, D., Tang, Y., Yang, Z., Cao, Y., Gu, Q.: Closing the generalization gap of adaptive gradient methods in training deep neural networks. arXiv preprint arXiv:1806.06763 (2018)
16. Hu, E.J., et al.: LoRA: low-rank adaptation of large language models. arXiv preprint arXiv:2106.09685 (2021)

17. Wu, J., Gan, W., Chen, Z., Wan, S., Lin, H.: AI-generated content (AIGC): a survey. arXiv preprint arXiv:2304.06632 (2023)
18. Radford, A., Wu, J., Child, R., Luan, D., Amodei, D., Sutskever, I.: Language models are unsupervised multitask learners. OpenAI blog **1**(8), 9 (2019)
19. Vaswani, A., Shazeer, N., Parmar, N., Uszkoreit, J., Jones, L., Gomez, N.: Attention is all you need. Adv. Neural Inf. Process. Syst. **30** (2017)
20. Devlin, J., Chang, M. W., Lee, K., Toutanova, K.: BERT: pre-training of deep bidirectional transformers for language understanding. arXiv preprint arXiv:1810.04805 (2018)
21. Wei, J., et al.: Finetuned language models are zero-shot learners. arXiv preprint arXiv:2109.01652 (2021)
22. Kaplan, J., et al.: Scaling laws for neural language models. arXiv preprint arXiv:2001.08361 (2020)
23. Azerbayev, Z., et al.: Llemma: an open language model for mathematics. arXiv preprint arXiv:2310.10631 (2023)
24. Villalobos, P., Sevilla, J., Heim, L., Besiroglu, T., Hobbhahn, M., Ho, A.: Will we run out of data? an analysis of the limits of scaling datasets in machine learning. arXiv preprint arXiv:2211.04325 (2022)
25. Mehmood, A., Natgunanathan, I., Xiang, Y., Hua, G., Guo, S.: Protection of big data privacy. IEEE Access **4**, 1821–1834 (2016)
26. Li, X., Huang, K., Yang, W., Wang, S., Zhang, Z.: On the convergence of FedAvg on non-iid data. arXiv preprint arXiv:1907.02189 (2019)
27. Kingma, D.P., Ba, J.: Adam: a method for stochastic optimization. arXiv preprint arXiv:1412.6980 (2014)
28. Fan, T., et al.: Fate-LLM: a industrial grade federated learning framework for large language models. arXiv preprint arXiv:2310.10049 (2023)
29. Zhang, J., et al.: Towards building the federatedGPT: federated instruction tuning. In: ICASSP 2024-2024 IEEE International Conference on Acoustics, Speech and Signal Processing (ICASSP), pp. 6915–6919. IEEE (2024)
30. Ye, R., et al.: OpenFedLLM: training large language models on decentralized private data via federated learning. arXiv preprint arXiv:2402.06954 (2024)
31. Taori, R., et al.: Stanford alpaca: an instruction-following llama model. (2023)
32. Hendrycks, D., et al.: Measuring massive multitask language understanding. arXiv preprint arXiv:2009.03300 (2020)
33. Zheng, L., et al.: Judging LLM-as-a-judge with MT-bench and chatbot arena. Adv. Neural Inf. Process. Syst. **36** (2024)
34. Cowen, T.: Introducing GPT-4o (2024)

Feature Augmented Meta-Learning on Domain Generalization for Evolving Malware Classification

Fangwei Wang[1], Yinhe Chen[1], Ruixin Song[2], Qingru Li[1], and Changguang Wang[1](✉)

[1] Key Laboratory of Network and Information Security of Hebei Province, College of Computer and Cyber Security, Hebei Normal University, Shijiazhuang 050024, China
{qingruli,wangcg}@hebtu.edu.cn

[2] College of Computer Science and Technology, Guizhou University, Guiyang 550025, China

Abstract. The proliferation of malware has resulted in substantial harm to various sectors and economies. Various deep learning-based malware classification methods have been suggested as a means of mitigating malware threats. These methods typically operate under the assumption of independent and identically distributed training and test data. However, this assumption becomes invalid with the evolving malware family. While domain adaptation models offer a potential solution to this issue, their implementation is hindered by the difficulty of collecting new malware variants. In order to address the previously mentioned problem, we suggest an image-based technique for categorizing malware families utilizing domain generalization. Initially, malware is transformed into gray-scale images that depict byte patterns of the malware. Subsequently, these gray-scale images are fed into a model incorporating convolutional block attention to extract features. Furthermore, data augmentation is implemented at the feature level to broaden the distribution of the source domain and enhance the model's generalization capabilities. Finally, meta-learning is utilized as a training approach to effectively extract domain-invariant representations. A series of experiments are conducted on the BIG2015 and BenchMFC-G1P1P2. The proposed method demonstrates a higher accuracy rate of 88.66% on the BIG2015 and 80.25% on the BenchMFC-G1P1P2, which is better than the existing methods.

Keywords: Malware Classification · Domain Shift · Domain Generalization · Feature Argumentation

1 Introduction

The proliferation of malware presents a substantial threat to a large number of internet users in the present day. According to AV-TEST [1], there was an excess of 110 million new instances of malware across all platforms in 2023, with over 70 million new cases specifically targeting Windows. These nefarious software

T. Zhu et al. (Eds.): ICA3PP 2024, LNCS 15252, pp. 204–223, 2025.
https://doi.org/10.1007/978-981-96-1528-5_14

programs engage in illicit actions, including the theft of user information and the destruction of user devices, leading to considerable economic and industrial losses across various sectors.

In the realm of cybersecurity, numerous models [2–5] are proposed which aim at identifying and categorizing various malware families as a means of safeguarding against potential threats. Nevertheless, certain limitations exist within these models, as evidenced by the assumption of independent and identical distribution within the dataset. Previous research has demonstrated that there is frequently a high degree of similarity between the same variant type samples of the same malware family [6]. In contrast, there is considerable variation in the texture of different malware families and between variant type samples of the same family. As illustrated in Fig. 1, the C2LOP family exhibits a high degree of similarity in the texture of samples belonging to the same variant. However, there is some variability between the g variant and the gen!g variant. In certain instances, it is necessary to categorize malware only at the family level, without specifying the types of variants within the family. In other words, the sole criterion for categorizing malware is whether it belongs to the same family, without regard to its variant types. In such instances, the variability between variants of the same family can present challenges for family classification. As malware families evolve, new variants with different textures from the old ones appear in the malware family, resulting in a shift in the distribution of the family. This phenomenon is also known as concept shift. The model will perform significantly below ideal design levels in the face of these shifted data. In Fig. 2, it can be observed that the evolving malware exhibits novel texture features that will alter the distribution of the malware. If the model is trained only on the original fareit malware, the model can be highly accurate against samples from the original fareit family, but the model's performance drops dramatically when confronted with concept shift caused by evolving malware. The greater the shift, the more pronounced the model's performance degradation will be [7].

Although some methods [8–11] based on unsupervised domain adaptation have been proposed, these methods require the collection of many malware samples. And current approaches [12,13] based on domain generalization do not discuss applications in malware family classification. So this study introduces a novel approach to domain generalization for malware family classification in order to address the aforementioned challenges. The original malware family samples, which have been labeled and collected, are considered to be the source domain, while the evolving malware family samples are the target domain. Consequently, the concept shift between the unevolved and evolving samples is regarded as the domain shift. Domain generalization facilitates the extraction of domain-invariant representations from complex source domains, thereby enhancing the model's performance on an unfamiliar target domain. Initially, the binary executable files of the malware are converted into gray-scale images. Subsequently, inspired by cross-validated committees [14] of ensemble learning, the grey-scale images were divided into two distinct source domains. We then train two feature extractors based on deep residual convolutional networks (RCN) to

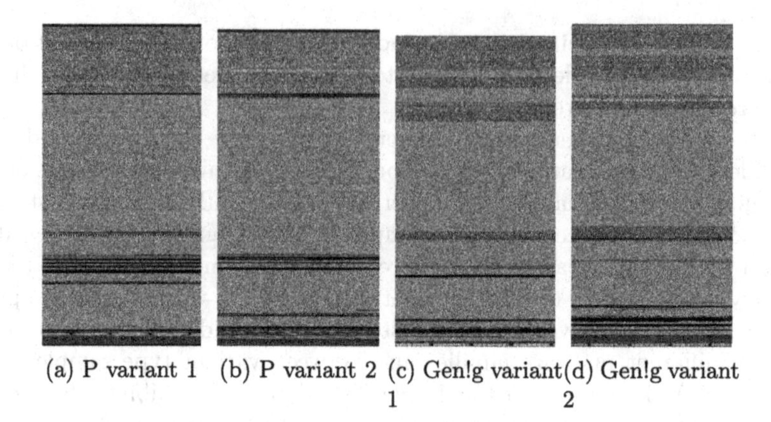

Fig. 1. C2LOP malware family in the Malimg.

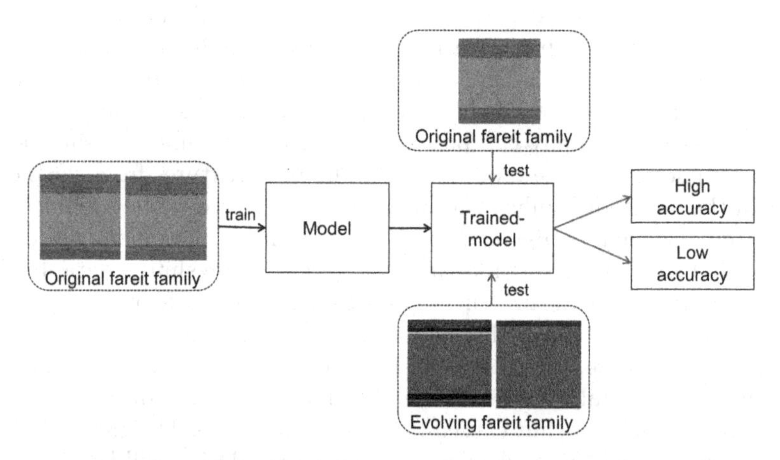

Fig. 2. Performance degradation caused by evolving malware.

extract features from the source domains. Within each feature extractor, the feature augmentation module MixSem proposed in this paper operates to locally amplify the high-level semantic features present in the malware images, resulting in the generation of a new distribution. At the same time, the MixStyle module proposed in [15] is inserted into the feature extractor to assist MixSem in generating more diverse distributions. The new distributions generated are more favourable for model extraction to domain invariant representations. Ultimately, the generalization capacity of the model is enhanced through the implementation of a meta-learning training strategy. The primary contributions of our study are outlined as follows:

1) This paper proposes a model that improves domain generalization. The MixStyle module is used to enhance low-level style features of images at the global level and the MixSem is used to enhance high-level semantic features

of images at the local level. Additionally, the CBAM (Convolutional Block Attention Module) [16] is used to extract and enhance important features more efficiently.
2) This study employs meta-learning techniques to improve the generalization capabilities of the model, and introduces a novel meta-learning loss according to the structure of the new model which could be more favourable for the extraction of the domain invariant representation.
3) A number of experiments on the BIG2015 and BenchMFC-G1P1P2 show that the proposed method effectively addresses domain shift caused by evolving malware. Also, there are fewer experiments based on the benchMFC-G1P1P2 dataset, so this paper can serve as a baseline for related research.

The structure is outlined as follows. Section 2 provides an overview of related work on image-based malware classification, malware classification based on domain adaptation, and domain generalization, followed by a detailed explanation of the proposed model in Sect. 3. Section 4 presents the results of an experiment that validates the effectiveness of the method, along with a discussion of its various components. Finally, Sect. 5 concludes this paper.

2 Related Work

Malware has inflicted substantial harm on the Internet, prompting a heightened focus on the classification of malware families as a critical area of research. The utilization of image-based techniques for malware analysis has become prevalent due to their intuitive nature and effectiveness. This section provides an overview of relevant studies, encompassing topics such as image-based malware classification, malware classification based on domain adaptation, and domain generalization.

2.1 Image-Based Malware Classification

The type of methods described is an end-to-end visualization technique for classifying malware based on texture features. Samples from the same malware family exhibit similar texture features, while samples from different families do not. Nataraj et al. [6] were the first to employ visual analysis techniques in malware classification, converting byte data from executable files into pixel representations and generating grayscale images. The utilization of deep learning for the classification of malware images has experienced substantial growth. In their study, Gibert et al. [2] employed convolutional neural networks (CNN) to extract features from gray-scale images of malware. The CNN architecture included three convolutional layers and a fully connected layer. Through experiments on malImg and BIG2015 datasets, they demonstrated the superior effectiveness of deep learning compared to conventional machine learning approaches for malware family classification tasks. Additionally, Kumar et al. [3] proposed that pre-trained models on IMAGENET can effectively transfer relevant knowledge to the domain of malware images. They utilized the pre-trained CNN for feature extraction and implemented early stopping methods to mitigate overfitting. This approach demonstrates rapid convergence and reduced computational demands, rendering it well-suited for environments with limited resources. In a related study, Rustam et al. [4] independently utilized ResNet50 and VGG16 for feature extraction, subsequently merging the extracted features and inputting them into a classifier composed of two identical machine learning classifiers. The proposed methodology yields more extensive and nuanced features capable of capturing the distinctive attributes of malware. Alanjani [5] proposed a deep learning network with dual attention, which improves the ability to capture malware signatures.

The aforementioned techniques for categorizing malware families operate under the assumption of independence and identical distribution, demonstrating high levels of accuracy. Nevertheless, the failure to account for potential shifts may result in a decline in model performance when the distributions differ.

2.2 Domain Adaptation-Based Malware Classification

Despite the availability of numerous high-accuracy models for classifying malware families, it is common practice to assume that the test and training sets are independent and identical. However, this assumption may not hold in real-world scenarios due to the shifting nature of target and source domains during malware evolution. Ma et al. [7] found that malware classification methods exhibit decreased performance when the training and test datasets are distributed dissimilarly.

Researchers have proposed several Unsupervised Domain Adaptation (UDA) strategies to tackle this issue. UDA can reduce the domain shift between source domain and target domain of malware under the condition of obtaining only malware samples in the target domain without obtaining labels. To reduce the shift between old malware and new malware, Wang et al. [8] utilized adversarial learning to mitigate domain discrepancies between source and target domains, and align class centers using pseudo-labels. In order to reduce the discrepancy between the target and source domains, Qi et al. [9] employed adversarial training, whereby pseudo-labels were assigned to novel unlabeled malware samples and incorporated into the training set in order to train the LightGBM model. Li et al. [10] utilize adaptive regularization transfer learning to harmonize the marginal and conditional distributions between the source and target domains. In a similar vein, Bhardwaj et al. [11] introduce the MD-ADA framework, employing a dynamic adversarial network to align distributions between target and source domains and predict the maliciousness of software.

Nevertheless, these UDA methods necessitate access to manually gathered target domain data, a process that is both labor-intensive and challenging.

2.3 Domain Generalization

Blanchard et al. [17] introduced the concept of Domain Generalization(DG). Domain generalization aims to train a model to perform well on unknown new distributions without access to knowledge of the new distributions. So it is a more challenging and relevant approach due to the lack of access to new distributions.

To improve the performance of domain generalisation, researchers have proposed many methods at the level of data enhancement or learning strategies. On the learning strategy level, methods such as MLDG [18], DAML [19] and DAEL [20] can effectively improve the generalisation ability of the model. At the data augmentation level, methods such as mixup [21], cutmix [22] and mixstyle [15] can improve model performance with very little resource. In addition, generative models such as DDAIG [23] have been used to generate data to improve the generalisation ability of the model. Domain generalization is gradually playing an important role in the field of malware classification. To address the adverse impact of bias among distinct malware type on the model, Rani et al. [12] trained the model using MLDG [18], thereby markedly enhancing the model's capacity to generalize to previously unidentified malware. To mitigate the impact of evolution on malware detection, Bosansky et al. [13] proposes a data augmentation approach that generates malware images at different times through conditional cycle GANs. This approach is then used to train the model using both the generated malware images and real malware images, thereby enhancing the model's generalization ability to future unknown malware. However, for the classification of evolving malware families, these methods have not yet been discussed.

To enhance the model's ability to generalize evolving malware families, particularly in cases where collecting such variants poses challenges, we introduce a novel approach based on domain generalization. This approach incorporates both stylistic and semantic features, considering both global and local information within the image. Our study implements distinct feature extractors for two distinct source domains to derive domain-invariant representations. Furthermore, we employ meta-learning as a training technique and introduce a unique meta-learning loss function to facilitate the extraction of domain-invariant representations by the model.

3 Proposed Methodology

Our study proposes a method based on domain generalization for malware family classification to mitigate the impact of domain shift on malware classification performance. The whole framework is illustrated in Fig. 3. To begin with, gray-scale images are generated from the malware binary files. Then, meta-learning is used to train a model. In this stage, to improve the performance of the model, the gray-scale images are randomly divided into two domains to train two different feature extractors named F_0 and F_1. The two source domains are labeled as source domain D_0 and D_1 respectively. The meta-training phase starts when the model receives samples from the source domains. The model enhances both low-level stylistic features and high-level semantic features of the image, thus generating a distribution D^{st} that enhances only the style and a distribution D^{se} that enhances both style and semantics. Then the stylistic loss $\mathcal{L}^{st}$ and semantic loss $\mathcal{L}^{se}$ are calculated, where $\mathcal{L}^{st}$ is the model's loss on D^{st} and $\mathcal{L}^{se}$ is the model's loss on D^{se}. These losses are combined to form the meta-training loss $\mathcal{L}^{mtr}$. Then the model is copied as a new model. The copied model is updated based on $\mathcal{L}^{mtr}$ through gradient descent, and the model parameters are updated from θ to $\theta^{'}$. This is followed by a meta-testing phase. The copied model processes new samples and calculates the losses $\mathcal{L}^{st'}$ and $\mathcal{L}^{se'}$, which are combined to form the meta-test loss $\mathcal{L}^{mte}$. A combination of $\mathcal{L}^{mtr}$ and $\mathcal{L}^{mte}$ is used to update the original model of parameters θ. The above training process is repeated continuously until the model converges. After training, a trained well model is obtained. When testing the samples from the target domain, the preprocessed samples are input directly into the trained model, and the category labels of samples are output.

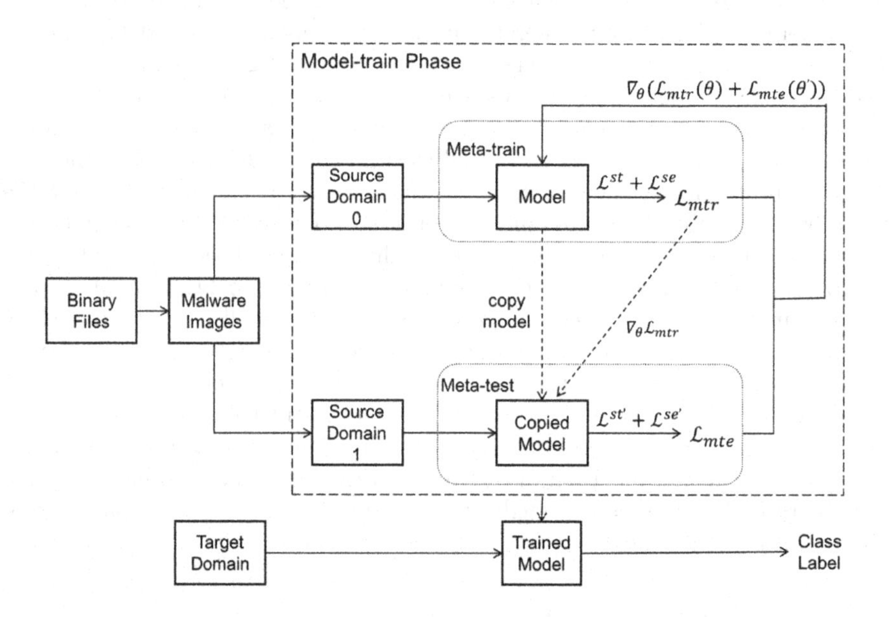

Fig. 3. The workflow of the proposed framework.

The architecture of the Model is illustrated in Fig. 4. It contains Feature Extractor F_0, Feature Extractor F_1 and a Classifier. Classifier is a fully connected layer. In meta-training, source domain D_0 serves as the meta-training domain for feature extractor F_0, while source domain D_1 serves as the meta-training domain for feature extractor F_1. In meta-testing, source domain D_1 serves as the meta-testing domain for feature extractor F_0, while source domain D_0 serves as the meta-testing domain for feature extractor F_1. The CBAM in the feature extractor enables the model to concentrate on significant channel and spatial features. The MixStyle and MixSem modules can produce more diverse distributions based on the source domain distribution. $\mathcal{L}_k^{st}$ is calculated from the feature output of the last residual convolution block in the feature extractor, and $\mathcal{L}_k^{se}$ is calculated from the feature output of the MixSem module.

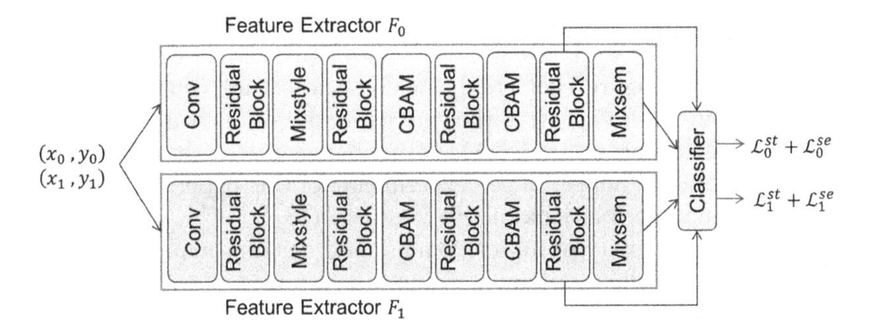

Fig. 4. The architecture of the proposed Model.

3.1 Data Preprocessing

During the preprocessing phase, the binary executable files are converted into gray-scale images. Each 8-bit binary data is transformed into a gray-scale value ranging from 0 to 255. The value 0 represents black in the image, while 255 represents white. Since the size of the malware is not consistent, the length of its binary sequence is also not same. So for each malware sample, we fix the width of the image and calculate the height according to Table 1. Then based on the calculated width and height the gray-scale values are stored in a two dimensional array in sequence. A gray-scale image is generated based on the gray-scale values in the array. Finally, in order to fit the input size of the ImageNet pre-trained model, we scale the gray-scale images in RGB dimension and then resize them to 224×224 pixels. Inspired by ensemble learning, we randomly divide the gray-scale images into two source domains thereby training two different deep residual networks to extract features from the gray-scale images.

Table 1. Image Width.

File Size	Image Width
<10 kB	32
10 kB–30 kB	64
30 kB–60 kB	128
60 kB–100 kB	256
100 kB–200 kB	384
200 kB–500 kB	512
500 kB–1000 kB	768
>1000 kB	1024

3.2 Model Structure and Training Strategy

Model Structure. Figure 3 depicts the model architecture, wherein RCNs serve as feature extractors for capturing the characteristics of malware images. Additionally, the model incorporates CBAM to emphasize crucial channel and spatial information within the image. In the experiments of this paper, the basic structure of the residual convolutional network is resnet18.

Enhancing the diversity of source domain data distribution is a key strategy for overcoming the challenge of generalizing to unfamiliar distributions. Numerous studies have utilized the existing source domain distribution at the feature level to create a novel distribution. Nevertheless, several augmentation techniques solely leverage either low-level stylistic information or high-level semantic information from the images. Consequently, the resultant new distribution lacks sufficient diversity, thereby constraining the model's capacity to generalize to unknown distributions. Additionally, these augmentation methods only enhance the global features of images. Malware images' local features, however, also reflect their evolutionary trend. Consequently, the proposed model augments data at both the low-level stylistic feature level and the high-level semantic feature level of every image in addition to enriching global and local levels.

MixStyle module is proposed in [15]. Using MixStyle, an image is normalized by its mean and variance, mixing the features of different images at the global level. As a plug-and-play module, in order to fully exploit the low-level features of the residual convolutional network to generate the unseen distribution, MixStyle module is inserted after the first residual convolutional block in this paper. However, MixStyle is unable to exploit the high-level features of RCNs, and to compensate for this shortcoming, we propose the MixSem.

At the high-level semantic feature level, we propose MixSem module based on cutmix [22] to facilitate the enhancement of high-level semantic features across different domains. Unlike cutmix, MixSem augments data at the feature level rather than at the image level. Based on the assumption that (z_0, y_0) and (z_1, y_1) represent the features that are extracted by feature extractors and labels in source domains D_0 and D_1. MixSem module enhances the features and the enhanced features and labels are denoted as $\widetilde{z}$ and $\widetilde{y}$, which are shown as follows:

$$\begin{aligned} \widetilde{z} &= rand\{m_0 z_0 + m_1 z_1, m_1 z_1 + m_0 z_0\}, \\ \widetilde{y} &= r y_0 + (1-r) y_1, \\ r &\sim beta(\alpha_{se}, \alpha_{se}), \end{aligned} \tag{1}$$

where $m \in \{0,1\}^L$ is a mask to select the local features from the feature vector z. The $rand\{\cdot\}$ represents a randomly selected element, and r follows a beta distribution. The length L of the mask is equal to the dimension of the feature vector. The position of 1 in the mask are determined by $O = (p_s, p_e)$, where

$$\begin{aligned} p_s &= Unif(0, L - rL), \\ p_e &= rL + p_s. \end{aligned} \tag{2}$$

In Eq. (2), $Unif$ represents a uniform distribution. In the mask, the value in the region O is 1 and the value in other regions is 0. Local features from different images are randomly combined and $\widetilde{z}$ is obtained. It should be noted that both the MixStyle and MixSem modules only augment the data during the training phase of the model. When testing the target domain, they will cease to function.

Meta-Training. Meta-training domains are used to train the model during the meta-training phase. Similar to [19], this paper expects the model to learn domain invariant representations from the augmented features.

We train two feature extractors separately. The input data for feature extractor $F_k(k \in \{0,1\}\})$, denoted as (X, Y), contains the sample (x_k, y_k) from this source domain D_k and the sample $(x_{|k-1|}, y_{|k-1|})$ from another source domain $D_{|k-1|}$. Assuming that f_k^{st} and $f_{|k-1|}^{st}$ represent the features extracted from (x_k, y_k) and $(x_{|k-1|}, y_{|k-1|})$ by the last convolutional block of the feature extractor F_k respectively, and $(f_k^{se}, y_k^{se}) = M(f_{st}^k, f_{st}^{|k-1|})$ represents the features and labels that have been augmented by the MixSem module. $G(\cdot)$ represents the classifier. This paper designs a new meta-training loss $\mathcal{L}_k^{mtr}$:

$$\begin{aligned}
\mathcal{L}_k^{st} &= \underset{(x_k, y_k) \sim D_k^{st}}{E}[-\sum_{m=0}^{|C|} (y_k)^m \log\left(G^m(F_k(x_k))\right)], \\
\mathcal{L}_k^{se} &= \underset{(f_k^{se}, y_k^{se}) \sim D_k^{se}}{E}[-\sum_{m=0}^{|C|} (y_k^{se})^m \log\left(G^m(f_k^{se})\right)], \\
\mathcal{L}_k^{mtr} &= \mathcal{L}_k^{st} + \mathcal{L}_k^{se},
\end{aligned} \tag{3}$$

where D_k^{st} represents the distribution of the meta-training domain D_k of F_k augmented by MixStyle, while D_k^{se} is the distribution of D_k^{se} enhanced by MixSem. The subscript m in the formula denotes the category probability, and $|C|$ denotes the total number of categories. $\mathcal{L}_k^{st}$ and $\mathcal{L}_k^{se}$ are the cross-entropy losses of the model over the D_k^{st} and D_k^{se}, respectively. $\mathcal{L}_k^{st}$ and $\mathcal{L}_k^{se}$ are summed as meta-training losses $\mathcal{L}_k^{mtr}$.

The parameters of the new model are updated using the meta-training loss $\mathcal{L}_k^{mtr}$ with a learning rate η , which is shown as Eq. (4).

$$\theta'_{k,G} = \theta_{k,G} - \eta \nabla_\theta \mathcal{L}_k^{mtr} \tag{4}$$

At this stage, the model with the original parameters $\theta_{k,G}$ are saved, and the model is then copied and updated with the new parameters $\theta'_{k,G}$.

Meta-Testing. Following the meta-training phase, we want to ascertain whether the parameters obtained from meta-training demonstrate a robust generalization ability for meta-testing domain. During the meta-testing phase, the model is replicated and parameters $\theta'_{k,G}$ are used as the parameters of the new model. Then, a new batch of source domain samples is inputted into the new model. Model performance is then tested in different distributions. It means that, in the mete-testing phase, the performance of feature extractor F_k on its meta-testing domain $D_{|k-1|}$ will be evaluated. In this paper, we design a new meta-testing loss, shown as Eq. (5):

$$\begin{aligned}
\mathcal{L}_k^{st'} &= \underset{(x_{|k-1|}, y_{|k-1|}) \sim D_{|k-1|}^{st}}{E} [-\sum_{m=0}^{|C|} (y_{|k-1|})^m \log (G^m(F_k(x_{|k-1|})))], \\
\mathcal{L}_k^{se'} &= \underset{(f_{|k-1|}^{se}, y_{|k-1|}^{se}) \sim D_{|k-1|}^{se}}{E} [-\sum_{m=0}^{|C|} (y_{|k-1|}^{se})^m \log (G^m(f_{|k-1|}^{se}))], \\
\mathcal{L}_k^{mte} &= \mathcal{L}_k^{st'} + \mathcal{L}_k^{se'}.
\end{aligned} \tag{5}$$

Equation (5) calculates the loss of the feature extractor F_k over the augmented meta-testing domain. If the model is able to learn a reasonable domain invariant representation, then the model will have less loss in both meta-training and meta-testing. In order to update the original parameters of the model, the meta-training loss $\mathcal{L}_k^{mtr}$ and the meta-testing loss $\mathcal{L}_k^{mte}$ are added together, shown as Eq. (6):

$$\theta_{k,G} = \theta_{k,G} - \beta \nabla_\theta (\mathcal{L}_k^{mtr} + \mathcal{L}_k^{mte}). \tag{6}$$

3.3 Model Training

The training process is as follows. The gray-scale images are initially divided into two source domains denoted as D_0 and D_1 to training two feature extractors with different parameters. During the meta-training phase, the model initial parameters are set to $\theta_{0,1,G}$. Samples from source domains D_0 and D_1 are fed into feature extractors. It should be noted that, in this phase, the feature extractor F_0 extracts the features of the augmented distribution on the source domain D_0. In a similar manner, the feature extractor F_1 is tasked with extracting only the features of the augmented distribution on domain D_1. The feature extractor F_k utilizes MixStyle and MixSem to enhance the images low-level stylistic and high-level semantic features on a global and local scale, respectively. This expands the distribution of the source domains. The features are extracted through a residual convolutional network with CBAM. These features include features f_k^{st} enhanced only style and features f_k^{se} enhanced both semantics and style. These features are fed into the classifier and the losses $\mathcal{L}_k^{st}$ and $\mathcal{L}_k^{se}$ are calculated. The losses are then summed to obtain the meta-training loss $\mathcal{L}_k^{mtr}$. Then the model is copied as a new model. $\mathcal{L}_k^{mtr}$ is used to update the copied model. After meta-training, the parameters of the copied model are $\theta_{0,1,G}^{'}$. During the meta-test phase, we feed another batch of samples from D_0 and D_1 into the copied models feature extractors. The feature extractor F_0 is tasked with extracting features from the augmented distribution of the meta-test domain D_1, while the feature extractor F_1 is employed to extract features from the augmented distribution of the meta-test domain D_0. The classifier computes losses $\mathcal{L}_k^{st'}$ and $\mathcal{L}_k^{se'}$ using features extracted by F_k. These losses are then summed to form the meta-test loss $\mathcal{L}_k^{mte}$. The original model parameters $\theta_{0,1,G}$ are updated using $\mathcal{L}_k^{mtr}$ and $\mathcal{L}_k^{mte}$. The model is meta-trained and meta-tested until convergence is achieved. The model training process is shown in Algorithm 1.

3.4 Testing on the Target Domain

After training the model with Algorithm 1, we obtain feature extractors F_0 and F_1 as well as the classifier G. When testing the target domain samples x_t, each feature extractor extracts the features and feeds them into the classifier separately to predict the results. Calculating the final prediction of x_t is accomplished by average voting, shown as Eq. (7):

$$y_t = \frac{1}{2}[G(F_0(x_t)) + G(F_1(x_t))]. \tag{7}$$

Algorithm 1. Training of our model

Input: source domain D_0 and D_1, meta learning rate η and sgd learning rate β, MixSem hyper-parameter α_{se}, MixStyle hyper-parameter α_{st}, epochs N
Output: model parameters $\theta_{0,1,G}$
Initialize: model parameters $\theta_{0,1,G}$

1: **for** epoch=1 to N **do**
2: Sample a batch of data $B^{mtr} = \{B_0^{mtr}, B_1^{mtr}\}$ from D_0 and D_1
3: **for** k=0 to 1 **do**
4: $B_k^{st} = (f_k^{st}, y_k) \leftarrow F_k(MixStyle(\alpha_{st}, B_k^{mtr}))$
5: $B_k^{se} = \{(f_k^{se}, y_k^{se})\} \leftarrow MixSem(\alpha_{se}, B_k^{st})$
6: $\mathcal{L}_k^{st} \leftarrow \{G(f_k^{st}), y_k\}$ $using$ B_k^{st}
7: $\mathcal{L}_k^{se} \leftarrow \{G(f_k^{se}), y_k^{se}\}$ $using$ B_k^{se}
8: $\mathcal{L}_k^{mtr} = \mathcal{L}_k^{st} + \mathcal{L}_k^{se}$
9: $\theta_{k,G}^{'} = \theta_{k,G} - \eta\nabla_\theta\mathcal{L}_k^{mtr}$
10: **end for**
11: Sample another batch of data $B^{mte} = \{B_0^{mte}, B_1^{mte}\}$ $from$ D_0 and D_1
12: **for** k=0 to 1 **do**
13: $B_{|k-1|}^{st'} = (f_{|k-1|}^{st'}, y_{|k-1|}) \leftarrow F_k^{'}(MixStyle(\alpha_{st}, B_{|k-1|}^{mte}))$
14: $B_{|k-1|}^{se'} = \{(f_{|k-1|}^{se'}, y_{|k-1|}^{se'})\} \leftarrow MixSem(\alpha_{se}, B_{|k-1|}^{st'})$
15: $\mathcal{L}_k^{st'} \leftarrow \{G(f_{|k-1|}^{st'}), y_{|k-1|}\}$ $using$ $B_{|k-1|}^{st'}$
16: $\mathcal{L}_k^{se'} \leftarrow \{G(f_{|k-1|}^{se'}), y_{|k-1|}^{se}\}$ $using$ $B_{|k-1|}^{se'}$
17: $\mathcal{L}_k^{mte} = \mathcal{L}_k^{st'} + \mathcal{L}_k^{se'}$
18: $\theta_{k,G} = \theta_{k,G} - \beta\nabla_\theta(\mathcal{L}_k^{mtr} + \mathcal{L}_k^{mte})$
19: **end for**
20: **end for**
21: return $\theta_{0,1,G}$

4 Experiment and Result Analysis

4.1 Experimental Settings

Dataset. The BIG2015 and benchMFC-G1P1P2 are used for related experiments. There are 21,741 samples in BIG2015 [24], which are divided into nine families. A total of 10,868 samples have been labeled. Each sample includes asm and bytes files. The experiments use the bytes files of the labeled samples to generate gray-scale images. Due to the presence of non-hexadecimal characters in 8 labeled samples, they could not be converted into images. Therefore, a total of 10860 labeled samples are used.

Since the BIG2015 dataset does not flag whether the malware has evolved or not, but by looking at the dataset we find that there are samples with large texture differences in some malware families, and we believe that these texture differences are caused by the evolution of the malware family. Therefore, in this paper, we refer to [25] for the idea of dividing the dataset and use the k-means algorithm to cluster each malware family in the BIG2015, so that the samples that have a large difference in texture compared to the majority of the samples are filtered out as the target domain and the rest of the samples are used as the source domain. Under this division method, there is a large distribution shift between the source domain (SD) and target domain (TD). In our experiments, we select 80% of the source domain samples as the training set and 20% as the validation set. Table 2 presents the sample information from the BIG2015 dataset used in this paper.

Table 2. BIG-2015 dataset.

No.	Family	SD Samples	TD Samples	Total Samples
1	Ramnit	1207	326	1533
2	Lollipop	2062	416	2478
3	Kelihos_ver3	2593	349	2942
4	Vundo	279	196	475
5	Simda	40	2	42
6	Traceur	624	127	751
7	Kelihos_ver1	184	214	398
8	Obfuscator.ACY	968	260	1228
9	Gatak	830	183	1013

BenchchMFC [26] is a dataset designed for the investigation of malware family classification under the influence of concept shift. The dataset encompasses 223,650 instances of malware between January 2012 and December 2022, with 526 distinct malware families being classified. Importantly, the dataset marks whether evolution is occurring in the malware family. Meanwhile, in order to test the performance of the model under concept shift due to malware family evolution, [26] took some samples from the dataset labeling unevolved malware families as G1P1 and evolving malware families as G1P2. The details of G1P1 and G1P2 in BenchMFC dataset are shown in Table 3. In this paper, we also use G1P1 and G1P2 for the related experiments, and use G1P1 as the source domain and G1P2 as the target domain. However, the division of training, validation and test sets in this paper is slightly different from [26]. [26] used only 20% of the samples in G1P2 as a test set, whereas we used the 100% G1P2 set to more fully test the model's ability to generalize the offset data. We train the model on the training set and select the best-performing weights on the validation set as the final weights of the model, and then we test the relevant evaluation metrics of the model on the test set.

Table 3. BenchMFC-G1P1P2 dataset.

No.	Family	Malware Samples	Evolving Samples	Total Samples
1	simda	500	500	1000
2	grandcrab	500	500	1000
3	yuner	500	500	1000
4	hotbar	500	500	1000
5	fareit	500	500	1000
6	zbot	500	500	1000
7	upatre	500	500	1000
8	parite	500	500	1000

Implementation Details. The SGD optimizer with cosine annealing is used in all models, along with a learning rate of 0.0009, momentum of 0.9, and weight decay of 0.0001. A batch size of 24 is used. PyTorch 2.1 and Python 3.10 is used for the experiments. The configuration of hardware is as follows: CPU: 7/Intel(R) Xeon(R) Silver 4314 CPU @ 2.40 GHz, GPU: NVIDIA GeForce RTX 3080. In this paper, the grid search is used to search for the optimal hyper-parameters within a certain range. The tuning range of hyper-parameters and the results of optimal hyper-parameters are shown in Table 4.

Table 4. Hyper-parameters used for the proposed model.

Hype-parameter	*Optimal value*	*Tuning range*
Meta learning rate η	0.008	(0.01,0.009,0.008,0.007,0.006)
SGD learning rate β	0.0009	(0.001, 0.0009, 0.0008, 0.0007, 0.0006, 0.0005)
MixSem α_{se}	0.6	(0.5, 0.6, 0.7, 0.8)
MixStyle α_{st}	0.1	(0.1, 0.2, 0.3)

Evaluation Metrics. Accuracy, macro-precision, macro-recall, and macro-F1 score are measures by which we evaluate the proposed model. The samples are classified according to their true and predicted labels as true positive (TP), false positive (FP), true negative (TN), and false negative (FN).

TP measures the number of samples with both true and predicted positive labels, and they are correctly classified.

FP measures the number of samples that have negative true labels but positive predicted labels, and they are classified incorrectly.

TN is the number of samples having negative true labels and negative predicted labels, and thus being classified correctly.

FN is the number of samples having positive true labels but negative predicted labels, and thus being classified incorrectly.

The *Accuracy* is expressed as the ratio between correctly predicted samples and the total sample:

$$Accuracy = \frac{TP + TN}{TP + FP + TN + FN} \tag{8}$$

As a percentage of correctly predicted positive samples to all predicted positive samples, *Precision* can be considered:

$$Precision = \frac{TP}{TP + FP} \tag{9}$$

As a percentage of correctly predicted positive samples to all positive samples, *Recall* can be considered:

$$Recall = \frac{TP}{TP + FN} \tag{10}$$

As a result of weighing precision and recall, $F1$ scores are calculated as follows:

$$F1 = 2 \times \frac{Recall \times Precision}{Recall + Precision} \tag{11}$$

Calculating *Precision*, *Recall*, and $F1$ scores for each malware family, we calculate an average of these metrics to determine macro-precision, macro-*Recall*, and macro-$F1$.

4.2 Comparison with Other Methods

This paper evaluates the performance of various models using the aforementioned datasets and evaluation metrics. Due to the lack of malware family classification methods based on domain generalization, the paper uses several malware classification models and object recognition models based on domain generalization as comparative experiments. We make reference to the original text of the comparison models with a view to selecting appropriate hyper-parameters. Subsequently, the models are retrained or fine-tuned on the datasets. It should be noted that for the BenchMFC-G1P1P2 dataset, the classification results of the MLP in [26] are directly used as the baseline in this paper.

- CNN [2]: The malware images are classified using a CNN with three convolutional layers.
- DTMIC [3]: A new model using VGG16 as a feature extractor is proposed. The model loads parameters pretrained on the ImageNet dataset and uses the early stop method to prevent overfitting. The hyperparameter patience of the early stopping method is set to 7 in the experiments of this paper.
- BiModel [4]: The paper employs VGG16 and Resnet50 as feature extractors, and fuses the extracted features before feeding them into a bi-classifier for classification. Two SVMs are used as classifiers in the experiments.
- MLDG [18]: One domain per training epoch is randomly selected as the meta-testing domain, while the remainder serves as the meta-training domain. This study employs the MLDG method, which uses resnet18 as a feature extractor, for comparative experiments.
- Vrex [27]: This paper employs Vrex loss function and uses resnet18 as a feature extractor for experiments.
- MixStyle [15]: As a module, MixStyle can be inserted into different convolutional blocks of a convolutional neural network. In the comparison experiments, this paper uses resnet18 as the model and inserts MixStyle behind the 1st and 2nd residual blocks.

Table 5. Performance comparison on the BIG2015.

Method	*Accuracy*	Macro-*Recall*	Macro-*Precision*	Macro-*F1*
CNN [2]	79.55	66.39	69.91	67.04
DTMIC [3]	75.11	59.99	65.25	59.46
BiModel [4]	81.91	66.66	75.29	69.13
MLDG [18]	80.66	66.31	72.79	68.84
Vrex [27]	84.56	70.46	76.44	71.72
Mixstyle [15]	84.52	71.84	75.79	72.24
Ours	**88.66**	**75.05**	**79.97**	**75.57**

Table 6. Performance comparison on the BenchMFC-G1P1P2.

Method	*Accuracy*	Macro-*Recall*	Macro-*Precision*	Macro-*F1*
MLP [26]	76.56	76.56	79.56	76.60
CNN [2]	67.28	67.28	73.62	66.83
DTMIC [3]	78.12	78.13	81.49	78.12
BiModel [4]	73.75	73.75	77.70	73.48
MLDG [18]	76.15	76.15	79.66	76.54
Vrex [27]	74.08	74.08	77.55	74.43
Mixstyle [15]	77.18	77.17	78.22	77.36
Ours	**80.25**	**80.21**	**83.08**	**80.10**

The performance comparison in Tables 5 and 6 demonstrates that our method performs well. With the BIG2015, our method produces significant improvements over other methods in accuracy, macro-precision, and macro-$F1$ scores, achieving 88.66%, 75.05%, 79.97%, and 75.57%, respectively. The results in Table 6 show that our method has a good ability to generalize to evolving malware families. Our method achieves an accuracy of 80.25%, macro recall of 80.21%, macro precision of 83.08% and macro $F1$ score of 80.10%. The evaluation metrics outperform other methods on this dataset. Figure 5 shows the confusion matrix of MLP [26] and the proposed method in this paper on the benchMFC-G1P1P2 dataset. From the confusion matrix, we can see that our method is only slightly less accurate than MLP on the three families of fareit, gandcrab and upatre, but its generalization ability to the simda family is much better than MLP, and its accuracy in the parite and zbot families is slightly better than MLP. Taken together, our method outperforms MLP on the BenchMFC-G1P1P2 dataset.

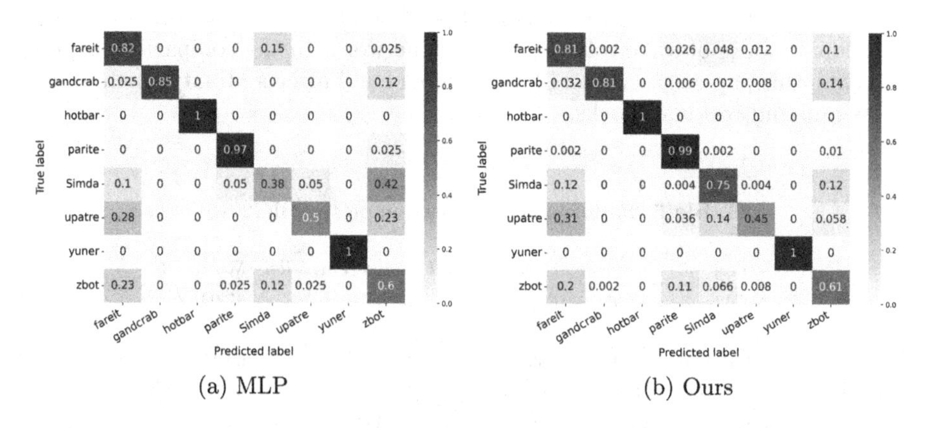

(a) MLP (b) Ours

Fig. 5. Confusion matrix of MLP and ours on the BenchMFC.

4.3 Ablation Study

An ablation experiment is conducted on feature augmentation, attention mechanisms, and meta-learning in order to study the function of each module in depth. The baseline model consists of only two resnet18 feature extractors and a classifier, and meta-learning is not used as a training strategy. Our ablation experimental results are listed in Table 7. D_s^{aug} represents the use of MixStyle and MixSem in the model, while $D_s^{aug-mixup}$ represents the use of Dir-mixup proposed in [19] to replace MixSem. D_s^{cbam} represents the use of CBAM, and $Meta$ represents the use of meta-learning. The results in the Table 7 show that the proposed method is significantly more accurate than the baseline model. Table 7 also shows that the model accuracy is improved when the MixStyle and MixSem modules are used, which indicates that a diverse data distribution enhances the model generalization ability. However, changing MixSem to Dir-mixup results in a slight decrease in accuracy, which suggests that strengthening the model ability to analyze malware local images contributes to its ability to generalize to unseen distributions. The use of the CBAM improves accuracy compared to without it, which indicates attention mechanism is helpful for extracting domain invariant features. Models with the meta-learning strategy achieves higher accuracy rates, which suggests meta-learning has the ability to transfer knowledge across domains or tasks.

Table 7. The results of the ablation experiments on the BIG2015.

D_s^{aug}	$D_s^{aug-Dirmixup}$	D_s^{cbam}	**Meta**	***Accuracy***
				85.24
	✓	✓	✓	87.75
✓		✓		85.87
✓			✓	87.60
		✓	✓	86.93
✓		✓	✓	**88.66**

5 Conclusion

This paper introduces a approach based on domain generalization for classifying malware. We utilize gray-scale images as input and divide them into two distinct source domains. With these two source domains, we can train two deep residual networks with different parameters as feature extractors. In addition, the model is developed to create a novel distribution derived from the source domain distribution and to extract domain-invariant representations from these domains. Our method demonstrates efficacy in classifying malware families, as evidenced by experimental results. The results of multiple evaluation metrics demonstrate that our method for classifying malware families is more effective than other methods when evolution occurs within the malware family. Meanwhile, our methodology eliminates the need for gathering target domain samples, distinguishing it from malware detection or classification models that rely on domain adaptation.

Notwithstanding the favourable outcome on the dataset, it is evident that some malware instances remain misclassified. It is therefore evident that further research is required in the area of malware family classification under shift conditions caused by evolving malware. Furthermore, the proposed method continues to depend on labeled data. It is therefore important to consider how unlabeled samples can be used to enhance the generalization ability of malware in future research. Additionally, we will delve into the classification of malware families for Android and Linux systems.

Acknowledgements. This research was funded by the National Natural Science Foundation of China under Grant 62462012, and Science and Technology Program of Hebei under Grant 22567606H.

References

1. Malware Statistics [EB/OL]. https://portal.av-atlas.org/malware/statistics. Accessed 20 May 2024
2. Gibert, D., Mateu, C., Planes, J., Vicens, R.: Using convolutional neural networks for classification of malware represented as images. J. Comput. Virol. Hack. Tech. **15**(2), 15–28 (2019)
3. Kumar, S., Janet, B.: Dtmic: deep transfer learning for malware image classification. J. Inform. Secur. Appl. **64**(103063), 1–18 (2022)
4. Rustam, F., Ashraf, I., Jurcut, A.D., Bashir, A.K., Zikria, Y.B.: Malware detection using image representation of malware data and transfer learning. J. Paral. Dist. Comput. **172**(7), 32–50 (2023)
5. Alandjani, G.: Securing edge devices: malware classification with dual-attention deep network. Appl. Sci. **14**(11), 4645 (2024)
6. Nataraj, L., Karthikeyan, S., Jacob, G., Manjunath, B.S.: Malware images: visualization and automatic classification. In: Proceedings of the 8th International Symposium on Visualization for Cyber Security, pp. 1–7 (2011)
7. Ma, Y., Liu, S., Jiang, J., Chen, G., Li, K.: A comprehensive study on learning-based PE malware family classification methods. In: Proceedings of the 29th ACM Joint Meeting on European Software Engineering Conference and Symposium on the Foundations of Software Engineering, pp. 1314–1325 (2021)
8. Wang, F., Chai, G., Li, Q., Wang, C.: An efficient deep unsupervised domain adaptation for unknown malware detection. Symmetry **14**(2), 296 (2022)
9. Qi, P., Wang, W., Zhu, L., Ng, S.K.: Unsupervised domain adaptation for static malware detection based on gradient boosting trees. In: Proceedings of the 30th ACM International Conference on Information & Knowledge Management, pp. 1457–1466 (2021)
10. Li, H., Chen, Z., Spolaor, R., Yan, Q., Zhao, C., Yang, B.: Dart: detecting unseen malware variants using adaptation regularization transfer learning. In: ICC 2019-2019 IEEE International Conference on Communications (ICC), pp. 1–6 (2019)
11. Bhardwaj, S., Li, A., Dave, M., Bertino, E.: Overcoming the lack of labeled data: training malware detection models using adversarial domain adaptation. Comput. Secur. **140**(5), 103769 (2024)
12. Rani, N., Mishra, A., Kumar, R., Ghosh, S., Shukla, S.K., Bagade, P.: A generalized unknown malware classification. In: International Conference on Security and Privacy in Communication Systems, pp. 793–806 (2022)
13. Bosansky, B., Hospodkova, L., Najman, M., Rigaki, M., Babayeva, E., Lisy, V.: Counteracting concept drift by learning with future malware predictions. arXiv preprint arXiv:2404.09352 (2024)
14. Dietterich T G.: Ensemble methods in machine learning. In: International Workshop on Multiple Classifier Systems, pp. 1–15 (2000)
15. Zhou, K., Yang, Y., Qiao, Y., Xiang, T.: Mixstyle neural networks for domain generalization and adaptation. Int. J. Comput. Vis. **132**(3), 822–836 (2024)
16. Woo, S., Park, J., Lee, J.Y., Kweon, I.S.: Cbam: convolutional block attention module. In: Proceedings of the European Conference on Computer Vision (ECCV), pp. 3–19 (2018)

17. Blanchard, G., Lee, G., Scott, C.: Generalizing from several related classification tasks to a new unlabeled sample. In: Advances in Neural Information Processing Systems (NIPS'11), pp. 2178–2186 (2011)
18. Li, D., Yang, Y., Song, Y., Hospedales, T.: Learning to generalize: meta-learning for domain generalization. In: Proceedings of the AAAI Conference on Artificial Intelligence, pp. 11–25 (2018)
19. Shu, Y., Cao, Z., Wang, C., Wang, J., Long, M.: Open domain generalization with domain-augmented meta-learning. In: Proceedings of the IEEE/CVF Conference on Computer Vision and Pattern Recognition, pp. 9624–9633 (2021)
20. Zhou, K., Yang, Y., Qiao, Y., Xiang, T.: Domain adaptive ensemble learning. IEEE Trans. Image Proc. **30**(4), 8008–8018 (2021)
21. Zhang, H., Cisse, M., Dauphin, Y.N., Lopez-Paz, D.: Mixup: beyond empirical risk minimization. In: International Conference on Learning Representations, pp. 1–13 (2018)
22. Yun, S., Han, D., Oh S.J. , Chun, S., Choe, J., Yoo, Y.: Cutmix: regularization strategy to train strong classifiers with localizable features. In: Proceedings of the IEEE/CVF International Conference on Computer Vision, pp. 6023–6032 (2019)
23. Zhou, K., Yang, Y., Hospedales, T., Xiang, T.: Deep domain-adversarial image generation for domain generalisation. In: Proceedings of the AAAI Conference on Artificial Intelligence, pp. 13025–13032 (2020)
24. Ronen, R., Radu, M., Feuerstein, C., Yom-Tov, E., Ahmadi, M.: Microsoft malware classification challenge. arXiv preprint arXiv:1802.10135 (2018)
25. Chen, Y., et al.: Achieving domain generalization for underwater object detection by domain mixup and contrastive learning. Neurocomputing **528**(7), 20–34 (2023)
26. Jiang, Y., Li, G., Li, S., Guo, Y.: Benchmfc: a benchmark dataset for trustworthy malware family classification under concept drift. Comput. Secur. **139**(8), 103706 (2024)
27. Krueger, D., et al.: Out-of-distribution generalization via risk extrapolation (rex). In: International Conference on Machine Learning, pp. 5815–5826 (2021)

Coordinated Multi-regional Logistics Path Planning: A Broad Reinforcement Learning Framework

Shengwei Li[1], Congcong Zhu[2](✉), and Zeping Tong[1]

[1] Wuhan University of Science and Technology, Wuhan 430065, Hubei, China
jasmine_li@wust.edu.cn
[2] School of Data Science, City University of Macau, Macau, China
cczhu@cityu.edu.mo

Abstract. In the context of coordinated multi-regional logistics, a sophisticated path planning approach is essential for optimizing operational efficiency. The dynamic path planning problem, characterized by its complexity and the need for real-time decision-making, presents a significant challenge. Traditional methods, often relying on heuristics, can fall short in providing the most effective solutions, particularly when faced with large-scale data sets. To address these challenges, we introduce a novel framework that employs broad reinforcement learning for coordinated multi-regional logistics path planning. Our algorithm is fortified by a pre-training phase, which enhances its adaptability across diverse scenarios and enables it to swiftly identify near-optimal solutions even with modest data sets. Through extensive experimentation with enterprise and Solomon data sets, our framework has demonstrated superior performance over established algorithms such as MAVFA, MADQN, and NSGA. While computationally intensive compared to heuristic methods, especially with an increase in order volume, our approach consistently outperforms in scenarios prioritizing optimal solution discovery. The implications of this research extend to a wide array of multi-agent sequential decision-making problems, suggesting potential for future exploration and application of our framework in various coordinated logistics endeavors.

Keywords: Path Planning · Broad Reinforcement Learning · Optimal Solution

1 Introduction

Logistics path planning, integral to supply chain management, encompasses decisions on the most efficient transportation modes, timing, and costs for the transport of goods from their origin to destination. Pivotal to supply chain efficiency [1] and profitability [2], its significance lies in reducing costs, enhancing service quality, mitigating risks [3], and promoting environmental sustainability [4].

T. Zhu et al. (Eds.): ICA3PP 2024, LNCS 15252, pp. 224–237, 2025.
https://doi.org/10.1007/978-981-96-1528-5_15

Effective planning significantly minimizes transportation expenses and bolsters resource utilization [5]; optimized routing streamlines delivery times, thereby elevating customer satisfaction. As computing technology advances, the domain of path optimization has undergone transformative developments. Traditional optimization techniques [6], once confined by computational constraints [7], have evolved with the advent of high-speed networks and deep learning technologies. Contemporary path optimization systems now address complex and dynamic logistical challenges. By leveraging offline training and distributed computational resources [8], these systems assimilate and analyze substantial logistics data in real time, thereby facilitating more precise and efficacious path planning. Nonetheless, an exigent challenge remains: ensuring solution quality while minimizing response times-a critical technological imperative for optimizing system performance.

Usually, the multi-regional dynamic path planning problem can be decomposed into two key sub-problems: how to determine the priority order of transportation tasks [9] and how to dynamically allocate these tasks to transportation resources in different regions [10]. The core challenge is how to dispatch all tasks in the transportation network to appropriate transportation resources in real time, and strive to minimize the total delay and cost of the whole transportation process. The problem of multi-regional dynamic path planning has become a NP-hard problem because it involves extensive geographical distribution and complex task interaction. [11] With the increase of transportation area and the number of tasks, the number of scheduling schemes to be considered increases sharply, which leads to the exponential expansion of search space.

Due to its significance, considerable effort has been channeled into the study of multi-regional dynamic path planning, yielding a multitude of heuristic methods [12]. Algorithms such as NSGA, a heuristic approach, swiftly yield feasible solutions but are constrained in their capacity for global optimization and responsiveness to dynamic shifts. To transcend these limitations, reinforcement learning has been integrated into dynamic path planning to refine path selection through experiential learning. However, current reinforcement learning-based algorithms often grapple with high computational demands and necessitate simplification of complex problem models [13].

Confronting these challenges, we introduce a novel framework termed Broad Reinforcement Learning, tailored to address multi-regional dynamic path planning. This framework distinguishes itself from traditional reinforcement learning by employing adaptive pre-training and expansive strategy learning to more adeptly manage large-scale, multi-regional, and highly dynamic path planning scenarios. It adeptly captures intricate inter-regional interactions and real-time fluctuations, concurrently optimizing the allocation and scheduling of transportation resources with computational efficiency. Empirical results substantiate the efficacy of this innovative approach.

The main contributions of this paper are delineated as follows:

- We propose a multi-region path planning solution grounded in a broad reinforcement learning algorithm. The model undergoes pre-training to fine-tune

its parameters, enabling the algorithm to conduct extensive offline searches through node data, thereby ultimately reducing the total program completion time.
- We define an effective network architecture and iterative methodology designed to encapsulate the nuances of complex inter-regional interactions and real-time fluctuations. This approach circumvents the algorithmic issues of time-consuming or unattainable convergence that may arise from excessively large state spaces.
- Extensive simulations and practical experiments have been conducted. Our algorithm has been compared against the Multi-Agent Value Function Approximation (MAVFA), Multi-Agent Deep Q-Networks (MADQN), and heuristic NSGA algorithms across a spectrum of types and parameters.

The remainder of this paper is organized as follows: Sect. 2 introduces the related work of solving multi-regional path optimization. Section 3 introduces the basic concepts and models of path optimization. Section 4 describes the process of solving the optimal path of Broad reinforcement Learning. Section 5 provides experimental results and comparative analyses. Finally, Sect. 6 summarizes the contributions and conclusions of this paper.

2 Related Works

Research into wage determination within multi-logistics center systems [14], typically falls into two primary methodologies: (1) wage publication, where the intermediary platform sets wages; and (2) wage negotiation, where wages are established through a collective bidding process among workers. In both scenarios, there exists a policy transparency yet income uncertainty, potentially leading workers to exploit the policy to mitigate personal risk, to the detriment of the system's overall performance. Consequently, the optimization of wage and task allocation has become a pivotal issue, with the discovery of a unified and publishable criterion being a central concern.

Previous scholarly work has delved into the online matching of controllable rewards alongside competition probabilities for resources that are either reusable or non-reusable, as well as the co-optimization of wages and resource-user matching facilitated by matching platforms. This bears resemblance to our investigation, which also initiates wage determination from the distribution process. However, our system incorporates temporal dynamics, eschewing a discrete time division, thereby ensuring the platform's long-term benefits remain consistent.

Furthermore, certain studies have crafted mechanisms for wage determination and task allocation on crowdsourcing platforms predicated on advance inquiry [15]. Such mechanisms enable workers to accurately convey task costs, facilitating the allocation of wages and tasks. Yet, the practice of advance inquiry does not align with real-world conditions [16]. We employ reinforcement learning to preemptively condition the environment [17], effectively circumventing this discrepancy and aligning more closely with actual circumstances to fulfill operational requirements.

Given that the Vehicle Routing Problem (VRP) [18]is recognized as an NP-hard problem, all its variations are considered to belong to the same class of challenges. As a result, the academic community has often turned to the creation of metaheuristics to address VRPs. [19] In light of the traditional Vehicle Routing Problem with Order Delivery (VRPOD) [20], Table 1 presents a comparative analysis of recent research trends, highlighting both the congruencies and divergences with the research presented in this paper.

3 Path Planning Problem

In the context of multi-regional dynamic path planning, we encounter a network of regions, numbered by $|I|$, each necessitating systematic monitoring or service delivery. Each region is conceptualized as an autonomous agent capable of sophisticated decision-making processes. Specifically, region i encompasses $|Ii|$ operational units that are tasked with covering $|Ji|$ designated operational zones within its territory, with operations extending over $|T|$ time intervals. At the commencement of operations, a pre-established itinerary assigns a specific route and time allocation to each agent.

3.1 Scheduling

Each delivery schedule is intricately designed to outline the sequence of delivery areas to be serviced (routing) and the precise temporal windows for these services (timing). This is akin to a comprehensive academic timetable, with the essential addition of transit time considerations between service areas. At the onset of operations, each fleet is endowed with an initial delivery schedule, denoted as $\delta_i(0)$. We presuppose that an initial collective schedule is pre-established and independently computed.

3.2 Order Dynamics

A new order, denoted as ω_k, emerges at decision epoch k and is characterized by a tuple: $\langle \omega_i^k, \omega_j^k, \omega_t^k, \omega_s^k \rangle$. In this context, ω_i^k signifies the region from which the order originates, $\omega_j^k \in J_{\omega_i^k}$ pinpoints the customer's location, $\omega_t^k \in T$ marks the time at which the order is received, and ω_s^k indicates the service time required for order fulfillment.

3.3 The Existence and Effectiveness of Rewards

Empirical research shows that the obvious existence of rewards can increase employees' satisfaction and trust. We define that the reward exists as a function of the cumulative effective time period allocated to each distribution area service during distribution. Each area j needs at least a time period of service. We introduce an existential utility function for distribution orders, labeled $R_p(j)$, in which the utility of the extra service time exceeding the requirement of the longest response time decreases exponentially as a parameter β_p(as shown in Eq. 1). This reflects the diminishing marginal utility of response time exceeding the minimum requirements in improving customer perception. The σ_j represents the cumulative rewards of all teams in the distribution area J in a given cross-regional distribution plan $\delta(k)$.

$$R_p(j) = \min\left(\sigma_j, Q_j\right) + 1_A \times \sum_{i=1}^{\sigma_j - Q_j} i \times e^{-\beta_p i}$$
$$\text{where } \mathrm{t}_A = \begin{cases} 0, \sigma_j - Q_j \leq 0 \\ 1, \sigma_j - Q_j > 0 \end{cases} \tag{1}$$

We introduce a performance index $f_d(\delta(k))$ to evaluate the effectiveness of a given distribution plan $\delta(k)$ in ensuring the existence of services.

$$f_d(\delta) = \frac{\sum_{j \in J} R_p(j)}{|T| \times |I|} \tag{2}$$

3.4 Response Time

The time for processing an order in the decision-making period k, called, refers to the sum of the time interval $(x_k^t - \omega_k^t)$ between the appearance of an order and the completion of its execution and the transit time from its current position to the accident coordinates. When $\tau_k \leq \tau_{\text{target}}$, it is recognized as a successful response. It is stipulated that all scheduled calls should be responded within τ_{max}.

$$\tau_k = \left(x_k^t - \omega_k^t\right) + d\left(j_{t'}^{x_k^m}, \omega_k^j\right)$$
$$\text{where } x_k^t - \omega_k^t \leq \tau_{\max}, t' = x_k^t \tag{3}$$

3.5 Objective Function

At each order input node k, the goal is to determine an optimal path x_k^*, which can not only obtain immediate benefits, but also explain the expected future benefits brought by upcoming dynamic events. This is encapsulated by the estimate function and expressed as $\hat{V}\left(S_k^x\right)$.

$$x_k^* = \underset{x_k \in X(S_k)}{\operatorname{argmax}} \left\{R\left(S_k, x\right) + \gamma \hat{V}\left(S_k^x\right)\right\} \tag{4}$$

4 Methodology

Upon examination of the aforementioned issues, we proceed under the assumption that the collective value function can be decomposed into constituent agent-specific value functions. Leveraging this premise, we have architected a framework aimed at addressing the issue at hand. The framework operates on the principle that each region acts as an autonomous agent, and upon the introduction of an order into the system, it is disseminated accordingly. Should one agent receive an order along with its associated reward, a corresponding decrement is incurred by the other agents. For a scenario involving $|N|$ orders, this results in $|N|$ distinct selection mechanisms. Upon attaining an optimal configuration for all entities, they are collectively fed into a mixed network layer designed to resolve the route valuation. Subsequently, these inputs are directed towards a Broad learning network, which refines the nodal activations and their associated weights, alongside the intrinsic values, and propels them towards the output layer. Throughout the Broad learning network, there is an integration of manually derived regional sector attributes alongside parameter settings that have been honed through pre-training (Fig. 1).

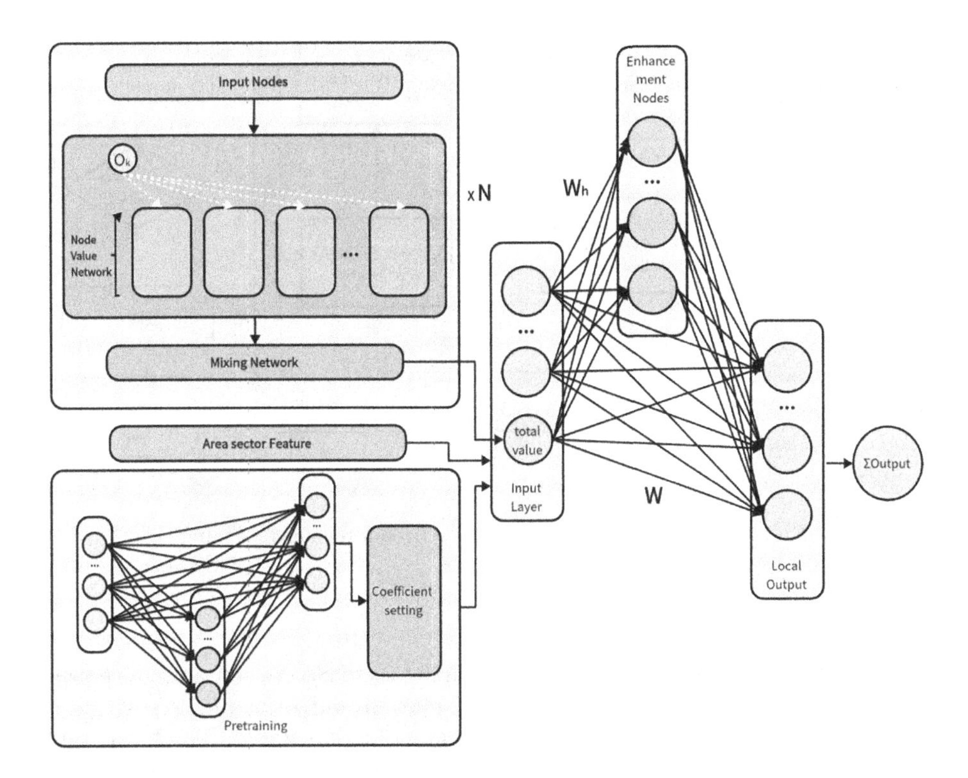

Fig. 1. A Broad Reinforcement Learning Framework.

4.1 Reinforcement Learning

Utilizing a temporal sequence diagram [21], we devise the reinforcement learning model's state and reward update mechanisms. Here, the initial node encapsulates the starting condition at the planning phase's onset. Descendant nodes delineate potential subsequent states accessible post the execution of subsequent actions, indicating possible allocation maneuvers. This study presents an innovative framework of extensive reinforcement learning, segmenting the path planning endeavor into quartile phases in accordance with distributed operations, as illustrated in Fig. 2.

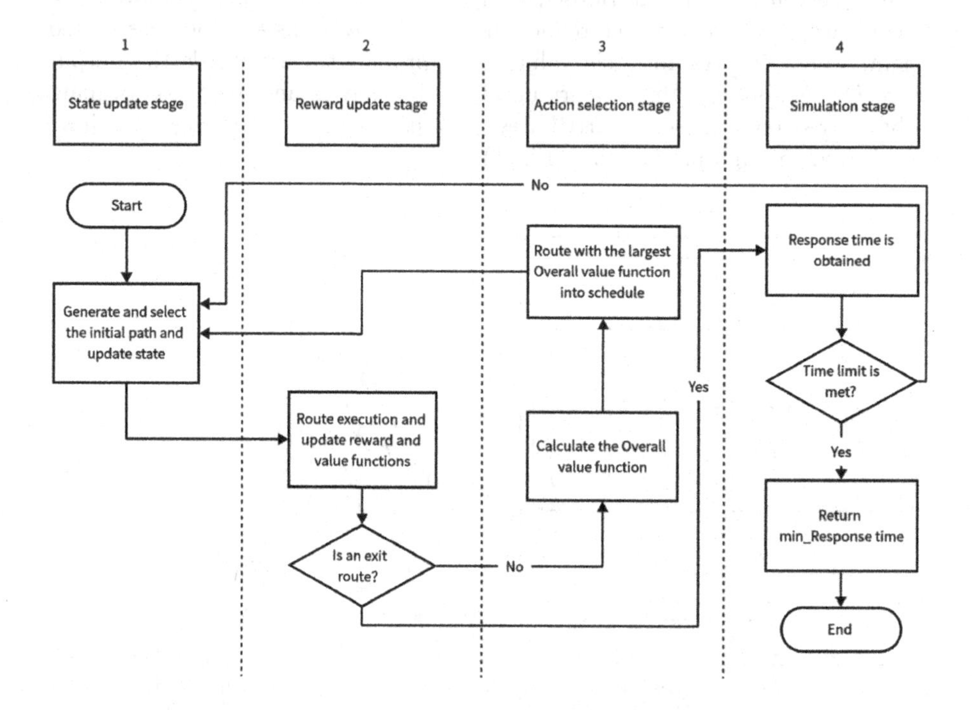

Fig. 2. Path update process in broad reinforcement learning.

State. planning state S_k and the post-planning state S_k^x. S_k contains the essential information necessary for making distribution and re-planning path decisions. States can be delineated into local and global categories. The local state pertains to information specific to an individual agent, whereas the global state encompasses information shared among agents, potentially facilitating coordination. Each local post-planning state, $S_{k,i}$ is articulated as the following tuple:

$$\left\langle \delta_i(k), \frac{\sigma_i(k)}{Q_j}, f_{util}\left(\delta_i(k)\right) \right\rangle \tag{5}$$

Action. X_k refers to the action of assigning a region/agent to send a fleet to execute a group of orders during the decision-making period, and updating the joint path planning table of all agents. X_k represented as the following tuples: $x_k^i, x_k^m, x_k^t, \gamma'(k)\rangle$ where $x_k^i \in I$ is the region/agent assigned to respond to the order. $x_k^m \in I_{x_k^i}$ is the regional motorcade assigned to the corresponding order, $x_k^t \in T$ is the time period to start responding to the delivery order after the task is assigned, and not the result of the joint path planning table obtained after the action X_k is carried out.

5 Experiment

We evaluated our framework in an actual scenario, which comes from the data of express delivery enterprises in the city where we live. There are $|j|=17$ distribution sectors in total. According to the actual situation, the number of crossing areas for any order does not exceed 4, and the response time and value function of the algorithm will show different complexity with the change of the total number of orders. Based on the particularity of the problem, the data of this experiment are generated by sharing data from some enterprises and public Solomon data sets.

5.1 Experimental Setup and Design

We divide the experiment into two stages, and verify the effectiveness of the framework through comparative analysis in many aspects. In order to make a fairer comparison, the parameters of unrelated variables such as rewards in the same group of algorithms are the same.

The following is the framework involved in this section

- BRL. This is the framework we put forward, in which B refers to breadth learning and RL refers to reinforcement learning.
- MAVFA. [22] This is a multi-agent value approximation method using explicit communication mechanism and heuristic algorithm.
- MADQN. [23] This is a framework method using depth Q-learning based on two-stage method.
- NSGA. [24] This is a multi-objective heuristic algorithm using non-dominated sorting and fitness sharing strategy.

Phase1. We assess the efficacy of our framework within the spectrum of optimal and suboptimal solution scenarios, alongside two contemporary and efficacious algorithms that leverage a reinforcement learning framework predicated on depth-based strategies.

Our evaluation encompasses a comprehensive suite of 800 training simulations, each of which is tailored with a unique configuration of initial and terminal order coordinates, alongside an array of dynamic events that emerge within a predefined spatial domain. The performance of our approach is quantified by the aggregate success rate across all orders-reflecting the proportion of dynamic orders addressed within the allocated timeframe-and the success rate within the most challenging agents. In terms of response latency, our evaluation metrics focus on the temporal expenditure for each decision-making instance, encompassing both route formulation and the updating of the value function.

Given the stochastic nature of the queries and the initialization of pre-training parameters, we present our findings as the mean outcome and its corresponding 95% confidence interval, thereby providing a robust statistical representation of our results.

Phase2. Regarding the response time, we use the experimental results of the latest and most effective framework to analyze and compare with the experimental results of large and small examples. 30–100 orders are used for evaluation in small examples, and 300–800 orders are used for comparative analysis in large examples. Two widely recognized metrics are employed to evaluate the framework's performance across various dimensions, and the framework is subjected to a side-by-side comparison with numerous well-established algorithms.

5.2 Performance Metrics

To evaluate the performance of path planning algorithms, we employed two metrics.

Response Time Ratio. The complexity of order conditions and the original routes can significantly vary the impact of different orders on the overall response time, thereby complicating the effective comparison of algorithmic performance. To address this challenge, we introduce the concept of the response time ratio. The response time ratio is defined as the ratio of the actual response time to the ideal response time derived from a predetermined route.

$$\text{Response time ratio} = \frac{\sum_{n_i \in CP_{\text{MIN}}} \min_{p_j \in Q} \{w_{i,j}\} - \text{makespan}}{\text{makespan}}. \tag{6}$$

The numerator of the response time ratio index represents the actual value minus the cumulative value of the minimum response time of all critical path tasks, that is, the distance from the response time to the ideal minimum scheduling time. Therefore, the response time ratio after regularization is always less than 1, because the scheduling time is usually longer than the ideal limit. Compared with the larger response time ratio, the smaller response time ratio indicates better algorithm performance.

Efficiency. The efficiency of parallel processing is gauged by the speedup factor, which is calculated by comparing the aggregate time expenditure of executing tasks sequentially to the overhead associated with employing parallel execution techniques.

$$\text{Speedup} = \frac{\min_{p_j \in Q} \left\{\sum_{n_i \in V} w_{i,j}\right\}}{\text{makespan}} \tag{7}$$

The calculation of the speedup index involves taking the minimum response time required for processing all routes sequentially on a solitary processor and dividing it by the response time achieved through the algorithm's application. An elevated acceleration metric indicates superior parallel execution capabilities of the scheduling mechanism. Efficiency is quantified by the quotient of the achieved speedup and the count of processors engaged (denoted as p^*). This metric mirrors the extent of processor exploitation within the scheduling paradigm, with a more substantial efficiency figure denoting a heightened level of processor engagement.

$$\text{Efficiency} = \text{Speedup} / p^* \tag{8}$$

5.3 Experimental Results and Discussion

Phase1. When contrasted with three prevailing solution architectures, our approach yields statistically significant decisions that enhance both the aggregate success rate and the performance rate of the least successful agent, as detailed in Table 1. Our framework's overall success rate surpasses that of MADQN-the current premier framework-showing an improvement of 2.523% on this metric. This improvement translates to an additional 20 orders per round within the context of an 800-order scenario (which correlates to approximately 400 units of product, as per enterprise data). While there exists the potential for enhancing convergence velocity when matched against heuristic methods, the overall rate of convergence coupled with the solution quality has witnessed marked enhancement.

Table 1. Performance table of frame algorithm on optimal solution and worst solution.

Model	Success Rate	Time Per Decision(s)
BRL	**(O) 65.0 ± 0.4% (W) 52.4 ± 0.8%**	**23.4 ± 2.0**
MAVFA	(O) 62.2 ± 0.6% (W) 51.8 ± 1.0%	26.5 ± 1.6
MADQN	(O) 63.4 ± 0.7% (W) 50.7 ± 0.2%	25.6 ± 2.9
NSGA	(O) 49.4 ± 0.5% (W) 31.7 ± 0.2%	20.6 ± 6.2

Phase2. Our method achieves stability after approximately 500 training sessions, with the number of iterations being directly proportional to the execution duration. We evaluate the efficiency of our framework relative to existing, widely-used frameworks in terms of response time ratio and overall efficiency.

As depicted in Fig. 3, a comparison of large and small instances reveals that our framework exhibits more pronounced advantages under conditions of smaller order volumes. This advantage leverages the principle of horizontal data enhancement inherent in broad learning. In scenarios with constrained data, our framework facilitates more expeditious responses than deep learning methods, which rely on multi-level complex networks and substantial data volumes, and it also delivers superior solution quality. When compared to the heuristic algorithm NSGA, which has a similar response time ratio, our framework demonstrates superior performance in solution quality.

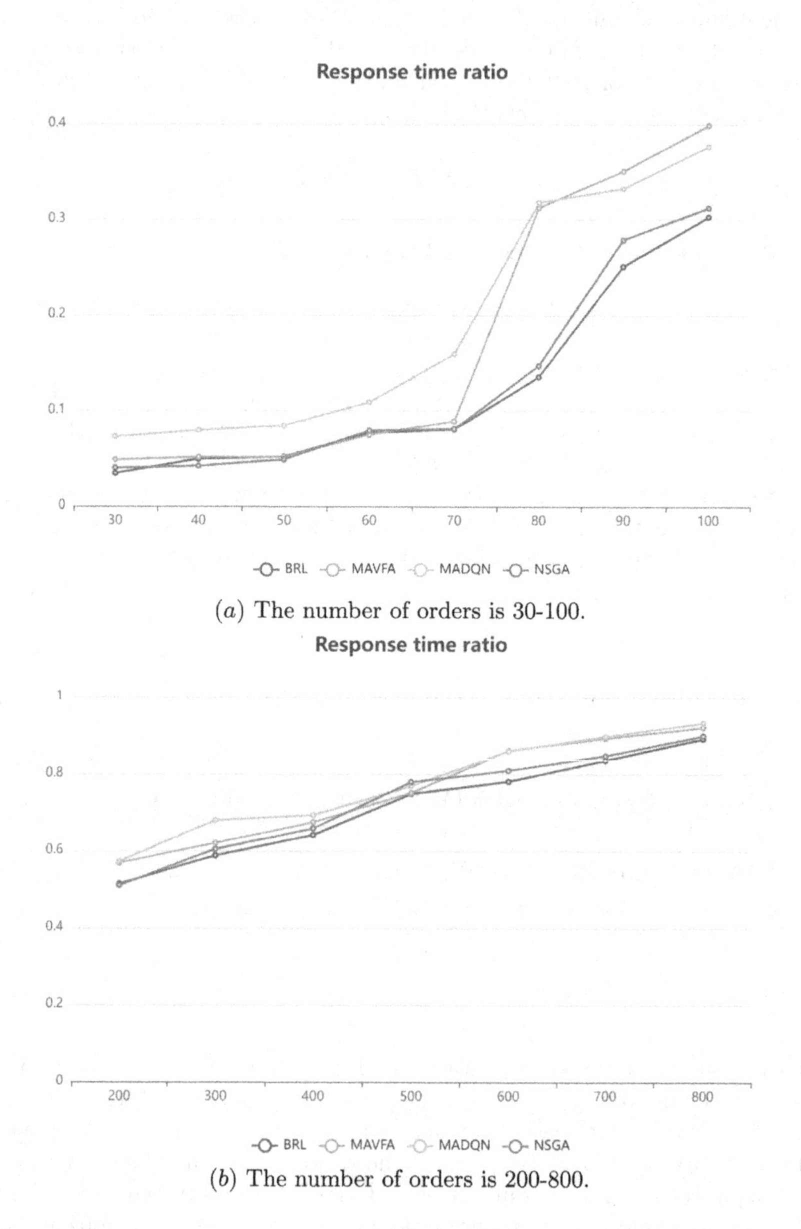

(a) The number of orders is 30-100.

(b) The number of orders is 200-800.

Fig. 3. The correlation between the response time ratio and the order number.

Figure 4 illustrates that as the number of processors increases, the efficiency values of all algorithms tend to decline gradually. Nonetheless, for any specified number of processors, the BRL framework consistently outperforms other frameworks in terms of efficiency. For instance, with four processors, the efficiency of BRL is recorded at 0.949, contrasting with the efficiency values of MAVFA, MADQN, and NSGA, which stand at 0.893, 0.869, and 0.831, respectively. Consequently, the BRL framework realizes an efficiency enhancement of 6.270%.

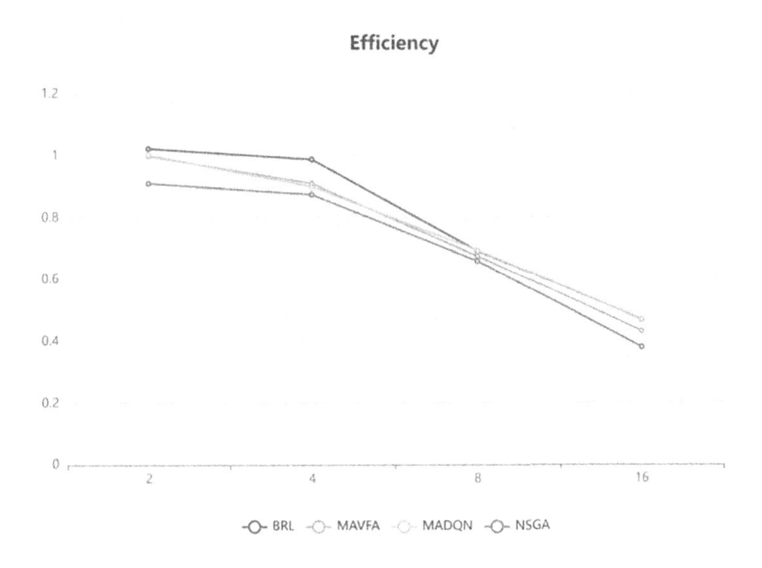

Fig. 4. The correlation between the efficiency and the processor number.

6 Conclusion

We poposed a broad reinforcement learning framework adept at tackling the complexities of multi-regional dynamic path planning, a quintessential challenge within the realm of multi-agent sequential pathfinding with intricate action schemes. Our algorithm is underpinned by a Broad learning framework that undergoes pre-training, thereby bolstering its overall adaptability. Concurrently, it manages to expedite response times and secures optimal solutions, particularly within the confines of modest data sets. This strategy is specifically crafted to mitigate the issue of non-convergence due to vast search spaces. Empirical evidence drawn from enterprise-level data and the publicly available Solomon data sets corroborates the superiority of our framework over prevalent routing algorithms such as MAVFA, MADQN, and NSGA. Nonetheless, when compared to heuristic approaches, our algorithm requires a more substantial execution timeframe, with search complexity increasing exponentially as order-related data grows. Despite this, it demonstrates superior performance in scenarios requiring the identification of optimal solutions.

Prospectively, ample avenues exist for the further assessment and assimilation of the concepts introduced herein. Our methodology holds promise for the evaluation of diverse multi-agent sequential decision-making dilemmas, necessitating the investigation of network architectures tailored to specific environmental contexts. Engaging in a comparative analysis of our approach with established cooperative Multi-Agent Reinforcement Learning (MARL) methods presents an intriguing prospect.

References

1. Di Puglia Pugliese, L., Ferone, D., Festa, P., Guerriero, F., Macrina, G.: Solution approaches for the vehicle routing problem with occasional drivers and time windows. Optim. Methods Softw. **37**(4), 1384–1414 (2022)
2. Feng, L., et al.: Solving generalized vehicle routing problem with occasional drivers via evolutionary multitasking. IEEE Trans. Cybern. **51**(6), 3171–3184 (2019)
3. Yıldız, B.: Express package routing problem with occasional couriers. Transport. Res. Part C: Emerg. Technol. **123**, 102994 (2021)
4. Macrina, G., Pugliese, L.D.P., Guerriero, F., Laporte, G.: Crowd-shipping with time windows and transshipment nodes. Comput. Oper. Res. **113**, 104806 (2020)
5. Foerster, J., Farquhar, G., Afouras, T., Nardelli, N., Whiteson, S.: Counterfactual multi-agent policy gradients. In: Proceedings of the AAAI Conference on Artificial Intelligence, vol. 32 (2018)
6. Ma, Y., et al.: A hierarchical reinforcement learning based optimization framework for large-scale dynamic pickup and delivery problems. Adv. Neural. Inf. Process. Syst. **34**, 23609–23620 (2021)
7. Dahle, L., Andersson, H., Christiansen, M., Speranza, M.G.: The pickup and delivery problem with time windows and occasional drivers. Comput. Oper. Res. **109**, 122–133 (2019)
8. Los, J., Schulte, F., Gansterer, M., Hartl, R.F., Spaan, M.T.J., Negenborn, R.R.: Decentralized combinatorial auctions for dynamic and large-scale collaborative vehicle routing. In: Lalla-Ruiz, E., Mes, M., Voß, S. (eds.) ICCL 2020. LNCS, vol. 12433, pp. 215–230. Springer, Cham (2020). https://doi.org/10.1007/978-3-030-59747-4_14
9. Lei, K., et al.: A multi-action deep reinforcement learning framework for flexible job-shop scheduling problem. Expert Syst. Appl. **205**, 117796 (2022)
10. Defersha, F.M., Rooyani, D.: An efficient two-stage genetic algorithm for a flexible job-shop scheduling problem with sequence dependent attached/detached setup, machine release date and lag-time. Comput. Ind. Eng. **147**, 106605 (2020)
11. Zhu, C., Ye, D., Zhu, T., Zhou, W.: Time-optimal and privacy preserving route planning for carpool policy. World Wide Web **25**(3), 1151–1168 (2022)
12. Vincent, F.Y., Jodiawan, P., Redi, A.P.: Crowd-shipping problem with time windows, transshipment nodes, and delivery options. Transport. Res. Part E: Logist. Transport. Rev. **157**, 102545 (2022)
13. Zhu, C., Ye, D., Huo, H., Zhou, W., Zhu, T.: A location-based advising method in teacher-student frameworks. Knowl.-Based Syst. **285**, 111333 (2024)

14. Hikima, Y., Akagi, Y., Kim, H., Asami, T.: An improved approximation algorithm for wage determination and online task allocation in crowd-sourcing. In: Proceedings of the AAAI Conference on Artificial Intelligence, vol. 37, pp. 3977–3986 (2023)
15. Anari, N., Goel, G., Nikzad, A.: Mechanism design for crowdsourcing: an optimal 1-1/e competitive budget-feasible mechanism for large markets. In: 2014 IEEE 55th Annual Symposium on Foundations of Computer Science, pp. 266–275. IEEE (2014)
16. Zhu, C., Ye, D., Zhu, T., Zhou, W.: Location-based real-time updated advising method for traffic signal control. IEEE Internet Things J. (2023)
17. Chen, Y., et al.: Can sophisticated dispatching strategy acquired by reinforcement learning?-a case study in dynamic courier dispatching system. arXiv preprint arXiv:1903.02716 (2019)
18. Vincent, F.Y., Aloina, G., Jodiawan, P., Gunawan, A., Huang, T.C.: The vehicle routing problem with simultaneous pickup and delivery and occasional drivers. Expert Syst. Appl. **214**, 119118 (2023)
19. Zhu, C., Cheng, Z., Ye, D., Hussain, F.K., Zhu, T., Zhou, W.: Time-driven and privacy-preserving navigation model for vehicle-to-vehicle communication systems. IEEE Trans. Veh. Technol. (2023)
20. Archetti, C., Guerriero, F., Macrina, G.: The online vehicle routing problem with occasional drivers. Comput. Oper. Res. **127**, 105144 (2021)
21. Shu, L., et al.: Smart dag task scheduling based on mcts method of multi-strategy learning. In: Tari, Z., Li, K., Wu, H. (eds.) Algorithms and Architectures for Parallel Processing, pp. 224–242. Springer, Singapore (2024). https://doi.org/10.1007/978-981-97-0834-5_14
22. Joe, W., Lau, H.C.: Learning to send reinforcements: coordinating multi-agent dynamic police patrol dispatching and rescheduling via reinforcement learning. In: Elkind, E. (ed.) Proceedings of the Thirty-Second International Joint Conference on Artificial Intelligence, IJCAI-23, pp. 153–161. International Joint Conferences on Artificial Intelligence Organization (2023). https://doi.org/10.24963/ijcai.2023/18
23. Chen, X., Ulmer, M.W., Thomas, B.W.: Deep q-learning for same-day delivery with vehicles and drones. Eur. J. Oper. Res. **298**(3), 939–952 (2022)
24. Duan, P., Yu, Z., Gao, K., Meng, L., Han, Y., Ye, F.: Solving the multi-objective path planning problem for mobile robot using an improved NSGA-II algorithm. Swarm Evol. Comput. **87**, 101576 (2024)

Path Optimization Method Under UAV Charging Scheduling Network

Tingting Yang[1], Yiqian Wang[1], Jie Zhu[1,2(✉)], Shuyu Chang[1], and Haiping Huang[1,3(✉)]

[1] Nanjing University of Posts and Telecommunications, Nanjing 210003, China
[2] State Key Laboratory Chinese of Computer Architecture, Institute of Computing Technology, Academy of Sciences, Beijing 100864, China
zhujie@njupt.edu.cn
[3] Jiangsu High Technology Research, Key Laboratory for Wireless Sensor Networks, Nanjing 210003, China
hhp@njupt.edu.cn

Abstract. Unmanned Aerial Vehicles (UAVs) have emerged as integral components in logistics systems, where their potential for efficient delivery services is being explored. However, the limited battery capacity of UAVs poses a significant challenge for long-distance delivery applications. In this paper, we investigate a hybrid problem involving UAV trajectory planning and charging scheduling for long-distance logistics systems with a charging network. In the problem under study, multiple UAVs are assigned long-distance delivery tasks. A UAV can stop at any charging station in the charging network to charge if it is in a low-energy state. An Ant Colony Optimization-based UAV Trajectory Planning and Charging Scheduling (ACO-TPCS) algorithm is proposed to minimize task completion time by optimizing UAV trajectory and charging plans. The main framework of the ACO-TPCS algorithm consists of several key components, including reachable graph construction, candidate flight path generation, pheromone matrix construction, ant generation, feasible solution generation, and pheromone update method. All these components are delicately designed. Through extensive experimentation and comparison with baseline algorithms, we demonstrate the effectiveness of the ACO-TPCS algorithm in addressing the problem under study.

Keywords: UAVs · trajectory planning · charging scheduling · charging network · Ant Colony Optimization (ACO)

1 Introduction

Unmanned Aerial Vehicles (UAV) have been widely adopted in many fields, such as public safety [1], forestry [2], and agriculture [3]. There are many advantages

T. Yang and Y. Wang—Co-first authors.

T. Zhu et al. (Eds.): ICA3PP 2024, LNCS 15252, pp. 238–253, 2025.
https://doi.org/10.1007/978-981-96-1528-5_16

to employing UAVs. For example, they can respond quickly in many real-time applications, they can operate with clean energy and the traffic congestion does not affect them. Due to these advantages, UAV delivery has attracted a lot of attention in the logistics industry, and more and more logistics enterprises have been pre-researching UAV-based delivery applications.

However, the limited battery capacity of UAVs severely limits some applications that require long-distance flying. Without battery recharging, a UAV can conduct continuous flight for at most 3km to 33km [4] depending on the battery type. To implement long-distance delivery services, a UAV under a low energy level should be recharged at quick battery-charging platforms during the path to its destination. Multiple delivery stations equipped with quick battery-charging platforms can form a charging network to provide long-distance delivery services.

However, there are two major challenges for the logistics system with charging network to be fully operational. (i) Trajectory plan is critical to the system under study. Each UAV has a limited battery capacity. When and where to recharge should be determined before a UAV runs out of energy. It is possible that a UAV needs to be recharged multiple times. Therefore, a delicately designed trajectory plan with the shortest path can greatly reduce the completion time of tasks. (ii) Some stations may be on the key positions of the charging network. Severe congestion may occur when multiple UAVs visit these charging stations, which may lead to long waiting time inside the stations. Therefore, charging schedule in each station is the key problem to reduce the waiting time of UAVs, which will further lead to less completion time of tasks.

In this paper, we investigate the trajectory planning and the charging scheduling problem in the UAV-enabled long-distance delivery system with the charging network. In the considered problem, the delivery stations equipped with charging platforms are distributed in a given area. There are many long-distance delivery tasks with different original and destination stations. Each task is assigned to a UAV. We need to determine the trajectory plan for each UAV. For UAVs that are assigned to the same stations to recharge, the charging scheduling plan should also be determined at each station. The paper aims to obtain feasible and optimal solutions for all tasks with the objective of minimal completion time for all tasks.

The main contributions of this paper are summarized as follows:

1) A trajectory planning and charging scheduling problem with a charging network is modeled considering the energy consumption of UAVs and the multi-tasking characteristics of the system.
2) We have designed an algorithm for alleviating congestion at charging platforms. The algorithm addresses the challenge of minimizing the waiting time of UAVs at key charging stations, thereby optimizing the overall efficiency of the delivery system.
3) An improved shortest path algorithm is introduced to solve the problem of trajectory planning at the bottom of the system. Taking into account factors such as limited battery capacity and the need for multiple recharges, the method specializes in trajectory planning for each UAV.

4) An improved algorithmic framework for Ant Colony Optimization (ACO) is proposed. The framework is a higher-level approach for coordinating and optimizing the trajectory planning and charging scheduling processes across the delivery network. By leveraging the collective intelligence of artificial ants, the framework aims to efficiently explore and utilize the solution space, ultimately achieving optimal solutions and minimizing task completion time.

The rest of the paper is organized as follows. A review of existing related work on UAV applications is presented in Sect. 2. Section 3 describes the problem. In Sect. 4, the proposed algorithm for the problem is presented. To show the effectiveness and efficiency of our proposal, the comparison with other baseline algorithms is conducted in Sect. 5, followed by the conclusions in Sect. 6.

2 Related Work

The related work for the problem under study can be classified into three categories: (1) distribution tools, (2) UAV delivery path problem and (3) charging facilities.

Regarding distribution tools, many scholars have chosen to use trucks and drones for coordinated delivery. For example, Cho et al. [5] established accurate electric van and UAV power models and battery models. Then, a heuristic delivery scheduling algorithm was proposed to determine the delivery scheduling of electric vans and UAVs, and it was demonstrated that the method can successfully reduce the delivery cost. A novel collaborative Pareto ACO optimization algorithm was proposed by Das et al. [6] considering a vehicle routing problem with time windows and synchronized UAVs. Independent delivery with UAVs is another option. Chen et al. [7] explored solutions to enable UAVs to fly autonomously in mixed indoor and outdoor environments. They came out with a novel Internet of Things (IoT)-based drone delivery system that can avoid problems associated with drone collisions, and through simulation, the system can achieve a high user-defined flight success rate. Park et al. [8] designed a multi-UAV cooperative algorithm based on a novel communication network (CommNet) to achieve a robust and reliable multi-UAV autonomous aerial parcel delivery service, to maximize the total number of parcels delivered and the efficient use of energy.

Many scholars have studied the UAV delivery path problem, in which the UAV trajectory planning problem is crucial for optimizing the flight path. Huang et al. [9] use UAVs for last-mile parcel delivery. Consider using drone stations to change or charge batteries for drones. They are focusing on flight planning issues intending to minimize the total flight time from the warehouse to the customer. The problem studied by Huang et al. [10] is the round-trip routing problem, i.e., finding the shortest path in time in a time-dependent network under delivery deadline constraints and energy budget constraints. Du et al. [11] studied the influence of wind speed and direction on the flight state of UAVs, established the relevant parameters affected by wind conditions and their resolution methods, and deeply studied the logistics UAV path planning problem.

To meet the demand for UAVs to deliver goods over long distances, some academics have considered establishing charging facilities to extend the range of UAV deliveries. Fixed charging stations, are fixed in location and are permanent facilities that provide high-power charging and can fully charge drones in a short period. Li et al. [12] introduced a wireless static charger (WSC) in a UAV network and proposed an interruption-free wireless rechargeable UAV network (WRUN) model in which UAVs can be charged wirelessly without returning to the charging platform, and the WSC is timed to turn on and release energy only during the charging time period. The goal is to minimize energy wastage in the WSC without the UAV running out of energy. Fan et al. [13] investigated the UAV routing problem in the presence of multiple charging stations (URPMCS) and proposed a deep reinforcement learning-based approach. The objective is to minimize the total distance flown by UAVs during traffic monitoring. Qin et al. [14] analyses wireless charging infrastructure sharing strategies in UAV and EV networks. They considered a scenario where UAVs charge at an EV charging station and pay a shared fee.

Although there have been some studies on UAV trajectory planning and charging scheduling, there are still insufficient studies on integrating these two aspects to optimize the overall logistics system. We propose an ant colony optimization-based UAV trajectory planning and charging scheduling algorithm (ACO-TPCS), which aims to solve the trajectory planning and charging scheduling problems simultaneously in order to minimize the task completion time.

3 Problem Description

Table 1 gives the notations defined for the problem description.

In the considered problem, there are m delivery stations distributed in a given area. The location of these stations are denoted as $L_1, \ldots, L_m$. There are n packages distributed among these stations. The delivery of a package is taken as a task. A task t_i can be described by a 3-tuple $< l_i, d_i, w_i >$, where l_i is the original station, d_i is the destination station and w_i is the package weight. We have $l_i, d_i \in \{L_j | j = 1, \ldots, m\}$. Each task t_i is assigned to a specific UAV u_i $(i = 1, \ldots, n)$, and u_i is ready at station l_i.

The average flying speed of u_i is set to v_i. Each UAV is fully charged at its original station. According to the study in [15], the energy consumption model of UAVs involves two factors: the battery capacity and the average amperage draw. Usually, if the battery of a UAV is under a very low level, e.g., 20%, it is likely to cause permanent damage. Therefore, we define that the maximum available capacity of the battery is E_{max} which is about 80% of the actual battery capacity. Suppose the residual energy of UAV at time t is $R_i(t)$, we have $R_i(0) = E_{max}$. The average amperage draw A_i defines the average energy consumption of UAV u_i for each time unit. It mainly depends on the weight of the UAV and the carried package. We define that the motor ability of a UAV draws α amp in order to produce ϖ grams of thrust. In the paper, we assume the weights and motor abilities of all UAVs are the same. The weight of a UAV is w_{uav}. Then A_i can be computed by Eq. (1).

Table 1. Notations.

Notation	Definition
m	Number of delivery stations
L_j	Location of the jth station
n	Number of delivery tasks
t_i	i^{th} task
w_i	Package weight of t_i
l_i	Initial location of t_i
d_i	Destination of t_i
u_i	UAV that u_i is assigned to
v_i	Average flying speed of u_i
w_{uav}	Weight of a UAV
A_i	Average energy consumption of u_i for each time unit
α, ϖ	Parameters for computing the average amperage draw
E_{max}	Maximum available capacity of battery
$R_i(t)$	Residual energy of u_i at time t
$\Delta E(\Delta t)$	Energy consumption during time Δt
K	Number of chargers in a charging station
β	Gained energy per time unit on a charging platform
$\Delta G(\Delta t)$	Energy gained during time Δt
$f_i = \{f_{i,1}, f_{i,2}, \ldots, f_{i,N_i}\}$	Flight plan of u_i
$f_{i,h}$	h^{th}Sub-flight plan of f_i
N_i	Sub-flight number of f_i
$l_{i,h-1}$	Start station of sub-flight $f_{i,h}$
$l_{i,h}$	End station of sub-flight $f_{i,h}$
$D(l, l')$	Distance between locations l and l'
$a_{i,h}$	Arrive time of u_i at $l_{i,h}$
$ch_{i,h}$	Charger assigned to u_i at $l_{i,h}$
$s_{i,h}$	Start charging time of u_i at $l_{i,h}$
$c_{i,h}$	Charging time of u_i at $l_{i,h}$
$e_{i,h}$	End charging time of u_i at $l_{i,h}$
$S = \{f_1, f_2, \ldots, f_n\}$	Schedule
$T(S)$	Completion time of S

$$A_i = \frac{(w_i + w_{uav}) \times \alpha}{\varpi} \tag{1}$$

If a UAV starts to fly at time t and the duration of the continuous flight is Δt, then the energy consumption $\Delta E(\Delta t)$ is defined by Eq. (2). The residual energy after the flight is defined by Eq. (3).

$$\Delta E(\Delta t) = A_i \times \Delta t \tag{2}$$

$$R(t + \Delta t) = R(t) - \Delta E(\Delta t) \geq 0 \tag{3}$$

Since the energy capacity of UAVs are limited, it is necessary to provide charging platforms for them. There is a quick battery-charging platform provisioned in each delivery station. For each charging platform, K chargers are installed, which means at most K UAVs can be simultaneously charged on each platform. When a UAV is under a low battery level, it should fly to a nearby station and recharge on the corresponding charging platform. The gained energy per time unit on the charging platform is β. If a UAV starts to recharge at time t and the charging duration is Δt, then the gained energy $\Delta G(\Delta t)$ is defined by Eq. (4). The residual energy after charging is defined by Eq. (5).

$$\Delta G(\Delta t) = \beta \times \Delta t \tag{4}$$

$$R(t + \Delta t) = R(t) + \Delta G(\Delta t) \leq E_{max} \tag{5}$$

In order to accomplish task t_i, UAV u_i either flies directly to the destination station or visits some stations in the middle of the flight to recharge. The flight plan f_i of u_i can be represented as a sequence of sub-flight plan $f_{i,h}$, i.e., $f_i = (f_{i,1}, f_{i,2}, \cdots, f_{i,N_i})$. N_i is the number of sub-flights. A sub flight plan $f_{i,h}$ is a 7-tuple $< l_{i,h-1}, l_{i,h}, a_{i,h}, ch_{i,h}, s_{i,h}, c_{i,h}, e_{i,h} >$. $l_{i,h-1}$ and $l_{i,h}$ are the start and end station of the sub-flight $f_{i,h}$, respectively. The direct distance between $l_{i,h-1}$ and $l_{i,h}$ is defined as $D(l_{i,h-1}, l_{i,h})$. Apparently, $l_{i,0} = l_i$ in the first sub-flight $f_{i,1}$ and $l_{i,N_i} = d_i$ in the last sub-flight f_{i,N_i}. $a_{i,h}$ is the time that u_i arrives at $l_{i,h}$. $ch_{i,h}$ is the charger assigned to the UAV. $c_{i,h}$ is the charging time. $s_{i,h}$ and $e_{i,h}$ are the start charging and end charging times of u_i at $l_{i,h}$, respectively. $e_{i,h}$ also denotes the time when u_i leaves the station $l_{i,h}$.

The schedule solution S for the problem under study consists of the flight plans of all tasks, i.e., $S = \{f_1, f_2, \cdots, f_n\}$. The schedule is feasible if and only if the following constraints are satisfied.

- Non-overlapping on charger constraint makes sure that only one UAV can be charged per charger at any given time. The constraint can be defined by Eq. (6). Equation (6) indicates that if two UAVs are assigned to the same charger, the charging time of them cannot overlap.
- Non-preemption constraint means that when the UAV is charging, it cannot be interrupted. The constraint can be defined by Eq. (7).

- Electricity constraint ensures that the remaining energy of a UAV cannot be negative at any time and cannot exceed the maximum available capacity as well. The constraint can be defined by Eqs. (8)–(11). Initially, the UAV starts with a full charge (Eq. (8)). The residual energy of u_i when it arrives at station $l_{i,h}$ is defined as Eq. (9). During the waiting time for a charger, there is no energy consumption (Eq. (10)). The residual energy of u_i when it stops charging is computed by Eq. (11). Equation (11) also indicates that the residual energy cannot exceed the maximum available capacity.

$$\min\{e_{i,h}, e_{i',h'}\} \geq \max\{s_{i,h}, s_{i',h'}\} \quad \forall ch_{i,h} = ch_{i',h'} \tag{6}$$

$$c_{i,h} = e_{i,h} - s_{i,h} \tag{7}$$

$$R(0) = E_{max} \tag{8}$$

$$R(a_{i,h}) = R(e_{i,h-1}) - A_i \times \frac{D(l_{i,h-1}, l_{i,h})}{v_i} \geq 0 \tag{9}$$

$$R(s_{i,h}) = R(a_{i,h}) \tag{10}$$

$$R(e_{i,h}) = R(s_{i,h}) + \beta \times c_{i,h} \leq E_{max} \tag{11}$$

Given a feasible S, its completion time $T(S)$ can be computed by Eq. (12). The paper aims to obtain the optimal schedule with the minimum completion time of all tasks.

$$T(S) = \min \sum_{i=1}^{n} e_{i,N_i} \tag{12}$$

4 Proposed Method

In this paper, we investigate a hybrid problem involving both UAV trajectory planning and charging scheduling in a logistics system with a charging network. An ant colony optimization-based UAV trajectory planning and charging scheduling (ACO-TPCS) algorithm is proposed for the problem under study. In this section, we first introduce the main framework of the proposal, then followed by the details of key components.

4.1 Main Framework

Algorithm 1 describes the main framework of the proposed ACO-TPCS.

In Algorithm 1, we first construct the reachable graph for each task (Line 2). For each task, a set of candidate flight paths is generated (Line 3). An initial pheromone matrix is constructed (Line 4). Then an ACO-based iterative process is repeated until the iteration number I exceeds the upper bound I_{max}. In each

Algorithm 1: Main Framework of ACO-TPCS

```
1  for i = 1, 2, ..., n do
2      Perform Reachable Graph Construction Method to construct the reachable
         graph G_i for task t_i;
3      Compute candidate flight path set F_i^can for t_i by Candidate Flight Path
         Generation Method based on graph G_i;
4  Call Pheromone Matrix Construction to construct an initial pheromone matrix
     M = (τ_{i,i'})_{n×n};
5  I ← 0;
6  S* ← NULL, T* ← +∞;/* S* is the global best solution                    */
7  while I ≤ I_max do
8      I ← I + 1;
9      S' ← NULL, T' ← +∞; /* S' is the local best solution of current
           iteration                                                          */
10     for q = 1, ..., n_a do
11         Call Ant Generation Method to generate an ant π based on the
             pheromone matrix M;
12         Call Feasible Solution Generation Method to generate a feasible solution
             S based on the ant π and candidate flight paths of all tasks;
13         if T(S) < T(S') then
14             S' ← S, T' ← T(S);
15         if T(S) < T(S*) then
16             S* ← S, T* ← T(S);
17     Perform Pheromone Update Method to update pheromone matrix M based
         on S';
18 return S*;
```

iteration, we generate n_a ants. For each ant, the Feasible Solution Generation Method is performed to generate a feasible solution based on the pheromone matrix (Line 12). The local and global best solutions are updated accordingly (Lines 13–16). At the end of each iteration, the pheromone matrix is updated based on the local best solution obtained in the current iteration (Line 17). Finally, the global best solution is returned (Line 18).

As can be seen, the proposed algorithm consists of several major components: (1) Reachable Graph Construction Method to obtain the reachable graph; (2) Candidate Flight path Generation Method to generate candidate flight paths; (3)Pheromone Matrix Construction to define the pheromone matrix; (4) Ant Generation Method to generate an ant; (5) Feasible Solution Generation Method to compute feasible solution for a given ant; (6) Pheromone Update Method to update pheromone matrix. The details of the components are given in the following sub-sections.

4.2 Reachable Graph Construction Method

Reachable Graph Construction Method is designed to generate a reachable graph for each task. Given the flying speed v_i and package weight w_i, the maximum distance D_i^{max} that u_i can conduct can be computed by Eq. (13). A reachable graph G_i consists of a vertex set $\mathbb{V}_i$ and an edge set $\mathbb{E}_i$, where each vertex represents a station and any edge $(L, L') \in \mathbb{E}_i$ must have $D(L, L') \leq D_i^{max}$ satisfied.

$$D_i^{max} = \frac{E_{max}}{A_i} \times v_i \tag{13}$$

The main process of the Reachable Graph Construction Method is shown in Algorithm 2. ε_i is the weight of an edge (L, L') which is equal to the flying time Δt_{fly} from L to L' plus the charging time Δt_{ch} at L'. We assume u_i will be fully charged at every station.

Algorithm 2: Reachable Graph Construction Method $RGCM(t_i)$

---

1 $\mathbb{V}_i \leftarrow \varnothing$, $\mathbb{E}_i \leftarrow \varnothing$;
2 Compute D_i^{max} by Eq. (13);
3 **for** $\forall L, L' \in \{L_1, \ldots, L_m\}$ *and* $L \neq L'$ **do**
4   **if** $D(L, L') \leq D_i^{max}$ **then**
5         $\mathbb{V}_i \leftarrow \mathbb{V}_i \cup \{L, L'\}$;
6         $\Delta t_{fly} \leftarrow \frac{D(L,L')}{v_i}$;
7   Compute the energy consumption $\Delta E(\Delta t)$ by Eq. (2);
8         $\Delta t_{ch} \leftarrow \frac{\Delta E(\Delta t)}{\beta}$;
9         $\varepsilon \leftarrow \Delta t_{fly} + \Delta t_{ch}$;
10         $\mathbb{E}_i \leftarrow \mathbb{E}_i \cup \{(L, L', \varepsilon)\}$;

11 $G_i \leftarrow < \mathbb{V}_i, \mathbb{E}_i >$;
12 **return** G_i

4.3 Candidate Flight Path Generation Method

Given the reachable graph G_i, we generate at most $\mathbb{K}$ candidate flight paths for t_i. Algorithm 3 gives the details of the method. The first candidate path $\mathcal{L}$ of t_i is the shortest path from l_i to d_i obtained by directly performing the Dijkstra's algorithm on G_i (Line 2). The rest candidates are generated based on $\mathcal{L}$ (Lines 4–11).

4.4 Pheromone Matrix Construction Method

Pheromone Matrix Construction Method initializes the pheromone matrix as a matrix of size $n \times n$, where n is the number of tasks in the problem. Initially, all elements can be set to the same value, usually a small positive number indicating the initial pheromone level.

Algorithm 3: Candidate Flight Path Generation Method $CFPGM(G_i, l_i, d_i, \mathbb{K})$

```
1  F_i^can ← ∅;
2  Generate L = (l_[1], ..., l_[n_i]) by perform Dijkstra's algorithm on G_i ;
3  F_i^can ← F_i^can ∪ {L};
4  h ← 2 ;
5  while |F_i^can| < K do
6      Generate a graph G' by removing vertex l_[h] from G_i;
7      Generate L' by perform Dijkstra's algorithm on G';
8      F_i^can ← F_i^can ∪ {L'};
9      h ← h + 1;
10     if l_[h] = d_i then
11         Break ;
12 return F_i^can
```

4.5 Ant Generation Method

In the ant generation method, an ant is represented by a task sequence, which is an arrangement of tasks. Tasks are arranged in the order of their position in the given task sequence.

The ant generation process is shown in Algorithm 4. First, we initialize an empty task sequence π and a list of unselected tasks $\mathbb{T}$ (Lines 1–2). Next, the algorithm iterates from the first position, and for each position i, calculates the selection probability of each unselected task based on M (Line 4). Then, a random number r between 0 and 1 is generated for roulette selection (Line 6). During roulette, the selection probability for each task is calculated cumulatively and the task is selected based on the random number r. Once a task is selected, it is added to the task sequence π and removed from $\mathbb{T}$ (Lines 8–14). Finally, return the generated ant sequence π (Line 15).

4.6 Feasible Solution Generation Method

Algorithm 5 describes the method for generating feasible solutions.

In this method, two inputs need to be provided: the ant path π and the set of candidate flight paths $\mathbb{F}^{can}$ for all tasks. First, we traverse each mission in the ant path π. For each task, traverse its corresponding set of candidate flight paths (Lines 5–13). Then, for each candidate's flight path, calculate its total time, including flight time, charging time, and waiting time. The candidate flight path with the shortest total time is selected as the optimal solution for the current task (Lines 14–15). Eventually, we combine the optimal solutions of all tasks to form a feasible solution S and compute its total time $T(S)$ (Line 16).

Algorithm 4: Ant Generation Method $AGM(M, n)$

```
1  π ← ∅;
2  𝕋 ← {t_1, t_2, ..., t_n};
3  for i = 1, 2, ..., n do
4      Compute selection probability p_i for each task in 𝕋 using M;
5      Compute total probability p_sum as the sum of selection probabilities for all
         tasks;
6      Generate a random number r between 0 and 1;
7      p ← 0;
8      while t_i ∈ 𝕋 do
9          p ← p + p_i;
10         if p ≥ r then
11             π_i ← t_i;
12             π ← π ∪ {π_i};
13             𝕋 ← 𝕋 − {t_i};
14             Break;
15 return π;
```

Algorithm 5: Feasible Solution Generation Method $FSGM(\pi, \mathbb{F}^{can})$

```
1  S ← NULL;
2  T(S) ← 0;
3  n ← |π|;
4  for i = 1, 2, ..., n do
5      for k = 1, 2, ..., 𝕂 do
6          T_i ← ∞;
7          T_k^fly = D(l_{i−1}, l_i) / v_i;
8          compute T_k^charge by Eq. (7);
9          T_k^wait = s_{i,h} − a_{i,h};
10         T_k = T_k^fly + T_k^charge + T_k^wait;
11         if T_k < T_i then
12             T_i ← T_k;
13             f_i ← 𝔽^can_{i,k} ;
14     S ← S ∪ f_i ;
15     T(S) ← T(S) + T_i ;
16 return S, T(S)
```

4.7 Pheromone Update Method

There are two strategies for the pheromone update method. The first is a local update rule that performs a volatilization operation on the pheromone (Eq. (14)). Where ζ is the local pheromone volatilization factor. The pheromone on all paths dissipates partially, which means reducing the pheromone level in all paths, simulating the natural pheromone volatilization process. The purpose of volatilization is to prevent the pheromone level from becoming too high and causing the ants to concentrate too much on one path. The second one is the global update rule to increase the pheromone level in the paths. The pheromone level on the path is increased by Eqs. (15)–(17). The pheromone is increased according to the quality of the solutions searched by the ants. Only the ant that obtains the minimum task completion time in this iteration is allowed to leave pheromone on its resulting path.

$$\tau_{ii'} \leftarrow (1-\zeta)\tau_{ii'} \tag{14}$$

$$\rho = \begin{cases} e_{max}, & 0 < I \leq \frac{I_{max}}{3} \\ e_{min} \times (2 - \frac{I}{I_{max}}), & \frac{I_{max}}{3} < I \leq \frac{2I_{max}}{3} \\ e_{min}, & \frac{2I_{max}}{3} < I \leq I_{max} \end{cases} \tag{15}$$

$$\tau_{ii'} \leftarrow (1-\rho)\tau_{ii'} + \Delta\tau_{ii'} \tag{16}$$

$$\Delta\tau_{ii'} = \begin{cases} \frac{Q}{T(S')}, & \text{if } \pi_{i,i'} \in S' \\ 0, & else \end{cases} \tag{17}$$

5 Computational Experiments

This section compares the proposed algorithm with several baseline algorithms for the problem under study. We first give the metrics for evaluating the performance of the compared algorithms. Then the parameter settings for the task instances are described. Finally, the compared baseline algorithms are briefly described and the experimental results are given. All compared algorithms are coded in Java and run on a computer equipped with a M2 chip, 24GB RAM and 1TB SSD.

We measure the effectiveness of a solution in terms of the average relative percentage deviation. Given a solution S, the average relative percentage deviation is defined by Eq. (18). $T(S^*)$ is the shortest completion time obtained by all the compared algorithms in the same test instance. A smaller RPD indicates better performance.

$$RPD(S) = \frac{T(S) - T(S^*)}{T(S^*)} \times 100\% \tag{18}$$

The test instances were generated based on research investigating similar problems [15]. The parameters for generating test instances were set as follows.

- Three maps of different sizes Map_1, Map_2, and Map_3 are used to simulate the distribution of delivery stations. The sizes of these maps are 30 km × 30 km, 40 km × 40 km, and 50 km × 50 km, respectively.
- The number of tasks n is set to 60, 80 or 100. The number of delivery stations m is set to 20, 40 or 60. When generating the test instances, we randomly distribute the delivery stations in the given map. The origin and destination of each package are randomly distributed among different delivery stations.

- The average speed v of u is randomly set to [20,30) km/h, [30,40) km/h or [40,50) km/h.
- The package weight w is randomly set to [0,0.5] kg, [0.5,1] kg, [1,1.5] kg, or [1.5,2] kg.

There are $3 \times 3 \times 3 \times 3 \times 4 = 324$ instance combinations (three maps, $n \in \{60,80,100\}$, $m \in \{20,40,60\}$, $v \in \{$[20,30) km/h, [30,40) km/h,[40,50) km/h$\}$, $w \in \{$ [0,0.5] kg, [0.5,1] kg, [1,1.5] kg, [1.5,2] kg$\}$). For each instance combination, 10 testing instances are randomly generated. Therefore, there are $324 \times 10 = 3240$ testing instances in total.

The proposed ACO-TPCS is compared with two baseline algorithms: cloud-based drone navigation (CBDN) in [15], and the ant colony optimization (ACO) approach in [16]. The parameter settings of these algorithms are described below.

- In the CBDN, the number of candidate flight paths $\mathbb{K}$ is set to 3.
- In the ACO, the number of ants n_a is set equal to the number of tasks, the local pheromone volatilization factor ζ is set to 0.3, the initial pheromone level is set to 1, and the number of iterations I_{max} is limited up to 30. The total number of pheromones Q is set to 1000.

The compared three algorithms are performed on 3240 testing instances. Therefore, $3 \times 3240 = 9720$ experimental results are obtained.

Table 2. Experimental Results

Parameter	Value	ARPD (%)		
		CBDN	ACO	ACO-TPCS
n	60	16.45	2.22	0.02
	80	17.34	2.72	0.04
	100	18.39	3.22	0.04
m	20	20.63	3.70	0.06
	40	16.85	2.88	0.02
	60	14.71	1.57	0.02
v	[20–30)	11.31	1.98	0.08
	[30–40)	17.72	2.97	0.03
	[40–50)	22.17	3.08	0.08
w	[0–0.5)	16.26	2.62	0.02
	[0.5–1.0)	17.02	2.67	0.03
	[1.0–1.5)	17.91	2.74	0.05
	[1.5-2)	18.39	2.84	0.04
Map	Map1	23.35	3.16	0.01
	Map2	16.74	2.90	0.04
	Map3	12.09	2.09	0.05
Avg.		**17.33**	**2.71**	**0.04**

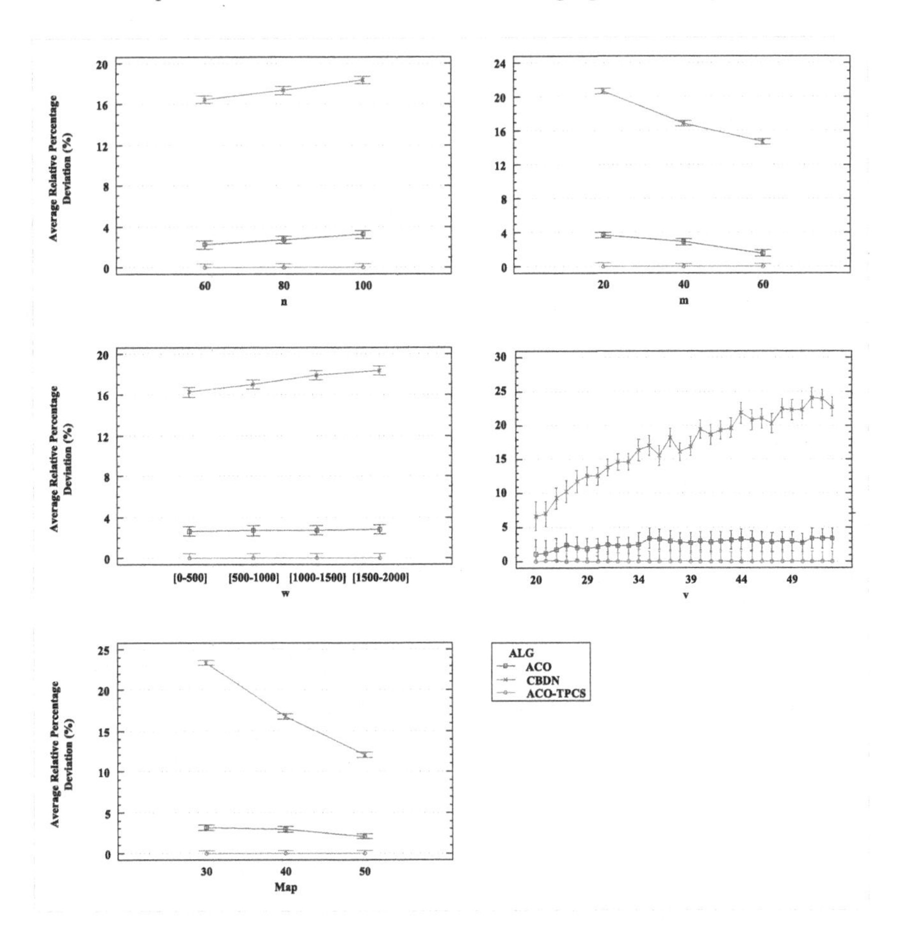

Fig. 1. Interactions between instance parameters and the compared algorithms with 95.0% Tukey HSD intervals

The average RPD (ARPD) of the compared algorithms is given in Table 2. In all cases, ACO-TPCS has the best performance on ARPD (0.04%). ACO is the second-best algorithm on ARPD with an average APRD of 2.71%. CBDN has the worst ARPD (17.33%). ACO-TPCS performs the best among all test instances, mainly due to its integrated trajectory planning and charge scheduling optimization capabilities, as well as the global search benefits from the pheromone update mechanism. The ACO algorithm, although it performs well in terms of path optimization, has a slightly lower overall performance due to the lack of charge scheduling optimization. The CBDN algorithm performs the worst in the complex task due to the lack of dynamic tuning and global optimization capabilities environment where it performs the worst.

Figure 1 depicts the interaction between the instance parameters and the comparison algorithm. From Fig. 1, we can see that ACO-TPCS is insensitive to parameter changes. We can conclude that ACO-TPCS is robust to the problem under study.

6 Conclusion

In this paper, the problem of trajectory planning and charging scheduling in a UAV tele-distribution system with a charging network. The problem involves no-overlap constraints, non-preemption constraints, and energy constraints on the chargers, and the objective is to minimize the completion time of all the tasks of the system. To address this problem, we propose an ant colony optimization-based algorithm for UAV trajectory planning and charging scheduling (ACO-TPCS). Our approach systematically solves the optimization problem of trajectory planning and charging scheduling by modeling the energy consumption characteristics of UAVs and the multitasking characteristics of the system. The proposed ACO-TPCS algorithm efficiently generates feasible and optimal solutions by iteratively optimizing UAV flight paths and charging schedules. The key components of the algorithm include reachability graph construction, candidate flight path generation, pheromone matrix initialization and updating, and methods to ensure solution feasibility and efficient pheromone updating. Experimental results show that the ACO-TPCS algorithm significantly reduces the overall task completion time by optimizing UAV trajectories and charging scheduling. The approach successfully mitigates congestion at key charging stations and ensures efficient energy management throughout the distribution network.

Future research directions could include further optimizing algorithms for trajectory planning and charging scheduling, considering real-time factors such as weather conditions and traffic patterns. Additionally, the integration of machine learning techniques could enhance the adaptability and efficiency of the delivery system in dynamic environments.

Acknowledgment. This work is sponsored by the State Key Laboratory of Computer Architecture (ICT, CAS) (Grant No. CARCHA202107), Jiangsu Province Universities Natural Science Research Major Project (Grant No. 23KJA520010), and the National Natural Science Foundations of China (Grant Nos. 62072252).

References

1. Khan, N., Ahmad, A., Wakeel, A., Kaleem, Z., Rashid, B., Khalid, W.: Efficient UAVs deployment and resource allocation in uav-relay assisted public safety networks for video transmission. IEEE Access **12**, 4561–4574 (2024). https://doi.org/10.1109/ACCESS.2024.3350138
2. Hao, Y., Widagdo, F.R.A., Liu, X., Liu, Y., Dong, L., Li, F.: A hierarchical region-merging algorithm for 3-D segmentation of individual trees using UAV-LiDAR point clouds. IEEE Trans. Geosci. Remote Sens. **60**, 1–16, Art no. 5701416 (2022). https://doi.org/10.1109/TGRS.2021.3121419
3. Huang, H., et al.: Object-based attention mechanism for color calibration of UAV remote sensing images in precision agriculture. IEEE Trans. Geosci. Remote Sens. **60**, 1–13, Art no. 4416013 (2022). https://doi.org/10.1109/TGRS.2022.3224580

4. How Far Can Drones Fly? Accessed 29 Nov 2018. https://3dinsider.com/drone-range/
5. Cho, Y.H., et al.: Multi-criteria coordinated electric vehicle-drone hybrid delivery service planning. IEEE Trans. Veh. Technol. **72**(5), 5892–5905 (2023). https://doi.org/10.1109/TVT.2022.3232799
6. Das, D.N., Sewani, R., Wang, J., Tiwari, M.K.: Synchronized truck and drone routing in package delivery logistics. IEEE Trans. Intell. Transp. Syst. **22**(9), 5772–5782 (2021). https://doi.org/10.1109/TITS.2020.2992549
7. Chen, K.-W., Xie, M.-R., Chen, Y.-M., Chu, T.-T., Lin, Y.-B.: DroneTalk: an internet-of-things-based drone system for last-mile drone delivery. IEEE Trans. Intell. Transp. Syst. **23**(9), 15204–15217 (2022). https://doi.org/10.1109/TITS.2021.3138432
8. Park, S., Park, C., Kim, J.: Learning-based cooperative mobility control for autonomous drone-delivery. IEEE Trans. Veh. Technol. **73**(4), 4870–4885 (2024). https://doi.org/10.1109/TVT.2023.3330460
9. Huang, C., Ming, Z., Huang, H.: Drone stations-aided beyond-battery-lifetime flight planning for parcel delivery. IEEE Trans. Autom. Sci. Eng. **20**(4), 2294–2304 (2023). https://doi.org/10.1109/TASE.2022.3213254
10. Huang, H., Savkin, A.V., Huang, C.: Round trip routing for energy-efficient drone delivery based on a public transportation network. IEEE Trans. Transport. Electrificat. **6**(3), 1368–1376 (2020). https://doi.org/10.1109/TTE.2020.3011682
11. Du, P., Shi, Y., Cao, H., Garg, S., Alrashoud, M., Shukla, P.K.: AI-enabled trajectory optimization of logistics UAVs with wind impacts in smart cities. IEEE Trans. Consum. Electron. **70**(1), 3885–3897 (2024). https://doi.org/10.1109/TCE.2024.3355061
12. Li, M., Liu, L., Gu, Y., Ding, Y., Wang, L.: Minimizing energy consumption in wireless rechargeable UAV networks. IEEE Internet Things J. **9**(5), 3522–3532 (2022). https://doi.org/10.1109/JIOT.2021.3097918
13. Fan, M., et al.: Deep reinforcement learning for UAV routing in the presence of multiple charging stations. IEEE Trans. Veh. Technol. **72**(5), 5732–5746 (2023). https://doi.org/10.1109/TVT.2022.3232607
14. Qin, Y., Kishk, M.A., Alouini, M.-S.: Performance analysis of charging infrastructure sharing in UAV and EV-involved networks. IEEE Trans. Veh. Technol. **72**(3), 3973–3988 (2023). https://doi.org/10.1109/TVT.2022.3219764
15. Kim, J., et al.: CBDN: cloud-based drone navigation for efficient battery charging in drone networks. IEEE Trans. Intell. Transport. Syst., 1-18 (2018)
16. Li, Y., Zhang, S., Chen, J., et al.: Multi-UAV cooperative mission assignment algorithm based on ACO method. In: 2020 International Conference on Computing, Networking and Communications (ICNC), pp. 304–308. IEEE (2020)

Cross-Modal Mask and Detail Alignment for Text-Based Person Retrieval

Ao Guo[1], Xuan Liu[1,2](✉), Xianggan Liu[3], Bingmeng Hu[1], Jie Yuan[1], and Chiawei Chu[4]

[1] Key Laboratory of Ethnic Language Intelligent Analysis and Security Governance, Ministry of Education, Minzu University of China, Beijing 100081, China
{guoao,liuxuan,hbm,ytgao92}@muc.edu.cn

[2] Hainan International College of Minzu University of China, Li'an International Education Innovation pilot Zone, Lingshui Li 572499, Hainan, China

[3] Natural Language Processing and Knowledge Graph Lab, School of Computer Science and Technology, Huazhong University of Science and Technology, Wuhan 430074, China
liuxianggan@msn.com

[4] City University of Macau, Macau, SAR, China
cwchu@cityu.edu.mo

Abstract. Text-based person retrieval primarily aims to retrieve the images of target persons represented by a given text query. In this task, how to effectively align images and text globally and locally is an important challenge. At the same time, since multiple modalities are involved, reducing the differences between different modalities is also an important challenge. Existing work has focused more on reconstructing an image rather than the semantic consistency between the image and text modalities. Therefore, we introduce the Cross-Modal Mask and Detail Alignment (CMDA) framework to address these challenges. Under this framework, our proposed Cross-Modal Mask Alignment module (CMA) semantically aligns features generated by supplementing randomly masked image/text with another modality text/image. Additionally, to narrow the gap between the image and text modalities, we designed the Cross-Modal Detail Alignment module (CDA), which establishes connections between images and texts and facilitates interactions between these two different modalities. Experimental results show that our model exhibits outstanding performance across multiple public datasets, *i.e.*, CUHK-PEDES, ICFG-PEDES, and RSTPReID.

Keywords: Text-based Person Retrieval · Multimodal · Cross-Model Retrieval · Artificial Intelligence

1 Introduction

In the backdrop of the rapid development of multimedia technology and the increasing awareness of safety, the field of text-based person retrieval has received

T. Zhu et al. (Eds.): ICA3PP 2024, LNCS 15252, pp. 254–269, 2025.
https://doi.org/10.1007/978-981-96-1528-5_17

more and more attention. In text description-based person retrieval, each image usually contains only one specific pedestrian, unlike in traditional cross-modal retrieval tasks that may include multiple object categories. Text descriptions provide more detailed information about the person, rather than merely summarizing the image's content. Since text queries are more user-friendly than image queries, text-based person retrieval is increasingly expected for various applications that benefit surveillance and public safety. However, because text-based person retrieval tasks involve both image and text modalities, and there is modal heterogeneity, it remains a challenging task. In cross-modal tasks, the main problem is caused by the inherent representational differences between vision and language.

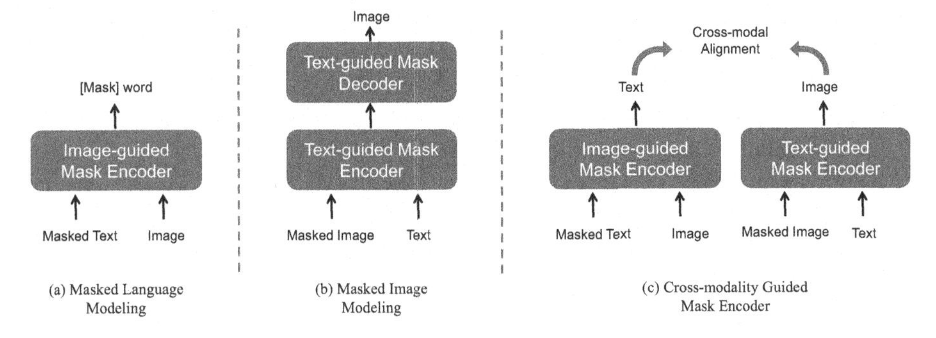

Fig. 1. Comparison with Various Mask Task Models. (a) For masked text, supplementary information from images retrieves the masked words. (b) For masked image tokens, images are reconstructed using a pre-trained decoder. (c) In our cross-modality mask alignment (CMA) method, we employ a Cross-Modality Guided Mask Encoder (CGME) to align both modalities, thereby eliminating the need for a pre-trained decoder and reducing the training burden.

In recent research work [2,4,22], models for image reconstruction require a decoder to perform the reconstruction. However, due to the limitations of pre-trained decoders, which need large amounts of training data and substantial training resources, and the fact that pre-training is targeted at a specific domain and cannot be fully applied to pedestrian tasks, we propose a cross-modal mask alignment module (CMA). This avoids the use of pre-trained decoders and instead directly acquires features, making the method more practical and convenient, as illustrated in Fig. 1(c). Furthermore, by completing masked images through text and masked texts through images, and aligning both, we enhance both the discriminability and the semantic consistency between images and texts. In previous research on text-based person retrieval, there not only was a lack of interaction between the two modalities but there was also a focus on local information without regard to global information. To address these issues, we propose the Cross-Modal Detail Alignment Module (CDA), which adds global features of another modality to local features, and to prevent global features from overshadowing important information in the local content of one modality, we weigh the local features of both modalities to capture important information. The experimental results show that our model achieves state-of-the-art (SOTA) results on the public datasets CUHKPEDES [11] and ICFG-PEDES [25] datasets for Rank-1, Rank-5, Rank-10, and MAP metrics; it also exceeds recent advanced research on the RSTPReid [26] dataset for Rank-5. In summary, the main contributions of this paper are as follows:

- We propose a Cross-Modal Mask Alignment Module (CMA), which uses masked images (texts) to enhance the discriminability of images (texts) and aligns the features completed by merging both modalities, thereby enhancing the semantic consistency between images and texts.
- We introduce a Cross-Modal Detail Alignment Module (CDA), which facilitates interactions between two different modalities by adding global text/image features to local features of another modality image/text, thereby reducing the differences between the two modalities.
- Experimental results show that our model achieves SOTA results on the CUHK-PEDES [11] and ICFG-PEDES [25] datasets for Rank-1, Rank-5, Rank-10, and MAP; it also exceeds recent advanced research on the RSTP-Reid [26] dataset for Rank-5.

2 Related Work

2.1 Text-Based Person Retrieval

First introduced by Li *et al.* [18] , the first benchmark dataset CUHK-PEDES [18] was proposed, along with a baseline model: the LSTM-based GNA-RNN. The primary challenge lies in effectively aligning image and textual features within a joint embedding space to facilitate rapid retrieval. Early works [5,17,18] utilized ResNet 50, VGG [27], and LSTM [23] to learn visual-textual modal representations, followed by using matching loss for alignment. For instance, CMPM [37] employed KL divergence to correlate representations across different modalities. Beyond aligning features, some studies also explored identity cues [17], aiding in learning discriminative representations. As text-based person retrieval necessitates fine-grained recognition of human figures, subsequent works began exploring global and local associations. Later research [6,24,37] enhanced the feature extraction backbones of ResNet 50/101 [13] and BERT [8] , designing new cross-modal matching losses for aligning global image-text features in a joint embedding space. Some studies [5] leveraged visual-textual similarity for part alignment. ViTAA [28] segmented the human body, associating visual and textual attributes through k-inverse sampling. Surbhi et al. [1] proposed creating semantically preserved embeddings through attribute prediction. Given the preprocessing required for visual and textual attributes, more research attempted fine-grained alignment exploration using attention mechanisms. PMA [15] introduced a posture-guided multi-granularity attention network. HGAN [38] segmented images into multi-scale stripes, using attention to select the alignment of the top k parts. Recent works [7,28,29,32,41] extensively adopted additional local feature learning branches, explicitly utilizing human segmentation, body parts, color information, and textual phrases.

Some studies [10,11,25,36] implicitly performed local feature learning through attention mechanisms. However, while these methods have been proven to provide better retrieval results than using global features alone, they also introduce additional computational complexity during image-text similarity inference. Other efforts include adversarial learning and relation modeling. TIMAM [24] learned modality-invariant feature representations using adversarial and cross-modal matching objectives. A-GANet [21] introduced textual and visual scene graphs composed of object attributes and relationships. In summary, most of the current work learns modal alignment by using local alignment. In our work, however, we use joint alignment of global and local content and reinforce the importance of local content through global information. Meanwhile, we study the visual and verbal alignment problems from different perspectives, especially the problem of generating incomplete information when there is an occlusion in the visual content or when the same textual content corresponds to different picture contents. The framework can solve these problems effectively while enhancing visual and verbal features with satisfactory performance.

2.2 Masked Image Modeling (MIM)

Masked Image Modeling (MIM) refers to the objective of pre-training MIM to recover masked patches of a given image so that the model captures rich contextual information. In BEiT [2], the approach involves tokenizing the original image and then randomly sampling patches for masking. However, the pre-training objective of BEIT is to restore the original visual tokens based on the corrupted image patches. There are various MIM strategies [2–4,22,34], all of which are formulated as the problem of reconstructing visual tokens (i.e., patches) that are randomly masked by unmasked tokens. Due to the significant semantic gap between modalities, reconstructing images from text is challenging, and the model's performance primarily depends on the visual tokenizer used. Although this method of reconstructing images by identifying masked tokens from tokenized images yields good results, it is highly dependent on the performance of the visual tokenizer, which poses limitations in its usage. In studies [2,4,22], models for image reconstruction require a decoder for reconstruction. However, due to the limitations of pre-trained decoders, which need large amounts of training data and substantial training resources, and the fact that pre-training is targeted at a specific domain and cannot be fully applied to pedestrian tasks, we designed a Cross-Modal Mask Alignment (CMA) module. This module employs a cross-attention mechanism to integrate masked images and complete text to reconstruct image features, reducing the difficulty of image reconstruction. Additionally, it introduces a method to integrate masked text and complete images to reconstruct text features. By aligning these two obtained features, the semantic consistency between images and text is enhanced, while more information is extracted. The comparison with various mask task models is shown in Fig. 1.

3 Proposed Method

In this section, we will show and illustrate the architecture of our proposed model. The specific model architecture is shown in Fig. 2 and the details of the model will be presented in the following subsections. We will mainly introduce two modules: (1) Cross-modal Mask Alignment (CMA) and (2) Cross-modal Detail Alignment (CDA) in Sects. 3.2 and 3.3.

3.1 Dual Encoders

We initialize our model with the complete CLIP image and text encoders to enhance its underlying cross-modal alignment capabilities.

Image Encoder. Given an input image I, we employ a CLIP pre-trained ViT model for image embedding. The process begins with visual tokenization to convert the image into a discrete token sequence, adding positional embeddings and appending an additional [CLS] token as an image-level representation at the beginning of the sequence. This is then fed into the CLIP pre-trained ViT model, where the relevance of each local image feature is modeled through L layers of Transformer blocks. Finally, a linear projection maps f_{cls}^{v} into a joint image-text embedding space. The output of the image encoder is represented as $f^{V} = \{f_{cls}^{v}, f_{1}^{v}, ..., f_{N}^{v}\}$, with f_{cls}^{v} serving as the global image representation f_{g}^{v}.

Text Encoder. Given the input text T, we utilize the CLIP text encoder to extract text representations. Following CLIP's approach, we first tokenize the input text description using Byte Pair Encoding (BPE) with a vocabulary size of 49152, which is case-insensitive. The text description begins with an [SOS] token at the start of the sequence and an [EOS] token at the end. Subsequently, the tokenized text is fed into a Transformer to obtain the output of the text encoder $f^{T} = \{f_{sos}^{t}, f_{1}^{t}, ..., f_{eos}^{t}\}$. Finally, the Transformer's output at the [EOS] token is linearly projected to the joint image-text embedding space at its top layer, obtaining the global text representation f_{eos}^{t}, which serves as the global text representation f_{g}^{t}.

3.2 Cross-Model Mask Alignment

This is essential to fully utilize global information and bridge the significant modal gap between vision and speech. While many current methods accomplish this by explicitly aligning local features between image and text, this paper presents a new approach. We employ a mask approach to implicitly explore the relationship between the two modalities and acquire global features that are both discriminative and information-enhancing.

In studies [2,4,14], models for image reconstruction require a decoder for reconstruction. However, due to the limitations of pre-trained decoders, which need large amounts of training data and substantial training resources, and the

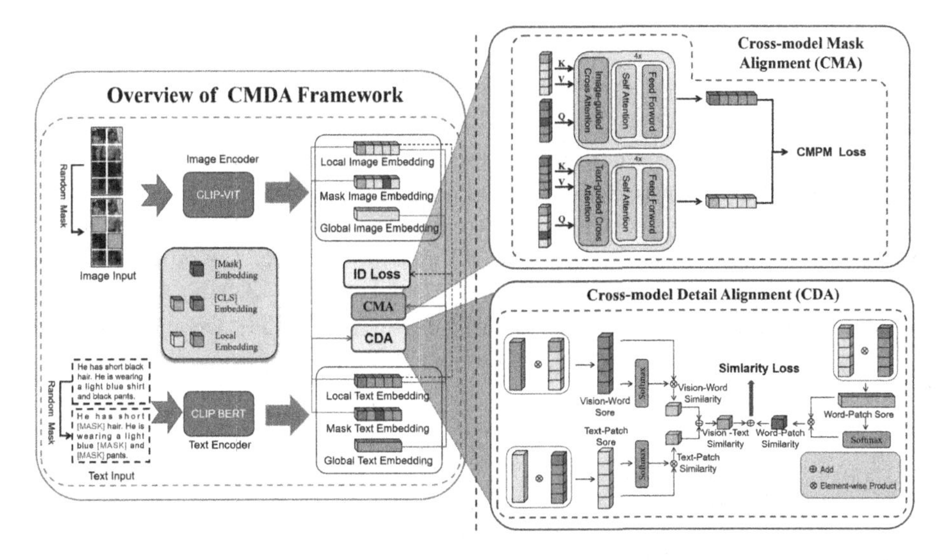

Fig. 2. The Overview of CMDA Framework. We will mainly introduce two modules: (1) Cross-modal Mask Alignment (CMA) and (2) Cross-modal Detail Alignment (CDA). Furthermore, we utilize an ID loss to consolidate feature representations of identical identities, leading to enhanced retrieval performance. In addition, we employ ID loss to aggregate feature representations of the same identity, which further improves retrieval performance.

fact that pre-training is targeted at specific domains and cannot be fully applied to pedestrian tasks, we propose training masked images and complete texts to obtain complete image features, supplementing the images with complete texts and establishing the connection between images and texts without relying on pre-trained decoders. We mainly use the Cross-modality Guided Mask Encoder (CGME) to guide and supplement one masked modality with another to obtain features. This encoder is specifically composed of a cross-attention mechanism and four transformer blocks.

Given an input image I, we randomly mask patches in the image and take the image features including the masked patches as I'. I' is passed through an image encoder to obtain the final hidden states $h^{v_m}i^N i = 1$ as the query (Q) for the cross-attention layer. The input text T is passed through a text encoder to obtain the final hidden states $h^{t}{}^{L}_{i\,i=1}$ as the key (K) and value (V) for the cross-attention layer. Through the Cross-modality Guided Mask Encoder (CGME), the contents of the two modalities interact, and the final image features $h^{V_M}i^N i = 1$ completed with integrated textual information are obtained. The specific implementation is shown in the following formula:

$$V_M = MCA_{V2T}(LN(h^{v_m}, h^t, h^t)), \tag{1}$$

where d is the dimension of the image patches, $LN(\cdot)$ represents layer normalization, and $MCA(\cdot)$ represents multi-head cross-attention.

$$\{h_i^{V_M}\}_{i=1}^N = CGME(V_M), \quad (2)$$

where $h^{V_M}i^Ni = 1$ are the image features completed with integrated textual information, and N is the number of input image patches.

Symmetrically, from text to image, the masked text features $h^{t_m}i^Li = 1$ are used as the query (Q) for the cross-attention layer, and the complete image features $h^vi^Ni = 1$ are used as the key (K) and value (V) for the cross-attention layer. Through the Cross-modality Guided Mask Encoder (CGME), the contents of the two modalities interact, and the text features $h^{T_M}i^Li = 1$ completed with the complete image are obtained. The specific implementation formula is as follows:

$$T_M = MCA_{T2V}(LN(h^{t_m}, h^v, h^v)), \quad (3)$$

where d is the dimension of the image patches, $LN(\cdot)$ represents layer normalization, and $MCA(\cdot)$ represents multi-head cross-attention.

$$\{h_i^{T_M}\}_{i=1}^L = CGME(T_M), \quad (4)$$

where $h^{T_M}i^Li = 1$ are the complete text features integrated with image information, and L is the length of the input text.

The specific definition of multi-head cross-attention $MCA(\cdot)$ is as follows:

$$MCA(Q, K, V) = softmax(\frac{QK^T}{\sqrt{d}})V, \quad (5)$$

where d is the dimension of the image patches. The text features $\{h_i^{T_M}\}_{i=1}^L$, enhanced with picture information, and the picture features $\{h_i^{V_M}\}_{i=1}^N$, enriched with text information, are utilized for learning visual-textual alignment through cross-modal projection matching (CMPM) loss [37].

Initially, we calculate the matching probability between the text feature $\{h_i^{T_M}\}_{i=1}^L$ of the supplementary image and the image feature $\{h_i^{V_M}\}_{i=1}^N$ of the supplementary text:

$$p_{i,j} = \frac{exp(h_i^{V_M} \cdot (\overline{h}_i^{T_M})^\top)}{\sum_{k=1}^B exp(h_i^{V_M} \cdot (\overline{h}_i^{T_M})^\top)} \quad (6)$$

Here, $\overline{h}_i^{T_M}$ represents the normalization of textual features, and $\overline{h}_i^{T_M}$ is defined as $\overline{h}_i^{T_M} = \frac{h_i^{T_M}}{\|h_i^{T_M}\|}$, while $p_{i,j}$ signifies the matching probability.

$$L_{V2T} = \frac{1}{B}\sum_{i=1}^B\sum_{j=1}^B (p_{i,j} \cdot \log \frac{p_{i,j}}{q_{i,j} + \epsilon})) \quad (7)$$

Here, B denotes a small batch, and ϵ is a minute value used to prevent numerical issues. $q_{i,j}$ denotes the normalized true matching probability, defined as follows:

$$q_{i,j} = \frac{y_{i,j}}{\sum_{k=1}^{B} y_{i,k}} \tag{8}$$

When $y_{i,j} = 1$, it indicates that the two different modes $\{h_i^{T_M}\}_{i=1}^{L}$ and $\{h_i^{V_M}\}_{i=1}^{N}$ represent features originating from the same individual.

The corresponding L_{T2V} can be introduced by inverting the above Eq. 7. Hence, the CMPM loss is computed as outlined below:

$$L_{cmpm} = L_{V2T} + L_{T2V} \tag{9}$$

3.3 Cross-Modal Detail Alignment

The main objective of text-based pedestrian retrieval tasks is cross-modal alignment. Therefore, we aim to extract information that is not only highly distinguishable but also shared across modalities. Previous research [5] has highlighted the lack of interaction between modalities and the tendency to focus solely on local information, neglecting global content. To address these issues, we propose a solution that incorporates global features from one modality into the local features of another. This fusion process combines image patches with global text features, emphasizing detailed image content, and vice versa, merging text words with global image features to enhance text details.

To explore the fine-grained correspondence between image and text, the fused patch features and word features are encouraged to be similar to each other. Additionally, to avoid losing crucial global information when integrating global features from one modality to another, we assign weights to the local features of both modalities. Specifically, the global text feature f_g^t is combined with the patch-level feature $\{f_i^v\}_{i=1}^{N}$ of the image, and the global image feature f_g^v is merged with the word-level feature $\{f_i^t\}_{i=1}^{L}$ of the text to compute similarity matrices Γ_{V-W} and Γ_{T-P}.

$$\Gamma_{V-W} = (f^t \cdot f_g^v)^T \tag{10}$$

$$\Gamma_{T-P} = f^v \cdot f_g^t \tag{11}$$

Softmax generates different weights during the aggregation process based on the scores of Γ_{V-W} and Γ_{T-P}, assigning higher weights to image chunks (words) related to the sentence (image). New similarities S_{V-W} and S_{T-P} are calculated using these weights, and the local feature similarity $S_{(V,T)}$ that incorporates global features from another modality is derived.

$$S_{V-W} = \sum_{i=1}^{L} \frac{exp(\Gamma_{V-W(1,i)})}{\sum_{j=1}^{L} exp(\Gamma_{V-W(1,j)})} \Gamma_{V-W(1,i)} \tag{12}$$

$$S_{T-P} = \sum_{i=1}^{N} \frac{exp(\Gamma_{T-P(i,1)})}{\sum_{j=1}^{N} exp(\Gamma_{T-P(j,1)})} \Gamma_{T-P(i,1)} \tag{13}$$

$$S_{(V,T)} = (S_{V-W} + S_{T-P})/2 \tag{14}$$

To enhance the focus on original details, we calculate the similarity between text word tokens $\{f_i^t\}_{i=1}^{L}$ and image patch features $\{f_i^v\}_{i=1}^{N}$ and weight the local information similarities using a softmax operation to obtain the local feature similarity $S_{(W,P)}$ that incorporates global features from another modality.

$$S_{(W,P)} = \sum_{i=1}^{N} \frac{exp(s(f_j^v, f_j^t))}{\sum_{j=1}^{N} exp(s(f_j^v, f_j^t))} s(f_j^v, f_j^t) \tag{15}$$

where $s(\cdot)$ calculates the similarity between text and image features as defined below:

$$s(I, T) = \frac{I \cdot T}{||I||||T||} \tag{16}$$

Here, I represents image features and T represents text features.

The final loss value is the average of the cross-modal weighted results combined with the multimodal local feature similarity.

$$L_{sim} = (S_{(V,T)} + S_{(W,P)})/2 \tag{17}$$

The ID loss is a softmax loss which classifies an image or text into distinct groups based on their identities. The model is trained by minimizing the following objective loss L:

$$L = L_{CMPM} + L_{sim} + L_{ID} \tag{18}$$

4 Experimental Results

4.1 Datasets

We introduce three benchmark datasets in text-based person retrieval. We evaluate our approach on three benchmark datasets, i.e., CUHK-PEDES [11], ICFG-PEDES [25], and RSTPReid [26]. The dataset statistics are shown in Table 1.

Table 1. Dataset statistics of CUHK-PEDES, ICFG-PEDES, and RSTPReid.

Datasets	IDs			Images			Text Descriptions		
	train	test	val	train	test	val	train	test	val
CUHK-PEDES [11]	11003	1000	100	34054	30474	3078	68126	6156	6158
ICFG-PEDES [25]	3102	1000	/	34674	19848	/	34674	19848	/
RSTPReid [26]	3701	200	200	18505	1000	1000	37010	2000	2000

CUHK-PEDES [11] is a commonly used database for cross-modal pedestrian retrieval tasks based on natural language descriptions, initially proposed by The Chinese University of Hong Kong. The pedestrian images in this dataset are sourced from five different pedestrian re-identification databases (CUHK03 [20], Market-1501 [39], SSM [33], VIPER [12], CUHK01 [19]), comprising 13,003 pedestrians with a total of 40,206 images. The dataset is divided using the original partitioning strategy, with the training set containing 34,054 images, 11,003 IDs, and 68,126 textual descriptions. The validation set includes 3,078 images, 1,000 IDs, and 6,158 textual descriptions; the test set consists of 3,074 images, 1,000 IDs, and 6,156 textual descriptions. CUHK-PEDES [11] is currently the most commonly used database for text-based pedestrian retrieval tasks.

RSTPReid [26], collected from MSMT [31], contains 20,505 images of 4,101 individuals. Each image is associated with a single corresponding textual description. The textual descriptions for each image comprise two sentences, with each sentence containing no fewer than 23 words. The dataset is divided into a training set and a test set, with the former containing 34,674 image-text pairs with 3,102 IDs, and the latter containing 19,848 image-text pairs with the remaining 1,000 IDs.

ICFG-PEDES [25] is also collected from MSMT 17 [31], encompassing 54,522 images of 4,102 individuals. Each image is accompanied by a descriptive sentence, averaging 37.2 words. The training and testing subsets include 34,674 image-text pairs (3,102 individuals) and 19,848 image-text pairs (1,000 individuals) from MSMT 17, respectively. Each image in these subsets is paired with a descriptive sentence, also averaging 37.2 words.

4.2 Evaluation Metrics

In all our experiments, we use Rank@k (k = 1, 5, 10) to evaluate performance, which measures the proportion of correct matches found within the top k retrieval results. Given a textual query description, all images in the gallery are ranked based on their similarity scores. A search is considered successful if the correct matching person image is found within the top k images. Additionally, to comprehensively assess performance, we also use the mean Average Precision (mAP) as an evaluation metric. The mAP quantifies the accuracy of retrieval by calculating the average precision at each correct match occurrence across all queries. Therefore, higher values of Rank-k and mAP indicate better retrieval performance.

4.3 Implementation Details

We conducted experiments using the PyTorch library on a single RTX4090 24GB GPU. The pre-trained image encoder is CLIP-ViT-B/16, and the pre-trained text encoder is the CLIP text transformer. The size of the input images was resized to 384×128. For all datasets, the maximum text length of the input sentences was set to 77. During training, our model was trained for 60 epochs using the Adam optimizer [9] and employed a linear warm-up strategy. The initial learning rates for the image and text backbones were set at 1e-5, with cosine learning rate decay.

4.4 Comparison with State-of-the-Art Methods

We compared our work with the most recent state-of-the-art studies on three existing datasets: CUHK-PEDES, ICFG-PEDES, and RSTPReid.The specific results are as follows Table 2, 3, 4.

Table 2. Performance comparisons with state-of-the-art methods on CUHK-PEDES datasets. '*' denotes that the model is trained under our environment.

Model	Ref	CUHK-PEDES			
		R@1	R@5	R@10	MAP
ViTAA [28]	ECCV20	54.92	75.18	82.90	51.60
SSAN [10]	arXiv21	61.37	80.15	86.73	–
LapsCore [32]	ICCV21	63.40	–	87.80	–
LBUL [30]	MM22	64.04	82.66	87.22	–
SAF [16]	ICASSP22	64.13	82.62	88.40	–
TIPCB [7]	Neuro22	64.26	83.19	89.10	–
AXM-Net [11]	MM22	64.44	80.52	86.77	58.73
LGUR [25]	MM22	65.25	83.12	89.00	–
IVT [26]	ECCVW22	65.59	83.11	89.21	–
CFine [35]	IEEE23	69.57	85.93	91.15	–
PLIP [42]	CoRR23	69.23	85.84	91.16	–
IRRA [14]*	CVPR23	72.09	88.84	93.25	65.99
CPCL [40]	CoRR24	70.03	87.28	91.78	63.19
Our Method	–	**73.21**	**88.92**	**93.29**	**66.14**

Performance Comparisons on CUHK-PEDES. Our Rank-1, Rank-5, Rank-10, and MAP outperform the recent state-of-the-art study IRRA* on CUHK-PEDES by + 1.12%, + 0.08%, + 0.04%, and + 0.1%, respectively, where IRRA* is the result tested in an environment consistent with our model. And it exceeds CFine [35] on the metric Rank-1 reaching +3.64% and exceeded CPCL by 3.19%.

Performance Comparisons on ICFG-PEDES. Our Rank-1, Rank-5, Rank-10, and MAP outperform the recent state-of-the-art study IRRA* by +0.37%, +0.13%, +0.04%, and +0.03%, respectively, on ICFG-PEDES, where IRRA* is the result tested in an environment consistent with our model. It also exceeded CFine [35] on the metric Rank-1 reaching +2.05% and exceeded CPCL by 0.28%.

Table 3. Performance comparisons with state-of-the-art methods on ICFG-PEDES datasets. The * denotes that the model is trained under our environment.

Model	Ref	ICFG-PEDES			
		R@1	R@5	R@10	MAP
LGUR [25]	MM22	59.20	75.32	81.73	–
IVT [26]	ECCVW22	56.04	73.60	80.22	–
CFine [35]	IEEE23	60.83	76.55	82.42	–
IRRA [14]*	CVPR23	62.51	79.59	85.17	38.13
CPCL [40]	CoRR24	62.60	79.07	84.46	36.16
Our Method	–	**62.88**	**79.72**	**85.21**	**38.16**

Table 4. Performance comparisons with state-of-the-art methods on RSTPReid datasets. The * denotes that the model is trained under our environment.

Model	Ref	RSTPReid			
		R@1	R@5	R@10	MAP
IVT [26]	ECCVW22	46.70	70.00	78.80	–
CFine [35]	IEEE23	50.55	72.50	81.60	–
IRRA [14]*	CVPR23	**59.45**	78.90	86.40	**47.53**
CPCL [40]	CoRR24	58.35	**81.05**	**87.65**	45.81
Our Method	–	58.20	79.90	86.25	47.11

Performance Comparisons on RSTPReid. Our Rank-5 outperforms the recent state-of-the-art study IRRA* by +1.00% on RSTPReid, where IRRA* is the result tested in an environment consistent with our model. But it also exceeds CFine [35] by +7.65% on the metric Rank-1 and exceeds CPCL on the indicator MAP by 1.30%. This result suggests that our model is more applicable to datasets containing larger character data, and it is also common to have a large amount of character data in real applications whereby our model is effective.

In summary, these results demonstrate the superiority and discriminative power of our proposed method.

4.5 Ablation Study

To better understand the contributions of each component in our framework, we conduct a comprehensive empirical analysis in this section. Specifically, the results of different modules of our framework on the CUHK-PEDES [11] datasets are shown in Table 5.

Ablations on Cross-model Mask Alignment. Integrating the cross-modal Mask Alignment Module (CMA) into the base model led to enhancements in Rank-1, Rank-10, and MAP by 0.28%, 0.04%, and 0.06% correspondingly. This

Table 5. Ablation study on each component of CMDA on CUHK-PEDES datasets.

No.	Methods	Components		CUHK-PEDES			
		CMA	CDA	R@1	R@5	R@10	MAP
0	baseline			72.09	88.84	93.25	65.99
1	+CMA	✓		72.37	88.84	93.29	66.05
2	+CDA		✓	72.66	**88.94**	**93.55**	66.11
3	Our Method	✓	✓	**73.21**	<u>88.92</u>	<u>93.29</u>	**66.14**

indicates a slight improvement over the baseline. The limited hardware capacity in the experimental setup might hinder the utilization of larger training batch sizes, thereby not fully leveraging the potential of mask images.

Ablations on Cross-model Detail Alignment. Adding the Cross-model Detail Alignment (CDA) module to the baseline model improves Rank-1, Rank-5, Rank-10, and MAP by 0.57%, 0.10%, 0.30%, and 0.12% respectively. This suggests that the CDA module enhances our model metrics more significantly than our cross-modal Mask alignment module, with superior performance in Rank-5 and Rank-10 compared to both modules combined. This result indicates that this module has a greater improvement on the results of each metric of our model compared to our Cross-modal Mask Alignment (CMA) module, and the results on Rank-5, and Rank-10 are better than the two modules together.

4.6 Qualitative Results

Figure 3 shows the top 10 retrieval results of our model for the given target image in the text query. As shown in Fig. 3, images with green frames are correct, those

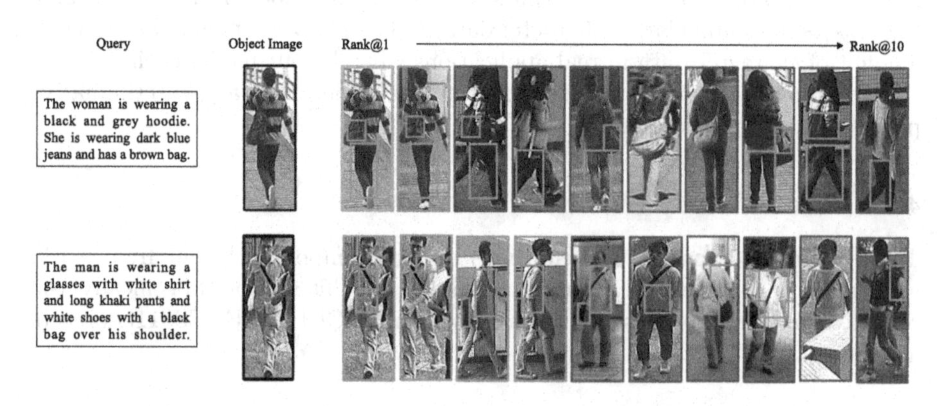

Fig. 3. Comparison of top-10 retrieved results on CUHK-PEDES for each text query. The image corresponds to the query text. Matched and mismatched images are marked with green, and red rectangles, respectively. (Color figure online)

with red frames are incorrect, and those with black frames correspond to the target pictures described in our text. The accuracy of the retrieval results, as demonstrated in Fig. 3, is primarily attributed to our Cross-modal Detail Alignment (CDA) module. This module effectively utilizes fine-grained discriminative clues to distinguish different pedestrians, which are shown in the orange and blue textual and image area boxes in Fig. 3.

5 Conclusion

In this study, we propose a Cross-Modal Masking Detail Alignment (CMDA) framework, which includes a Cross-Modal Masking Alignment module (CMA). This module applies masks to images and text, not only enhancing their discriminative power but also improving the semantic consistency between text and images. In addition, we also developed a cross-modal detail alignment module (CDA), which facilitates the interaction between text and image modalities and extracts more critical information. In three benchmarks for text-based person retrieval, our model achieved SOTA on the public datasets CUHK-PEDES and ICFG-PEDES, achieving top rankings in Rank-1, Rank-5, Rank-10, and MAP metrics, and it also reached the best performance in Rank-5 on the RSTPReid dataset.

Disclosure of Interests. The authors have no competing interests to declare that are relevant to the content of this article.

References

1. Aggarwal, S., Radhakrishnan, V.B., Chakraborty, A.: Text-based person search via attribute-aided matching. In: Proceedings of the IEEE/CVF Winter Conference on Applications of Computer Vision, pp. 2617–2625 (2020)
2. Bao, H., Dong, L., Piao, S., Wei, F.: Beit: bert pre-training of image transformers. arXiv preprint arXiv:2106.08254 (2021)
3. Cao, S., Xu, P., Clifton, D.A.: How to understand masked autoencoders. arXiv preprint arXiv:2202.03670 (2022)
4. Chen, J., et al.: Eve: Efficient vision-language pre-training with masked prediction and modality-aware moe. In: Proceedings of the AAAI Conference on Artificial Intelligence, vol. 38, pp. 1110–1119 (2024)
5. Chen, T., Xu, C., Luo, J.: Improving text-based person search by spatial matching and adaptive threshold. In: 2018 IEEE Winter Conference on Applications of Computer Vision (WACV), pp. 1879–1887. IEEE (2018)
6. Chen, Y., Huang, R., Chang, H., Tan, C., Xue, T., Ma, B.: Cross-modal knowledge adaptation for language-based person search. IEEE Trans. Image Process. **30**, 4057–4069 (2021)
7. Chen, Y., Zhang, G., Lu, Y., Wang, Z., Zheng, Y.: Tipcb: a simple but effective part-based convolutional baseline for text-based person search. Neurocomputing **494**, 171–181 (2022)

8. Devlin, J., Chang, M.W., Lee, K., Toutanova, K.: Bert: pre-training of deep bidirectional transformers for language understanding. arXiv preprint arXiv:1810.04805 (2018)
9. Diederik, P.K.: Adam: a method for stochastic optimization (2014)
10. Ding, Z., Ding, C., Shao, Z., Tao, D.: Semantically self-aligned network for text-to-image part-aware person re-identification. arXiv preprint arXiv:2107.12666 (2021)
11. Farooq, A., Awais, M., Kittler, J., Khalid, S.S.: Axm-net: implicit cross-modal feature alignment for person re-identification. In: Proceedings of the AAAI Conference on Artificial Intelligence, vol. 36, pp. 4477–4485 (2022)
12. Gray, D., Brennan, S., Tao, H.: Evaluating appearance models for recognition, reacquisition, and tracking. In: Proceedings of IEEE International Workshop on Performance Evaluation for Tracking and Surveillance (PETS), vol. 3, pp. 1–7 (2007)
13. He, K., Zhang, X., Ren, S., Sun, J.: Deep residual learning for image recognition. In: Proceedings of the IEEE Conference on Computer Vision and Pattern Recognition, pp. 770–778 (2016)
14. Jiang, D., Ye, M.: Cross-modal implicit relation reasoning and aligning for text-to-image person retrieval. In: Proceedings of the IEEE/CVF Conference on Computer Vision and Pattern Recognition, pp. 2787–2797 (2023)
15. Jing, Y., Si, C., Wang, J., Wang, W., Wang, L., Tan, T.: Pose-guided multi-granularity attention network for text-based person search. In: Proceedings of the AAAI Conference on Artificial Intelligence, vol. 34, pp. 11189–11196 (2020)
16. Li, S., Cao, M., Zhang, M.: Learning semantic-aligned feature representation for text-based person search. In: ICASSP 2022-2022 IEEE International Conference on Acoustics, Speech and Signal Processing (ICASSP), pp. 2724–2728. IEEE (2022)
17. Li, S., Xiao, T., Li, H., Yang, W., Wang, X.: Identity-aware textual-visual matching with latent co-attention. In: Proceedings of the IEEE International Conference on Computer Vision, pp. 1890–1899 (2017)
18. Li, S., Xiao, T., Li, H., Zhou, B., Yue, D., Wang, X.: Person search with natural language description. In: Proceedings of the IEEE Conference on Computer Vision and Pattern Recognition, pp. 1970–1979 (2017)
19. Li, W., Zhao, R., Wang, X.: Human reidentification with transferred metric learning. In: Lee, K.M., Matsushita, Y., Rehg, J.M., Hu, Z. (eds.) ACCV 2012. LNCS, vol. 7724, pp. 31–44. Springer, Heidelberg (2013). https://doi.org/10.1007/978-3-642-37331-2_3
20. Li, W., Zhao, R., Xiao, T., Wang, X.: Deepreid: deep filter pairing neural network for person re-identification. In: 2014 IEEE Conference on Computer Vision and Pattern Recognition, CVPR 2014, Columbus, OH, USA, 23–28 June 2014, pp. 152–159. IEEE Computer Society (2014). https://doi.org/10.1109/CVPR.2014.27
21. Liu, J., Zha, Z.J., Hong, R., Wang, M., Zhang, Y.: Deep adversarial graph attention convolution network for text-based person search. In: Proceedings of the 27th ACM International Conference on Multimedia, pp. 665–673 (2019)
22. Ma, Z., Xu, F., Liu, J., Yang, M., Guo, Q.: Sycoca: symmetrizing contrastive captioners with attentive masking for multimodal alignment. arXiv preprint arXiv:2401.02137 (2024)
23. Memory, L.S.T.: Long short-term memory. Neural Comput. **9**(8), 1735–1780 (2010)
24. Sarafianos, N., Xu, X., Kakadiaris, I.A.: Adversarial representation learning for text-to-image matching. In: Proceedings of the IEEE/CVF International Conference on Computer Vision, pp. 5814–5824 (2019)

25. Shao, Z., Zhang, X., Fang, M., Lin, Z., Wang, J., Ding, C.: Learning granularity-unified representations for text-to-image person re-identification. In: Proceedings of the 30th ACM International Conference on Multimedia, pp. 5566–5574 (2022)
26. Shu, X., et al.: See finer, see more: implicit modality alignment for text-based person retrieval. In: European Conference on Computer Vision, pp. 624–641. Springer, Heidelberg (2022). https://doi.org/10.1007/978-3-031-25072-9_42
27. Simonyan, K., Zisserman, A.: Very deep convolutional networks for large-scale image recognition. arXiv preprint arXiv:1409.1556 (2014)
28. Wang, Z., Fang, Z., Wang, J., Yang, Y.: *ViTAA*: visual-textual attributes alignment in person search by natural language. In: Vedaldi, A., Bischof, H., Brox, T., Frahm, J.-M. (eds.) ECCV 2020. LNCS, vol. 12357, pp. 402–420. Springer, Cham (2020). https://doi.org/10.1007/978-3-030-58610-2_24
29. Wang, Z., et al.: Caibc: capturing all-round information beyond color for text-based person retrieval. In: Proceedings of the 30th ACM International Conference on Multimedia, pp. 5314–5322 (2022)
30. Wang, Z., et al.: Look before you leap: improving text-based person retrieval by learning a consistent cross-modal common manifold. In: Proceedings of the 30th ACM International Conference on Multimedia, pp. 1984–1992 (2022)
31. Wei, L., Zhang, S., Gao, W., Tian, Q.: Person transfer gan to bridge domain gap for person re-identification. In: Proceedings of the IEEE Conference on Computer Vision and Pattern Recognition, pp. 79–88 (2018)
32. Wu, Y., Yan, Z., Han, X., Li, G., Zou, C., Cui, S.: Lapscore: language-guided person search via color reasoning. In: Proceedings of the IEEE/CVF International Conference on Computer Vision, pp. 1624–1633 (2021)
33. Xiao, T., Li, S., Wang, B., Lin, L., Wang, X.: End-to-end deep learning for person search. CoRR arxiv:1604.01850 (2016)
34. Xie, Z., et al.: Simmim: a simple framework for masked image modeling. In: Proceedings of the IEEE/CVF Conference on Computer Vision and Pattern Recognition, pp. 9653–9663 (2022)
35. Yan, S., Dong, N., Zhang, L., Tang, J.: Clip-driven fine-grained text-image person re-identification. IEEE Trans. Image Process. (2023)
36. Yan, S., Tang, H., Zhang, L., Tang, J.: Image-specific information suppression and implicit local alignment for text-based person search. IEEE Trans. Neural Netw. Learn. Syst. (2023)
37. Zhang, Y., Lu, H.: Deep cross-modal projection learning for image-text matching. In: Proceedings of the European Conference on Computer Vision (ECCV), pp. 686–701 (2018)
38. Zheng, K., Liu, W., Liu, J., Zha, Z.J., Mei, T.: Hierarchical gumbel attention network for text-based person search. In: Proceedings of the 28th ACM International Conference on Multimedia, pp. 3441–3449 (2020)
39. Zheng, L., Shen, L., Tian, L., Wang, S., Bu, J., Tian, Q.: Person re-identification meets image search. CoRR arxiv:1502.02171 (2015)
40. Zheng, Y., et al.: CPCL: cross-modal prototypical contrastive learning for weakly supervised text-based person re-identification. CoRR arxiv:2401.10011 (2024). https://doi.org/10.48550/ARXIV.2401.10011
41. Zhu, A., et al.: DSSL: deep surroundings-person separation learning for text-based person retrieval. In: Proceedings of the 29th ACM International Conference on Multimedia, pp. 209–217 (2021)
42. Zuo, J., Yu, C., Sang, N., Gao, C.: PLIP: language-image pre-training for person representation learning. CoRR arxiv:2305.08386 (2023). https://doi.org/10.48550/ARXIV.2305.08386

Prototype Enhancement for Few-Shot Point Cloud Semantic Segmentation

Zhengyao Li[1,2], Gengshen Wu[1], and Yi Liu[1,2](✉)

[1] Changzhou University, Changzhou 213164, China
liuyi0089@gmail.com
[2] City University of Macau, Macau 999078, China

Abstract. Few-shot point cloud semantic segmentation plays a fundamental role in the computer vision community since annotating point cloud data is quite time-consuming and labor-intensive. Current semantic segmentation methods employ few-shot learning to reduce dependence on labeled samples and enhance model generalization to new categories. Due to the complex 3D geometries of point clouds, significant feature variations exist even within the same category, meaning that a few training samples (support set) might not fully capture all category features. This discrepancy leads to differences in distribution between the support set and the samples used to evaluate the model (query set), impacting the effectiveness of traditional semantic segmentation approaches. In our paper, we employ a prototype enhancement strategy for few-shot point cloud semantic segmentation. Specifically, to align the prototype representation from the support set more closely with the query set, our framework proposes two modules to enhance the generated original prototype, we have developed a Cross Feature Enhancement module, which enhances support set features by reducing differences in terms of distribution of support and query sets. Moreover, we proposed a prototype correction module to refine the prototypes with the aim of matching query sets accurately. We conducted thorough experiments demonstrates the state-of-the-art performance of our model on publicly available benchmarks including S3DIS and ScanNet.

Keywords: Few-Shot learning · Point cloud · Semantic segmentation

1 Introduction

Point clouds are crucial in scene semantic segmentation tasks like autonomous driving [9,24,35] and robotics [1,11] for capturing accurate 3D geometric details. Despite advancements in fully supervised semantic segmentation algorithms [3,13,19], they demand extensive labeled data, which becomes increasingly challenging as point clouds grow denser due to advances in 3D acquisition technologies. Which need for extensive labeling severely limits the practical application of these algorithms. Alternatively, recent research has shifted towards few-shot learning framework for point cloud semantic segmentation.

T. Zhu et al. (Eds.): ICA3PP 2024, LNCS 15252, pp. 270–285, 2025.
https://doi.org/10.1007/978-981-96-1528-5_18

The task of Few-shot 3D Point Cloud Semantic Segmentation introduces a novel challenge by placing the conventional 3D point cloud semantic segmentation within a few-shot framework. Here, the model is equipped with the capability to segment new classes using only a small amount of supporting data. Zhao et al. [34] introduce the concept of attention-aware multi-prototype transductive inference (attMPTI) to address this challenge, marking the first attempt at few-shot point cloud semantic segmentation. However, attMPTI is hindered by its complexity and time-intensive nature, attributed to its utilization of multiple prototypes and the establishment of graph structures for few-shot point cloud semantic segmentation. As a result, attMPTI fails to achieve remarkable performance despite its efforts (Fig. 1).

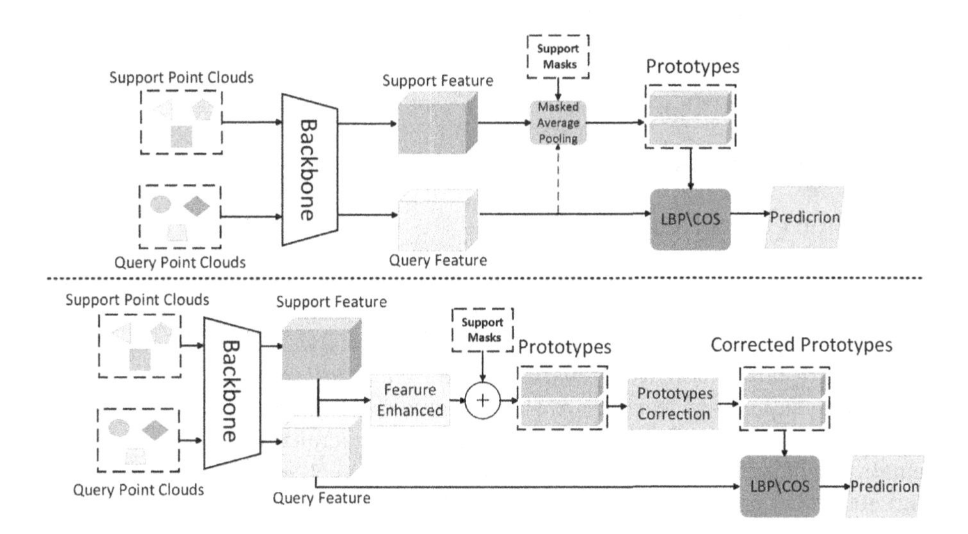

Fig. 1. Comparison of our method with previous methods, we further enhance the support set features and calibrate the prototypes generated using the optimized features.+: Element by element addition, LBP: non-parametric label propagation, COS: cosine similarity.

Building on these insights, we propose a framework to enhance the prototype of support features within the meta-learning strategy for few-shot semantic segmentation of point clouds. Specifically, the proposed model organizes data into labeled support sets and unlabeled query sets, using insights from the support set to guide the semantic segmentation of new categories. Firstly, to bridge the gap between support set prototypes and query set samples, we introduced the Cross Feature Enhancement (CFE) Module, which uses query set features to optimize and generate enhanced features, which reduces distance between the sets and improves the model's adaptability to new categories. Secondly, we design a Prototype Correction (PC) Module to refine support prototypes, which transfers prototypes to the query set domain by reducing the domain difference between the support set and the query set. To improve label prediction efficiency and counter the randomness of the Label Propagation Algorithm (LPA), we use a distance metric to assign labels based on the closest prototype similarity In summary, our primary contributions are:

- We propose an prototype enhancement network for few-shot point cloud semantic segmentation.
- We design a Cross-Feature Enhancement module to generate enhanced feature representations and reduce domain differences between support sets and query sets.
- We design a Prototype Correction module to enhance the support set features and optimize the generated prototypes to further correct for differences in feature distribution.
- We conducted thorough experiments to demonstrate state-of-the-art performance of our model on benchmark datasets including S3DIS and ScanNet.

2 Related Work

2.1 Point Cloud Recognition

Point cloud semantic segmentation has advanced from global feature-focused PointNet [17] to PointNet++ [18] which captures local details, and DGCNN [25] that uses graph convolution for better point analysis. As fully supervised methods are label-intensive, research has moved to less supervised approaches like weakly supervised learning [26], self-supervised learning [4], and perturbed self-distillation [32] to lessen labeling needs. Yet, generalizing to new categories remains a challenge, leading to approaches like few-shot semantic segmentation that efficiently process both existing and new, unlabeled point categories.

2.2 Few-Shot Semantic Segmentation

In the realm of few-shot semantic segmentation, recent methods such as OSLSM [20] and others [28] leverage techniques like dual-branch structures and dense feature comparison modules, albeit facing challenges such as overfitting. Drawing inspiration from Prototypical Networks [21], some approaches [31] employ masked average pooling and cosine similarity to link features between support and query images, while alternative strategies [12,27] generate multiple prototypes for enhanced semantic representation. In point cloud semantic segmentation, while some methods [34] utilize FPS and KNN for prototype creation and label spread, they may overlook feature distribution discrepancies. Our proposed method simplifies this process by forming a single global prototype per category from the support set and employing cosine similarity for similarity assessment with query samples.

2.3 Attention Mechanisms in Point Cloud Semantic Segmentation

Attention mechanisms have been widely explored and applied in the field of 3D point cloud semantic segmentation, PointNet++ [18] uses both local and global attention mechanisms to improve the expressiveness of point cloud features and hence the performance of semantic segmentation, Dynamic Graph Attention Network [36] utilises a dynamic graph attention mechanism that enables the network to dynamically adapt the graph structure to different task requirements based on the features of the supported samples. Point Attenti on Network [33] Focuses on introducing point-level attention mechanisms in point cloud data. It enables the network to focus on important points more efficiently by weighting each point and using these important points for semantic segmentation tasks. These works demonstrate the effectiveness and importance of applying attention mechanisms in semantic segmentation of 3D point clouds, and provide strong references and insights to further enhance the research and applications in this area.

3 Proposed Method

3.1 Problem Definition

Point cloud semantic segmentation involves assigning semantic labels to individual points. In this study, we utilize episode training [17], a technique commonly employed in few-shot learning, to train our model. We represent each training task using the C-way K-shot format typical in episode training, where each task comprises a support set and a query set of data. For instance, considering a 2-way 1-shot task, the support set can be described as:

$$S = \left\{ \left(S_p^1, M^1\right), \left(S_p^2, M^2\right) \right\}, \tag{1}$$

S_p^1 denotes samples from the first category, and M^1 denotes the corresponding binary mask for S_p^1. Additionally, the query set can be described as:

$$Q = \left\{ \left(Q_p^i, L^i\right) \right\}_{i=1}^{T}. \tag{2}$$

The dataset consists of T pairs of query points and their corresponding label and the objective of this task is to train the model using the provided support set S, comprising samples from two categories, in order to minimize errors in label predictions on the query set Q.

3.2 Architecture Overview

Our network framework employs metric learning for few-shot point cloud semantic segmentation, As shown in Fig. 2. It accepts inputs with dimensions corresponding to the number of categories, samples per category, feature dimension, and points per sample. Using a 2-way 1-shot example, features are extracted from support and query set point clouds with a DGCNN backbone.

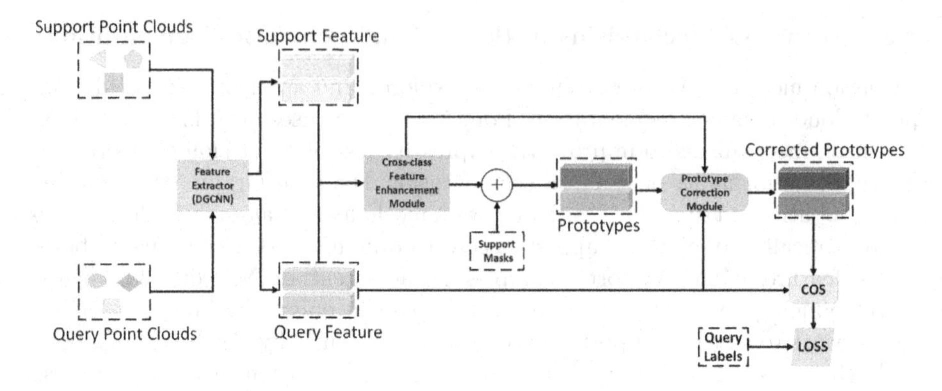

Fig. 2. Illustration of our proposed network framework (Take 2-way 1-shot for example. +: Element by element addition. "cos" denotes cosine similarity.)

Our Cross Feature Enhancement module boosts support features, which, along with support masks, are used to create prototypes. These prototypes are then optimized with prototype correction module and used to semantically label the query point cloud. Model performance is improved by minimizing the loss against ground truth labels.

3.3 Feature Embedding Network

The feature embedding module is a key part of our model, using the same architecture as AttMPTI [34] to extract unique features for each category, critical for prototype computation. It encodes a point cloud with C-dimensional features through three EdgeConvs [25], then uses MLPs to reduce it to a 256-dimensional space, and the formula is:

$$\mathbf{F}_{emb} = m_1(Concat(EC_{g1}(\mathbf{X}), EC_{g2}(\mathbf{X}))), \tag{3}$$

where EC_{g1}, EC_{g2} signifie EdgeConv, X represents a point cloud sample X with C-dimension features, m represents MLPs (Fig. 3).

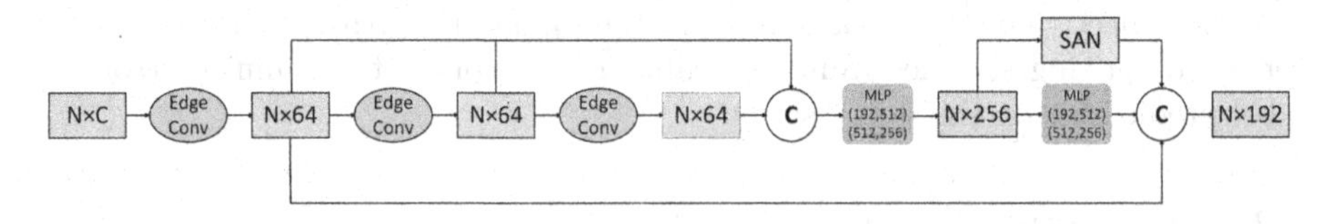

Fig. 3. The architecture of feature embedding module (C: Concatenation operation.). The basic feature extractor is trained in the pre-training process.)

These features are further processed by a self-attention module [29] and more MLPs to arrive at a 64-dimensional encoding, its expression is:

$$\mathbf{F}_{san} = \text{softmax}\left(m_2\left(\mathbf{F}_{emb}\right)^T \times m_3\left(\mathbf{F}_{emb}\right)\right) \times m_4\left(\mathbf{F}_{emb}\right), \tag{4}$$

where $\mathbf{F}_{san}$ denotes the output features of the SAN, m represents MLPs, and $\times$ denotes pointwise multiplication.

The final 192-dimensional output, F_{192}, combines these features with the first EdgeConv's output, creating a rich feature set for the point cloud, It's formulation can be expressed as:

$$\mathbf{F}_{out} = \text{Concat}(EC_{g1}(X), \mathbf{F}_{san}, m_5(\mathbf{F}_{emb})). \tag{5}$$

where $\mathbf{F}_{out}$ $\in$after the feature extractor, we finally get the 192-dimensional feature of the point cloud sample.

3.4 Cross Feature Enhancement Module

Acknowledging the feature distribution bias and overlap between support and query sets, we introduce a Cross Feature Enhancement Module. This module aligns features from both sets, inspired by techniques used in image semantic segmentation. For example, we process high-dimensional features from the same-class support and query sets through a feature embedding network, followed by global max-pooling for global statistics, and then feed them into a shared MLP. This not only economizes model parameters, avoiding separate MLPs, but also reinforces learning of co-occurring features. A softmax activation function post-MLP assigns feature importance at the channel level, The formula is as follows:

$$\begin{aligned} \mathbf{F}'_s &= \text{softmax}\left(m_6\left(\text{maxpooling}\left(\mathbf{F}_s\right)\right)\right), \\ \mathbf{F}'_q &= \text{softmax}\left(m_6\left(\text{maxpooling}\left(\mathbf{F}_q\right)\right)\right), \end{aligned} \tag{6}$$

where $\mathbf{F}_s$ and $\mathbf{F}_q$ represents the high-dimensional support set features and query set features obtained by the feature embedding network.

Subsequently, we perform an element-wise multiplication of feature mappings and then refine the support set features with this weighted representation.

$$\mathbf{F}_{com} = \left(\mathbf{F}'_s \times \mathbf{F}'_q\right) \times \mathbf{F}_s, \tag{7}$$

where $\times$ is the pointwise multiplication.

In scenarios with limited data, the disparity between the prototype derived from the support set and that from the entire dataset can weaken the model's ability to discriminate. To address this, we aim to close the feature gap between the support and query sets. Drawing from the domain adaptation field [23], we employ the mean-a first-order statistic-of datasets to signify their domain difference. By calculating the mean difference between the support and query sets, we measure their distance. We begin by determining the deviation of features

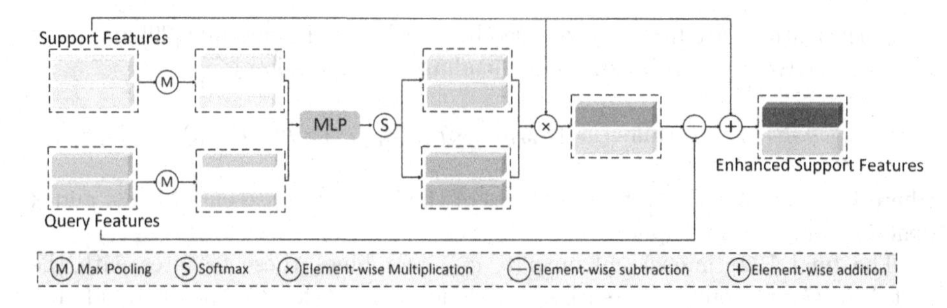

Fig. 4. The architecture of Cross Feature Enhancement Module.

between these sets, As shown in Fig. 4, taking a 2-way 1-shot task as an example, its expression is:

$$\begin{aligned} \mathbf{B}_{c1} &= \frac{1}{n_q}\sum_{j=1}^{n_q} \mathbf{f}_j^{q_1} - \frac{1}{n_s}\mathbf{F}_{com}^1, \\ \mathbf{B}_{c2} &= \frac{1}{n_q}\sum_{j=1}^{n_q} \mathbf{f}_j^{q_2} - \frac{1}{n_s}\mathbf{F}_{com}^2, \end{aligned} \tag{8}$$

where n_q denotes the number of points contained in the samples in the query set, and $\mathbf{f}_j^{q1}$ denotes the feature of the $j-th$ point in the first query set,$\mathbf{B}_{c1}$ denotes the feature distribution difference of the first category, $\mathbf{F}_{COM}^1$ denotes the enhanced feature of the first category, which focuses more on the co-occurring features in the support set and the query set. After obtaining the feature distribution deviation term for each category, we generate the enhanced support set features by adding the deviation terms to their corresponding support features:

$$\begin{aligned} \widetilde{\mathbf{F}}_{s1} &= \mathbf{F}_{s1} + \mathbf{B}_{c1}, \\ \widetilde{\mathbf{F}}_{s2} &= \mathbf{F}_{s2} + \mathbf{B}_{c2}, \end{aligned} \tag{9}$$

where $\widetilde{\mathbf{F}}_s$ denotes Enhanced support set features,F_s represents the raw support set features.

3.5 Prototype Correction Module

We generate prototypes for each category by averaging features using support masks for average pooling, normalizing by mask activations, and including a background prototype for non-category points. Our Prototype Correction module then aligns these prototypes with the query features, informed by query-support interactions, as shown in Fig. 5.

Prototypes for each category in the support set S, consisting of $S = \{(I_{n,k}^S, M_{n,k}^S)\}$ where k and n represent K-shot and N-way indices respectively, are generated using masked average pooling on corresponding features $\mathbf{F}_{n,k}^S \in \mathbb{R}^{N_p \times d}$.

Here, N_p is the number of points and d is the channel number. The prototype for category n is derived by averaging these features.

$$\mathbf{p}^n = \frac{1}{K}\sum_k \frac{\sum_x F_{S,x}^{n,k}\mathbb{q}(M_{S,x}^{n,k} = n)}{\sum_x \mathbb{q}(M_{S,x}^{n,k} = n)}, \tag{10}$$

For coordinates x ranging from 1 to N_p, the binary label indicator $\mathbb{q}(*)$ outputs 1 when $*$ is true. In addition to the target categories, a background prototype $\mathbf{p}_0$ is computed to represent points not belonging to any of the N target categories:

$$\mathbf{p}^0 = \frac{1}{NK}\sum_{n,k} \frac{\sum_x \mathbf{F}_{S,x}^{n,k}\mathbb{q}(M_{S,x}^{n,k} \notin \{1,...,N\})}{\sum_x \mathbb{q}(M_{S,x}^{n,k} \notin \{1,...,N\})}. \tag{11}$$

Now, our set of prototypes is represented as $\mathbf{P} = \{\mathbf{p}_0, \mathbf{p}_1, \ldots, \mathbf{p}_N\}$.

Detailedly, for a prototype $\mathbf{p}^i$ in $\mathbb{R}^{1\times d}$, ranging from 0 to N, and its corresponding enhanced support features, along with a query feature $\mathbf{F}_Q$ in $\mathbb{R}^{N\times d}$, we compute channel-wise attention, creating a projection attention matrix to adjust the prototype to align with the query feature distribution. The support features are initially averaged for each prototype:

$$\mathbf{F}_S^i = \begin{cases} \frac{1}{K}\sum_k \mathbf{F}_S^{n,k}, & i \in \{1,...,N\}, \\ \frac{1}{NK}\sum_{n,k} \mathbf{F}_S^{n,k}, & i = 0, \end{cases} \tag{12}$$

where $\mathbf{F}_S^i \in R^{N\times d}$ represent averaged support feature for prototype $\mathbf{P}^i$, Then the input to our PC is designed as:

$$\mathbf{Q} = \mathbf{F}_Q{}^\top \mathbf{W}_q, \mathbf{K} = \mathbf{F}_S^i{}^\top \mathbf{W}_k, \mathbf{V} = \mathbf{P}^i \mathbf{W}_v, \tag{13}$$

The matrices $\mathbf{W}_q$ and $\mathbf{W}_k$ are in $\mathbb{R}^{N\times N'}$, and $\mathbf{W}_v$ is in $\mathbb{R}^{d\times d}$, serving as adjustable parameters in fully connected layers to map features and prototypes into specific latent spaces. To enhance computational efficiency, N' is kept at or below N, leading to the derivation of $\mathbf{Q}$ and $\mathbf{K}$ matrices in $\mathbb{R}^{d\times N'}$, and the $\mathbf{V}$ matrix in $\mathbb{R}^{1\times d}$.

These matrices facilitate the transformer attention mechanism by multiplying the $\mathbf{Q}$ matrix with the transposed $\mathbf{K}$, This reflects the cross-correlation between different channels that reflect the characteristics of the support set and the query set, followed by applying a softmax layer to produce the channel attention map $\mathbf{Attn}$ in $\mathbb{R}^{d\times d}$, indicating the distribution of attention across channels:

$$\mathbf{Attn} = \text{softmax}(\frac{\mathbf{Q}\cdot\mathbf{K}^\top}{\sqrt{d}}), \tag{14}$$

The softmax function normalizes attention on a per-row basis. The resulting cross-attention map, $\mathbf{Attn}$, aligns query and support features across channels, guiding channel distribution. Ultimately, multiplying $\mathbf{Attn}$ by the transposed $\mathbf{V}$ refines the feature channel alignment from prototype to $\mathbf{Q}$:

$$\hat{\mathbf{P}}^i = \mathbf{W}_p(\mathbf{Attn}\cdot\mathbf{V}^\top)^\top, \tag{15}$$

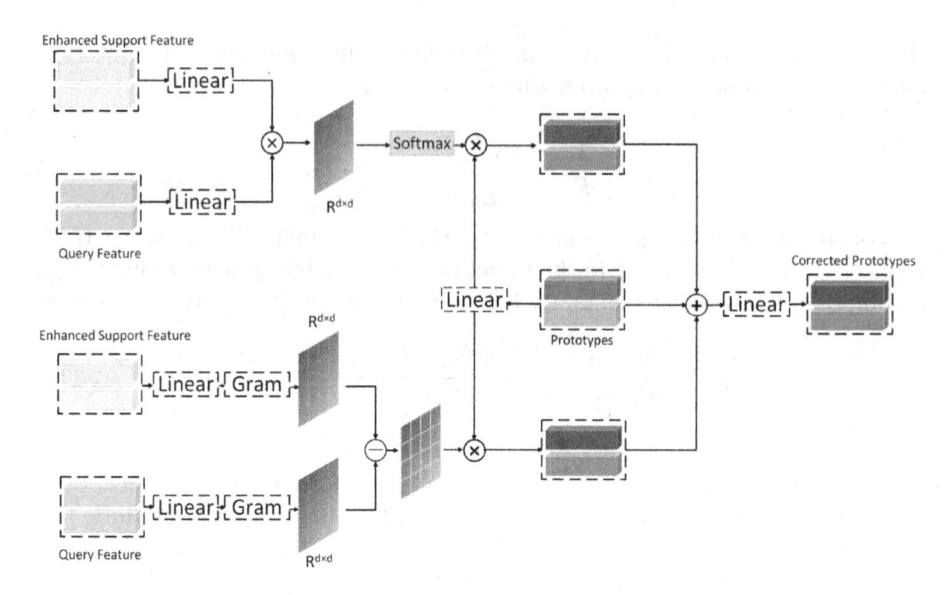

Fig. 5. The architecture of our proposed Prototype Correction module. Enhanced Support features and query features are extracted by DGCNN and Cross Feature Enhancement Module respectively)

different feature channels extracted from our hand-crafted filters are evidently not independent, as there are correlations between channels. Additionally, these correlations differ between the support-set and query-set domains. Therefore, we examine the difference in self-correlation between $\mathbf{F}_S$ and $\mathbf{F}_Q$, which also measures the domain gap between the support set and the query set. Typically, the Gram matrix of the feature vector is used to represent self-correlation [7,10]. Denoted as $\mathbf{G}^S$ and $\mathbf{G}^Q$, the Gram matrices for the support-set and query-set features are calculated as follows:

$$\mathbf{G}^S = \mathbf{F}_S^\top \mathbf{F}_S, \quad \mathbf{G}^Q = \mathbf{F}_Q^\top \mathbf{F}_Q, \tag{16}$$

both of these Gram matrices are symmetric. In addition, the difference between $\mathbf{G}^S$ and $\mathbf{G}^Q$ can also be a further measure of the domain gap between the support set and the query set. We extract this domain gap using a linear projection, $\mathbf{W}_g \in \mathbb{R}^{D\times 1}$, and utilize it to adjust the channels of $\mathbf{P}^i$:

$$\dot{\mathbf{P}}^i = \mathbf{P}^i \cdot [\mathrm{diag}((\mathbf{G}^Q - \mathbf{G}^S)\mathbf{W}_g)], \tag{17}$$

where diag$(\cdot)$ denotes diagonalization and $\dot{\mathbf{P}}^i$ represents the rectified prototypes through self-correlation.

Incorporating the original prototype as a residual connection promotes steadier training progression. We integrate the adjusted prototype with the original prototype to obtain the final prototype:

$$\mathbf{P}^* = \mathbf{P}^i + \dot{\mathbf{P}}^i + \hat{\mathbf{P}}^i. \tag{18}$$

Hence, when executing cosine similarity assessments between query features and these enhanced prototypes, channel-wise discrepancies are mitigated. This refinement is key to improving the outcome.

4 Experiments

4.1 Datasets and Data Processing

S3DIS dataset(Stanford large-scale 3D Indoor Spaces Dataset) [2] consists of 3D scans of 272 indoor rooms from 6 areas, captured with Matterport technology and annotated in 12 semantic categories plus clutter.

ScanNet [5] features 1, 513 point clouds from 707 indoor scenes, with points labeled in 20 semantic categories plus one for unannotated data.

Data Preprocessing: we use the method from [17] to manage the high point density in original room scans. Each room is segmented into 1 m × 1 m blocks on the XY plane, each containing 2,048 randomly chosen points. Consequently, S3DIS is segmented into 7,521 blocks, and ScanNet into 36,350 blocks. For few-shot learning, each dataset is split into two exclusive subsets based on the alphabetical order of semantic categories, The details of the category split are presented in Table 1. We designate one subset as the training set D_{train} and the other as the test set D_{test}. We evaluate our model through cross-validation between these subsets. For instance, in a 2-way 1-shot training episode, we select two classes from D_{train} and samples from these classes to create a support set S and a query set Q, along with their respective support masks M and labels L. In the 2-way 1-shot test phase, we conduct 100 episodes for each class combination from D_{test}, iterating through all possible pairs.

Table 1. List of category names for each subset of the S3DIS and ScanNet

Dataset	subset0		subset1	
S3DIS	beam	board	door	floor
	bookcase	celling	sofa	table
	chair	column	wall	window
ScanNet	bathtub	bed	otherfurniture	picture
	bookshelf	cabinet	refrigerator	show curtain
	chair	counter	sink	sofa
	curtain	desk	table	toilet
	door	floor	wall	window

4.2 Experimental Details

Experimental Environment: The implementation of our model is based on the Pytorch framework. All experiments are performed on one GPU (GeForce RTX 3090).

Pre−train: During pre-training, we append MLPs to the basic feature extractor, fix the neighbor number of sample points at 20, and employ the Adam optimizer. We use a batch size of 16, a learning rate of 0.001, and train for 100 epochs. After initializing the feature extractor with pre-trained weights, we train the full feature extractor using the Adam optimization algorithm [8]. The initial learning rate for the basic feature extractor is 0.0001, while for subsequent layers it is 0.001. We halve the learning rate every 5,000 iterations. Gaussian jittering and rotation around the z-axis augment both support and query sets.

Evaluation Metrics: For assessing the model in the point cloud semantic segmentation task, we employ mIoU (mean Intersection over Union) as the evaluation metric. The per-class Intersection over Union (IoU) is defined as

$$IoU = \frac{TP}{FN + FP + TP}, \tag{19}$$

In the few-shot setting, the mean IoU is computed by averaging across all classes in the testing set C_{unseen}, where TP, FN, and FP represent the counts of true positives, false negatives, and false positives, respectively.

Baseline: ProtoNet adopts the basic prototypical network [6,22] to the few-shot semantic segmentation task of the point cloud. It extracts the prototype representation by computing the mean of the sample features for each category in the support set. We adopt AttProtoNet as our baseline, This represents an upgraded iteration of ProtoNet, utilizing the feature extractor depicted in Fig. 2 as the sole modification.

4.3 Experimental Result

Result on S3DIS: In Table 2, we show the comparing results of our method with other methods on S3DIS. It is evident from the results that our method exhibits significant improvement over the baseline across all four N-way K-shot tasks, with an average mIoU increase of 10.15%. Even when compared with 2CBR [37], the current best-performing method, we outperform by an average of 5.59% across the four tasks. These enhancements demonstrate that our method effectively addresses the disparity in sample distribution between the support and query sets, guiding the model to learn common features among samples from both sets.

Result on ScanNet: Table 3 illustrates the semantic segmentation results of these methods on ScanNet. Notably, our method maintains highly competitive performance. In the 1-shot task, we exhibits an average improvement of 20.82% over the baseline. Furthermore, compared to 2CBR, our model demonstrates a notable improvement of 13.44% across all four N-way K-shot tasks.

Table 2. Comparison of results on S3DIS (%). SI: test on the I-th subset. AVG: average

Methods	2-way						3-way					
	1-shot			5-shot			1-shot			5-shot		
	S0	S1	AVG	S0	S1	AVG	S0	S1	AVG	S0	S1	AVG
Fine-tuning	36.34	38.79	37.57	56.49	56.99	56.74	30.05	32.19	21.12	46.88	47.57	47.23
ProtoNet	48.39	49.98	49.19	57.34	63.22	60.28	40.81	45.07	42.94	49.05	53.42	51.24
AttProtoNet	50.98	51.90	51.44	61.02	65.25	63.14	42.16	46.76	44.46	52.20	56.20	54.20
MPTI	52.27	51.48	51.88	58.93	60.56	59.75	44.27	46.92	45.60	51.74	48.57	50.16
AttMPTI	53.77	55.94	54.86	61.67	67.02	64.35	45.18	49.27	47.23	54.92	56.79	55.86
2CBR	55.89	61.99	58.94	63.55	67.51	65.53	46.51	53.91	50.21	55.51	58.07	56.79
Ours	**62.27**	**67.46**	**64.87**	**64.64**	**69.08**	**66.86**	**58.33**	**62.32**	**60.33**	**59.90**	**63.68**	**61.79**

Table 3. Comparison of results on ScanNet (%)

Methods	2-way						3-way					
	1-shot			5-shot			1-shot			5-shot		
	S0	S1	AVG	S0	S1	AVG	S0	S1	AVG	S0	S1	AVG
Fine-tuning	31.55	28.94	30.25	42.71	37.24	39.98	23.99	19.10	21.55	34.93	28.10	31.52
ProtoNet	33.93	30.95	32.44	45.34	42.01	43.68	28.47	26.13	27.30	37.36	34.98	36.17
AttProtoNet	37.99	34.67	36.33	52.18	46.89	49.54	32.08	28.96	30.52	44.49	39.45	41.97
MPTI	39.27	36.14	37.71	46.90	45.59	45.25	29.96	27.26	28.61	38.14	34.36	36.25
AttMPTI	42.55	40.83	41.69	54.00	50.32	52.16	35.23	30.72	32.98	46.76	40.80	43.77
2CBR	50.73	47.66	49.20	52.35	47.14	49.75	47.00	46.68	46.68	45.06	39.47	42.27
Ours	**56.82**	**57.17**	**56.99**	**59.88**	**62.60**	**61.24**	**60.66**	**62.63**	**61.65**	**63.76**	**59.75**	**61.76**

4.4 Ablation Experiments

To further showcase the effectiveness of each module, we conduct a series of ablation experiments under the 1-shot task and present the experimental results in Table 4 and Table 5. For each dataset, we compare four methods: the baseline, the method incorporating only CFE, the method incorporating only PC, and the complete our method.

Table 4. Ablation experiments of each module on S3DIS (%)

Methods	2-way			3-way		
	1-shot			1-shot		
	S0	S1	AVG	S0	S1	AVG
Baseline	50.98	51.90	51.44	42.16	46.76	44.46
+CFE	56.33	61.81	59.07	46.85	53.68	50.27
+PC	60.04	64.46	62.25	57.17	60.61	58.89
Ours	**62.27**	**67.46**	**64.87**	**58.33**	**62.32**	**66.86**

Table 5. Ablation experiments of each module on ScanNet (%)

Methods	2-way			3-way		
	1-shot			1-shot		
	S0	S1	AVG	S0	S1	AVG
Baseline	37.99	34.67	36.33	32.08	28.96	30.52
+CFE	52.44	49.78	51.11	46.32	46.19	46.26
+PC	53.94	53.58	53.76	56.61	58.89	57.75
Ours	**56.82**	**57.17**	**56.99**	**60.66**	**62.23**	**61.65**

Our ablation experiment demonstrate that experimental results without one of the two modules are not optimal, and that the best results are achieved when both modules are used simultaneously, which can show that our approach can significantly improve the distributional differences between the support set and the query set.

5 Conclusion

This paper introduces a point cloud few-shot semantic segmentation model based on single prototype learning, aiming to address the high cost associated with point cloud labeling. Simultaneously, to address the issue of sample distribution disparity between the support and query sets, we propose a CFE module can reduce the distance between the support and query sets. We also design a Prototype Correction module to map the prototypes extracted in Cross Feature Enhancement Module to the query feature space, which greatly improves the few-shot semantic segmentation performance. We assess the proposed approach on two widely used 3D point cloud semantic segmentation datasets, demonstrating new state-of-the-art performances with substantial improvements over prior methods. In future work, we will explore the potential of using the capsule network's [14,15] partial-whole hierarchy and capturing complex spatial relationships [16,30] in small sample point cloud segmentation.

Acknowledgement. This work was supported in part by the National Natural Science Foundation of Jiangsu Province under Grant BK20221379; in part by the CNPC-CZU Innovation Alliance, Changzhou University, under Grant CCIA2023-01; and in part by the Changzhou Leading Innovative Talent Introduction & Cultivation Project 20221460. It is also supported by the Science and Technology Development Fund, Macao SAR under Grant 0004/2023/ITP1.

References

1. Ahmed, S.M., Tan, Y.Z., Chew, C.M., Al Mamun, A., Wong, F.S.: Edge and corner detection for unorganized 3D point clouds with application to robotic welding. In: 2018 IEEE/RSJ International Conference on Intelligent Robots and Systems (IROS), pp. 7350–7355. IEEE (2018)
2. Armeni, I., et al.: 3D semantic parsing of large-scale indoor spaces. In: Proceedings of the IEEE Conference on Computer Vision and Pattern Recognition, pp. 1534–1543 (2016)
3. Chen, C., Qian, S., Fang, Q., Xu, C.: Hapgn: hierarchical attentive pooling graph network for point cloud segmentation. IEEE Trans. Multimedia **23**, 2335–2346 (2020)
4. Chen, Y., et al.: Shape self-correction for unsupervised point cloud understanding. In: Proceedings of the IEEE/CVF International Conference on Computer Vision, pp. 8382–8391 (2021)
5. Dai, A., Chang, A.X., Savva, M., Halber, M., Funkhouser, T., Nießner, M.: Scannet: richly-annotated 3D reconstructions of indoor scenes. In: Proceedings of the IEEE Conference on Computer Vision and Pattern Recognition, pp. 5828–5839 (2017)
6. Dong, N., Xing, E.P.: Few-shot semantic segmentation with prototype learning. In: BMVC, vol. 3, p. 4 (2018)
7. Jing, Y., Yang, Y., Feng, Z., Ye, J., Yu, Y., Song, M.: Neural style transfer: a review. IEEE Trans. Visual Comput. Graphics **26**(11), 3365–3385 (2019)
8. Kingma, D.P., Ba, J.: Adam: a method for stochastic optimization. arXiv preprint arXiv:1412.6980 (2014)
9. Krispel, G., Opitz, M., Waltner, G., Possegger, H., Bischof, H.: Fuseseg: lidar point cloud segmentation fusing multi-modal data. In: Proceedings of the IEEE/CVF Winter Conference on Applications of Computer Vision, pp. 1874–1883 (2020)
10. Lang, C., Cheng, G., Tu, B., Han, J.: Learning what not to segment: A new perspective on few-shot segmentation. In: Proceedings of the IEEE/CVF Conference on Computer Vision and Pattern Recognition, pp. 8057–8067 (2022)
11. Lewandowski, B., Liebner, J., Wengefeld, T., Müller, S., Gross, H.M.: Fast and robust 3D person detector and posture estimator for mobile robotic applications. In: 2019 International Conference on Robotics and Automation (ICRA), pp. 4869–4875. IEEE (2019)
12. Li, G., Jampani, V., Sevilla-Lara, L., Sun, D., Kim, J., Kim, J.: Adaptive prototype learning and allocation for few-shot segmentation. In: Proceedings of the IEEE/CVF Conference on Computer Vision and Pattern Recognition, pp. 8334–8343 (2021)
13. Liu, H., Guo, Y., Ma, Y., Lei, Y., Wen, G.: Semantic context encoding for accurate 3D point cloud segmentation. IEEE Trans. Multimedia **23**, 2045–2055 (2020)
14. Liu, Y., Cheng, D., Zhang, D., Xu, S., Han, J.: Capsule networks with residual pose routing. IEEE Trans. Neural Networks Learn. Syst. (2024)
15. Liu, Y., Zhang, D., Zhang, Q., Han, J.: Part-object relational visual saliency. IEEE Trans. Pattern Anal. Mach. Intell. **44**(7), 3688–3704 (2021)
16. Liu, Y., Zhang, Q., Zhang, D., Han, J.: Employing deep part-object relationships for salient object detection. In: Proceedings of the IEEE/CVF International Conference on Computer Vision, pp. 1232–1241 (2019)
17. Qi, C.R., Su, H., Mo, K., Guibas, L.J.: Pointnet: deep learning on point sets for 3D classification and segmentation. In: Proceedings of the IEEE Conference on Computer Vision and Pattern Recognition, pp. 652–660 (2017)

18. Qi, C.R., Yi, L., Su, H., Guibas, L.J.: Pointnet++: deep hierarchical feature learning on point sets in a metric space. In: Advances in Neural Information Processing Systems, vol. 30 (2017)
19. Qiu, S., Anwar, S., Barnes, N.: Geometric back-projection network for point cloud classification. IEEE Trans. Multimedia **24**, 1943–1955 (2021)
20. Shaban, A., Bansal, S., Liu, Z., Essa, I., Boots, B.: One-shot learning for semantic segmentation. arXiv preprint arXiv:1709.03410 (2017)
21. Snell, J., Swersky, K., Zemel, R.: Prototypical networks for few-shot learning. In: Advances in Neural Information Processing Systems, vol. 30 (2017)
22. Wang, K., Liew, J.H., Zou, Y., Zhou, D., Feng, J.: Panet: few-shot image semantic segmentation with prototype alignment. In: Proceedings of the IEEE/CVF International Conference on Computer Vision, pp. 9197–9206 (2019)
23. Wang, Y., Li, W., Dai, D., Van Gool, L.: Deep domain adaptation by geodesic distance minimization. In: Proceedings of the IEEE International Conference on Computer Vision Workshops, pp. 2651–2657 (2017)
24. Wang, Y., et al.: Pillar-based object detection for autonomous driving. In: Vedaldi, A., Bischof, H., Brox, T., Frahm, J.-M. (eds.) ECCV 2020. LNCS, vol. 12367, pp. 18–34. Springer, Cham (2020). https://doi.org/10.1007/978-3-030-58542-6_2
25. Wang, Y., Sun, Y., Liu, Z., Sarma, S.E., Bronstein, M.M., Solomon, J.M.: Dynamic graph CNN for learning on point clouds. ACM Trans. Graph. (TOG) **38**(5), 1–12 (2019)
26. Xu, S., Zhou, D., Fang, J., Yin, J., Bin, Z., Zhang, L.: Fusionpainting: multimodal fusion with adaptive attention for 3D object detection. In: 2021 IEEE International Intelligent Transportation Systems Conference (ITSC), pp. 3047–3054. IEEE (2021)
27. Yang, B., Liu, C., Li, B., Jiao, J., Ye, Q.: Prototype mixture models for few-shot semantic segmentation. In: Vedaldi, A., Bischof, H., Brox, T., Frahm, J.-M. (eds.) ECCV 2020. LNCS, vol. 12353, pp. 763–778. Springer, Cham (2020). https://doi.org/10.1007/978-3-030-58598-3_45
28. Zhang, C., Lin, G., Liu, F., Yao, R., Shen, C.: Canet: class-agnostic segmentation networks with iterative refinement and attentive few-shot learning. In: Proceedings of the IEEE/CVF Conference on Computer Vision and Pattern Recognition, pp. 5217–5226 (2019)
29. Zhang, H., Goodfellow, I., Metaxas, D., Odena, A.: Self-attention generative adversarial networks. In: International Conference on Machine Learning, pp. 7354–7363. PMLR (2019)
30. Zhang, Q., Duanmu, M., Luo, Y., Liu, Y., Han, J.: Engaging part-whole hierarchies and contrast cues for salient object detection. IEEE Trans. Circuits Syst. Video Technol. **32**(6), 3644–3658 (2021)
31. Zhang, X., Wei, Y., Yang, Y., Huang, T.S.: SG-one: similarity guidance network for one-shot semantic segmentation. IEEE Trans. Cybern. **50**(9), 3855–3865 (2020)
32. Zhang, Y., Qu, Y., Xie, Y., Li, Z., Zheng, S., Li, C.: Perturbed self-distillation: weakly supervised large-scale point cloud semantic segmentation. In: Proceedings of the IEEE/CVF International Conference on Computer Vision, pp. 15520–15528 (2021)
33. Zhao, H., Shi, S., Qi, X., Wang, X., Jia, J.: Point attention network for semantic segmentation of point clouds. In: Proceedings of the IEEE Conference on Computer Vision and Pattern Recognition (CVPR) (2019)
34. Zhao, N., Chua, T.S., Lee, G.H.: Few-shot 3D point cloud semantic segmentation. In: Proceedings of the IEEE/CVF Conference on Computer Vision and Pattern Recognition, pp. 8873–8882 (2021)

35. Zhou, D., et al.: Joint 3D instance segmentation and object detection for autonomous driving. In: Proceedings of the IEEE/CVF Conference on Computer Vision and Pattern Recognition, pp. 1839–1849 (2020)
36. Zhou, Y., Zhang, Z., Zha, H., You, J.: Dynamic graph attention network for few-shot 3D point cloud semantic segmentation. In: Proceedings of the AAAI Conference on Artificial Intelligence (AAAI) (2021)
37. Zhu, G., Zhou, Y., Yao, R., Zhu, H.: Cross-class bias rectification for point cloud few-shot segmentation. IEEE Trans. Multimedia (2023)

Towards Information Sharing Beetle Antennae Search Optimization

Xuan Liu[1,2], Chenyan Wang[1,2], Wenjian Liu[3], Lefeng Zhang[3], Xianggan Liu[4], and Yutong Gao[1,2](✉)

[1] Key Laboratory of Ethnic Language Intelligent Analysis and Security Governance of MOE, Minzu University of China, Beijing, China

[2] Hainan International College of Minzu University of China, Li'an International Education Innovation pilot Zone, Hainan, China

{liuxuan,23302121,ytgao92}@muc.edu.cn

[3] City University of Macau, Macau SAR, China

{andylau,lfzhang}@cityu.edu.mo

[4] Natural Language Processing and Knowledge Graph Lab, School of Computer Science and Technology, Huazhong University of Science and Technology, Wuhan, China

liuxianggan@msn.com

Abstract. The study of bioinformatics-based evolutionary computation has long been of significant interest within the scientific community. Beetle antennae search algorithm is widely used because of its lightweight, however, it lacks information sharing due to individual iteration in the algorithm. In this paper, we propose an innovative pheromone-based beetle antennae search algorithm, which evolves from a single iterative individual in BAS algorithm to multiple parallel iterative individuals and incorporates the pheromone sharing mechanism found in ant colony optimization algorithm. Applying the pheromone-based beetle antennae search algorithm to the virtual machine placement problem in cloud computing, we find that pheromone sharing mechanism allows the PB-BAS algorithm to exhibit superior optimization capabilities and effectively avoids convergence to local optimal. To verify the performance of the algorithm, we select other alternative algorithms and conduct a large number of comparative experiments under different experimental setups, the experimental results show the effectiveness and efficiency of our algorithm.

Keywords: Information Sharing · Beetle Antennae Search Algorithm · Evolutionary Computation

1 Introduction

In the field of biological information transmission, the study of pheromones occupies an extremely important position [1], especially in the exploration of the behavioral mechanisms of social insects. The sharing method based on

T. Zhu et al. (Eds.): ICA3PP 2024, LNCS 15252, pp. 286–295, 2025.
https://doi.org/10.1007/978-981-96-1528-5_19

pheromones enables ants to coordinate tasks such as searching for food, constructing nests, and defending against enemies without central control. It is precisely because of the efficiency of this biological information transmission that scientists have proposed the ant colony optimization (ACO) algorithm [7]. Drawing inspiration from the mechanism by which ants use pheromones to find and optimize paths, we propose to apply this pheromone sharing mechanism to enhance beetle antennae search (BAS) algorithm.

BAS is a novel heuristic bio-inspired optimization algorithm, which draws its concept from the observation and study of the foraging behavior of longhorn beetles [2]. Based on BAS, this paper proposes pheromone-based beetle antennae search (PB-BAS) algorithm, this development enhances the search capability by expanding the exploration space through concurrent operations. Additionally, it integrates the pheromone sharing mechanism that is a hallmark of the ACO algorithm, allowing for the exchange of information among the individual agents. Through this synergistic combination, the algorithm improves the likelihood of finding global optimal in complex optimization landscapes.

In order to evaluate the quality of the PB-BAS and compare with the proposed enhancements, we consider the mutation operation [3] in genetic algorithm (GA) to discretize PB-BAS, and apply it to the virtual machine placement (VMP) problem, which is a NP-hard problem. Experimental results by CloudSim tooklit shows the efficiency of our approach compared with other related algorithms.

Section 2 presents the related work with evolutionary computation optimization. Section 3 describes a BAS algorithm and proposed enhancements using pheromone sharing mechanism. Section 4 presents VMP problem model and applies PB-BAS to the VMP Problem. Section 5 presents the experimental results. Section 6 concludes the paper.

2 Related Works

A high-level comprehensive survey on evolutionary computation optimization has been discussed in the study [5]. Jing Liang provides an overview of various constrained multi-objective evolutionary algorithms, evaluating their advantages and drawbacks, and categorizing them accordingly.

The study [6] details the inception of the portia spider algorithm, a bio-inspired computational method derived from the elaborate hunting techniques of the portia spider. This species is distinguished by its advanced problem-solving skills and its ability to adapt and learn from experience, making it an ideal biological model for algorithm development. Marco Scianna propose to solve such a combinatorial optimization challenges with the AddACO algorithm [7], which an innovative variant of the Ant Colony Optimization. AddACO is designed to overcome two fundamental limitations commonly associated with traditional ACO algorithms.

Hongjian Li's contribution [8] is noteworthy for tackling the issue of cost efficiency in the deployment of communication-intensive and dynamically evolving web applications within Kubernetes environments. Li presents a cost-efficient scheduling algorithm, augmented by an Improved beetle antennae search-based method, which strategically co-locates network communication-intensive pods on the same node.

In order to efficient and effective placement of virtual machines into physical machines, which is an optimization problem that involves multiple constraints and objectives, the study [9] propose utilization based genetic algorithm (UBGA). UBGA focuses on machine utilization and node proximity to concurrently minimize resource wastage, network congestion, and energy consumption.

Unlike previous works, in this paper, we aim to improve the algorithm by incorporating the mechanism of inter-biological information transfer into the simple BAS algorithm. Therefore, PB-BAS algorithm is proposed and we apply the PB-BAS to the VMP problem. The experimental outcomes show the superiority of our proposed algorithm.

3 BAS Algorithm with Information Sharing

3.1 Simple BAS Algorithm

Within the realm of evolutionary optimization algorithms, there exists a category that takes its cues from the intricacies of natural processes [4]. This includes algorithms like ACO and GA. In this vein, BAS is designed to emulate the food-seeking behavior of beetles [8]. The underlying biological concept is that beetles rely on the intensity of food scents to guide its search. As depicted in the Fig. 1, the beetle's pair of elongated antennae are crucial in discerning the scent's intensity. When foraging, if the odor concentration on the left antenna is stronger than that on the right antenna, the beetle will move to the left and vice versa to the right. Through continuous adjustments based on the olfactory gradient, the beetle eventually finds food.

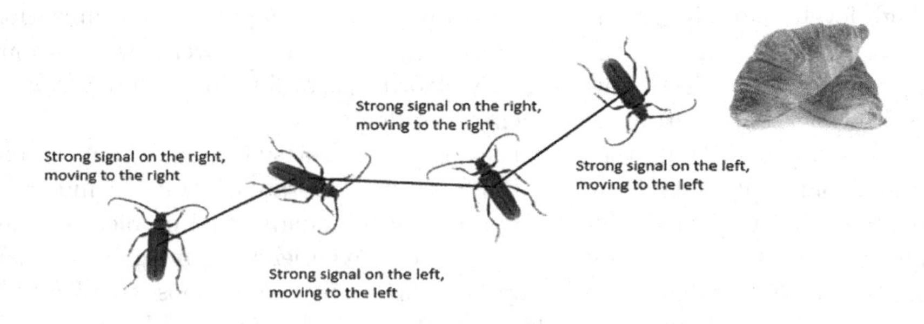

Fig. 1. Process of Searching for Food by Beetle

Inspiration can be drawn from the biological principles of beetle foraging. The smell of food is equivalent to the objective function. The two antennae of the beetle can collect the smell values of two points nearby itself. The purpose of the beetle is to find the location of the food, that is, the point with the highest global smell value. By imitating the behavior of longicorn beetles, we can conduct mathematical modeling and function optimization.

3.2 Enhancement of BAS with Information Sharing

Parallel Strategy. In the original BAS algorithm, using the optimization strategy of a single beetle makes it easy to implement. However, its reliance on a single-agent search strategy significantly raises the risk of premature convergence to local optima, potentially overlooking better solutions.

To mitigate this challenge and enhance the exploratory capabilities of the algorithm, this paper proposes the concept of parallelism. By deploying population of beetles that concurrently explore different areas of the solution space, the algorithm increases its chances of escaping local optima and locating more promising regions.

The Mechanism of Information Sharing. To ensure the reliability of information transfer, in each iteration we select the superior beetles; only these superior beetles have the authority to transmit information. Other beetles, driven by hunger and their inability to locate food, are compelled to rely on the guidance provided by the superior beetles to determine their subsequent foraging direction.

We consider the pheromone updating rule from the ACO algorithm and the position updating rule, which enables ants to determine their next target based on pheromone trails, and apply these rules to the PB-BAS framework. This integration endows beetle population with the capability of information dissemination, thereby enhancing the collective optimization prowess of the beetle swarm.

Discretization of PB-BAS Algorithm. Incorporating mutation operations from GA, we improve the original BAS algorithm, which was primarily designed for continuous optimization problems, to address the discrete optimization problem of VMP in cloud computing. Specifically, we abstract the optimization values obtained by the two antennae of the beetle as sequences for VMP. The antenna that detects a stronger food scent represents a more optimal virtual machine placement sequence.

The mutation operations [3] are mainly targeted at the superior beetles. Since superior beetles have already obtained superior solutions, they should narrow down their search range and mutation operations ensure beetles fine-tuned exploration. In this article, regarding the position update of superior beetles, we encode virtual resources using real number encoding, where the chromosome's length equals the number of VMs, denoted as m. Then, we randomly select two genes and exchange their positions to complete the mutation operation. Assuming $m = 6$, the mutation process is illustrated in Fig. 2:

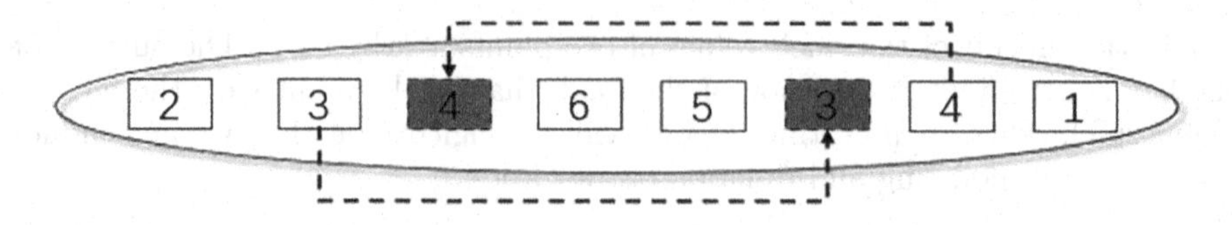

Fig. 2. Mutation Operation

4 Application of PB-BAS Algorithm to VMP Problem

4.1 VMP Problem Description

The VMP problem is a process of mapping from virtual machines (VMs) to physical machines (PMs) and is often described as a variant of the bin packing problem [11], which is a typical combinatorial optimization problem. Allocating VMs to as few PMs as possible can reduce energy consumption, but such allocation often leads to imbalanced resource utilization [12]. Therefore, we need to find an optimal solution for VM allocation that minimizes energy consumption while achieving resource load balancing.

Taking into account CPU and memory resources, we define the VMP problem as a constrained bi-objective optimization issue. To tackle this problem, we have developed models for both energy consumption and resource load balancing. Based on these models, we have constructed a bi-objective optimization function. The PB-BAS algorithm has been proposed to minimize the value of the joint-objective optimization function.

4.2 VMP Formulation

In this paper, we define n PMs and m VMs. Each PM $P_i\,(P_i \in P)$ has resource capacity $C_i^k = \{C_i^{cpu}, C_i^{ram}\}$. Similarly, each VM $V_j\,(V_j \in V)$ has resource demand $D_j^k = \{D_j^{cpu}, D_j^{ram}\}$. The resource utilization $U_i^k = \{U_i^{cpu}, U_i^{ram}\}$ of P_i is computed as the sum of the resource demand of the hosted VMs:

$$U_i^k = \sum_{j=1}^{m} D_j^k \ for\ \forall x_{ij} = 1 \tag{1}$$

where x is the placement matrix and is defined as follows:

$$x_{ij} = \begin{cases} 1 \text{ if } V_j \text{ is placed in } P_i \\ 0 \qquad \text{otherwise} \end{cases} \tag{2}$$

subsequently, the PM resource capacity constraint as shown in Eq. (3), and the Eq. (4) ensure that each VM is assigned to at most one PM:

$$\sum_{j=1}^{m} D_j^k x_{ij} \le C_i^k, \forall i \in \{1, \cdots, m\}, \forall k \in \{cpu, ram\} \tag{3}$$

$$\sum_{i=1}^{n} x_{ij} \leq 1, \forall j \in \{1, \cdots, m\} \tag{4}$$

Energy Consumption Model. We denote the energy consumption of the PM P_i as E_i. According to the most research [13], the CPU is the primary energy drain when compared to the other components. Consequently, the resource utilization of PMs are often represented by CPU workload. We draw upon empirical data derived from SPECpower benchmarking to inform our analysis. Table 1 presents the energy usage data for HP ProLiant G4 and G5 series servers under varying degrees of operational load.

Table 1. Energy consumption with different CPU utilization in watts

Server	0%	10%	20%	30%	40%	50%	60%	70%	80%	90%	100%
HP ProLiant G4	86	89.4	92.6	96	99.5	102	106	108	112	114	117
HP ProLiant G5	93.7	97	101	105	110	116	121	125	129	133	135

Resource Load Balance Model. Considering CPU and memory resources, we use W_i to measure the resource load imbalance degree for PM P_i as shown in the Eq. (5), where ε is a very small positive real number that is set to 0.0001.

$$W_i = \frac{|U_i^{cpu} - U_i^{ram}| + \varepsilon}{U_i^{cpu} + U_i^{ram}} \tag{5}$$

Problem Formulation. Based on resource load balance model and energy consumption model, the problem can be formulated as follows. Equations (6)–(7) are optimization objectives. Where E is the whole energy consumption and W represents all PMs resource load imbalance degree. λ is variable weight coefficient. We goal is minimize $f(X)$.

$$\min : f(X) = \lambda_1 E + \lambda_2 W \tag{6}$$

$$\text{s.t. Constraints (2)-(5)} \tag{7}$$

4.3 PB-BAS Algorithm Implementation

Pheromones and Heuristic Information. We adopt the pheromone strategy from ACO, utilizing a $m \times n$ matrix, denoted as τ, to represent the pheromone levels associated with the allocation of VMs to PMs. Where τ_{ij} represents the pheromones for placing VM j on PM i. The pheromone update rules as Eq. (8)–(9) shows.

$$\tau_{ij} = (1-\rho)\,\tau_{ij} + \rho\tau_0 \tag{8}$$

$$\tau_0 = \frac{1}{m\,(E_0 + W_0)} \tag{9}$$

where $\rho \in (0,1)$ is the pheromone decay parameter, τ_0 is the initial pheromone level, E_0 and W_0 are initialize energy consumption value and resource load imbalance degree value respectively.

We define a matrix $\eta_{m\times n}$ to record the heuristic information for the placement of VM to PM, with η_{ij} representing the heuristic information for placing VM j on PM i. In this paper, we consider resource load balancing and energy consumption simultaneously, therefore, heuristic information is defined as Eq. (10)–(12) shows.

$$\eta_{ij} = \eta_{ij}^{E} + \eta_{ij}^{W} \tag{10}$$

$$\eta_{ij}^{E} = \frac{1}{\varepsilon + E} \tag{11}$$

$$\eta_{ij}^{W} = \frac{1}{\varepsilon + W} \tag{12}$$

The Rules of VM Selection. Based on the ACO pseudorandom proportional state transition rule, we define the virtual machine selection rules as Eq. (13)–(14) shows.

$$j = \begin{cases} \max\{\alpha\tau_{ie} + (1-\alpha)\,\eta_{ie} | e \in \Omega_i\} & ,q \le q_0 \\ J & ,\text{otherwise} \end{cases} \tag{13}$$

$$pp_{ie} = \begin{cases} \frac{(\alpha\tau_{ig}+(1-\alpha)\eta_{ig})}{\sum_{g\in\Omega_i}(\alpha\tau_{ie}+(1-\alpha)\eta_{ie})} & ,e \in \Omega_i \\ 0 & ,\text{otherwise} \end{cases} \tag{14}$$

where $q \in [0,1]$ is a uniformly distributed random variable and $q_0 \in [0,1]$ is a parameter to determine relative importance of exploitation. α represents the relative importance of pheromones and heuristic information, Ω_i represents the set of PMs where VM can be placed. pp_{ie} is the probability of selecting VM e, and J is a random variable selected according to the probability distribution given in Eq. (14).

PB-BAS Description. The AVVMC algorithm pseudocode is shown in Algorithm 1. Lines 1 to 2 are initialization, we define G is number of iteration and N is number of beetles, then the optimization loop begins. PB-BAS orders all solutions by $f\,(X)$ to distinction between superior beetles and others, and then, both will enter different search processes.

Algorithm 1. VMP approach with PB-BAS

```
1: Initialize G,N
2: Initialize x,τ,η
3: for g = 1 to G do
4:    Order all solution
5:    Distinguish between superior beetles and others
6:    for n = 1 to N do
7:       if beetle n is superior beetle then
8:          Localized search by mutation operations
9:          if the right antenna's value is better than the left's then
10:            Move to the right
11:         else
12:            Move to the left
13:         end if
14:         pheromone update by equations (8)-(12)
15:      else
16:         Non-superior beetles update position by equations (13)-(14)
17:      end if
18:   end for
19:   Reorder all solution by the equations (6)-(7)
20: end for
21: return Best solution x and value
```

5 Experimental Evaluations

5.1 Experiment Settings

Our experimental setup includes 200 VMs and 200 PMs. Each host is equipped with two CPU cores, capable of 1880 or 2660 million instructions per second (MIPS), along with 4 GB of RAM and 1 TB of disk space. Regarding the VMs, we account for four distinct configurations, offering 500, 1000, 1500, and 2500 MIPS, with the quantity of each VM type being determined randomly.

In this paper, to evaluate the performance of the PB-BAS algorithm, we have conducted an extensive series of comparative experiments. These alternative algorithms are selected for comparisons in our experiments.

- RFF: RFF (random first-fit) algorithm randomly select a PM that can hold the target VM. So, RFF is normally used as a reference objective for other algorithms.
- ACO: ACO is a typical algorithm used for solving Combinatorial optimization problems which can be reduced to finding good paths through graphs. The algorithm was inspired by the behavior of ants in finding paths from their colony to food sources and back.
- GA: GA is an adaptive heuristic search algorithm based on the evolutionary ideas of natural selection and genetics. It generate solutions to the optimization problem using techniques inspired by natural evolution, such as inheritance, mutation, selection, and crossover.

– FFD: FFD (first fit decreasing) algorithm is a kind of classical greedy approximation algorithm, which prioritize allocating VMs that consume more resources.

5.2 Experimental Results and Evaluation

The optimization objective of VMP problem to be $f(X)$ (as shown in the Eqs. (6)–(7)) and we select the other four representative algorithms for comparisons in our experiments. Experiments are conducted at various iterations ($G = 50$, $G = 100$) and population sizes ($N = 20$, $N = 50$). Specifically, since the FFD algorithm and RFF algorithm do not have the concepts of population size and iteration, their values are directly used for comparison. We obtain the experimental results through simulations conducted using the CloudSim toolkit. The results of our experiments are shown in Fig. 3(a) to 3(d) and the experimental results show that PB-BAS improves performance by 45.28% compared to RFF, by 45.28% compared to ACS, and by 58.06% and 8.38% compared to the FFD and GA, respectively.

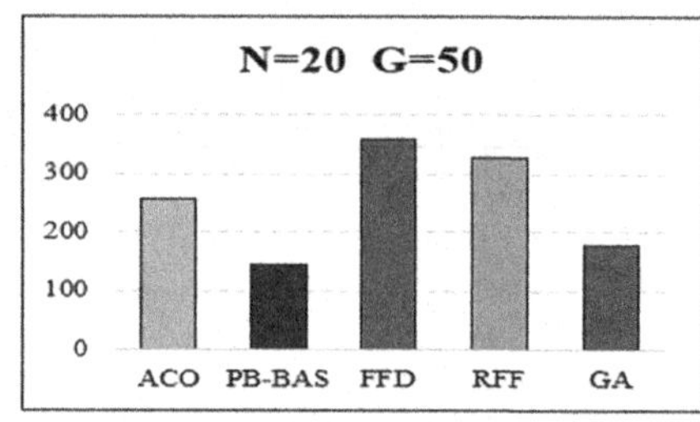

(a) function value of N=20,G=50

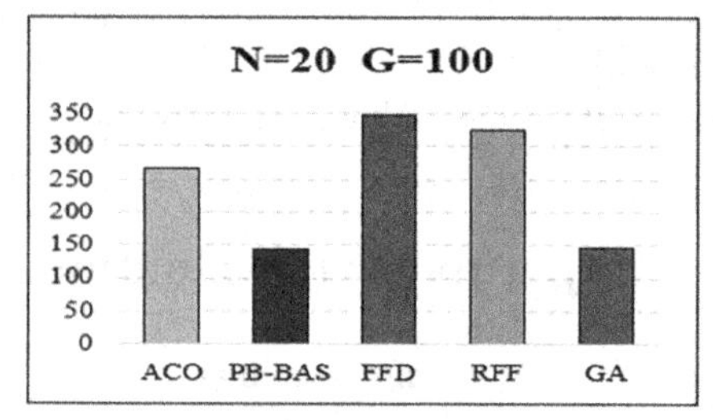

(b) function value of N=20,G=100

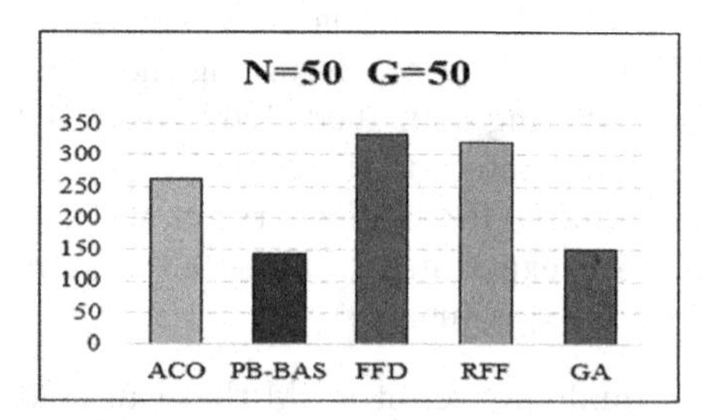

(c) function value of N=50,G=50

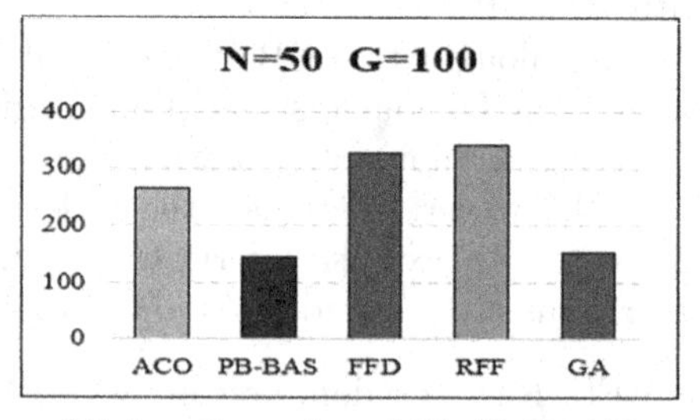

(d) function value of N=50,G=100

Fig. 3. Algorithms Comparison of ACO, PB-BAS, FFD, RFF and GA

6 Conclusion

In this paper, drawing inspiration from the process of information transmission among biological entities in nature, we innovatively propose the PB-BAS algorithm, which applies the mechanism of pheromone from ACO algorithm to the simple BAS algorithm. Then, we use the PB-BAS algorithm to solve the VMP problem, which is a typical combinatorial optimization issue in the field of cloud computing. To verify the performance of the PB-BAS algorithm, we conduct a comprehensive series of comparative experiments and executed simulations in a realistic cloud environment using the CloudSim simulator. Finally, experimental results show that our proposed algorithm outperforms other algorithms on the VMP problem.

References

1. Ivković, N., Kudelić, R., Golub, M.: Adjustable pheromone reinforcement strategies for problems with efficient heuristic information. Algorithms **16**(5), 251 (2023)
2. Wang, P., Li, G., Gao, Y.: A compensation method for gyroscope random drift based on unscented Kalman filter and support vector regression optimized by adaptive beetle antennae search algorithm. Appl. Intell. **53**, 4350–4365 (2023)
3. Chen, Q.-H., Wen, C.-Y.: Optimal resource allocation using genetic algorithm in container-based heterogeneous cloud. IEEE Access **12**, 7413–7429 (2024)
4. Farooq, H., Novikov, D., Juyal, A., Zelikovsky, A.: Genetic algorithm with evolutionary jumps. In: Guo, X., Mangul, S., Patterson, M., Zelikovsky, A. (eds.) ISBRA 2023. LNCS, vol. 14248, pp. 453–463. Springer, Singapore (2023). https://doi.org/10.1007/978-981-99-7074-2_36
5. Liang, J., et al.: A survey on evolutionary constrained multiobjective optimization. IEEE Trans. Evol. Computat. **27**(2), 201–221 (2023)
6. Pham, V.H.S., Nguyen Dang, N.T.: Portia spider algorithm: an evolutionary computation approach for engineering application. Artif. Intell. Rev. **57**(2), 24 (2024)
7. Scianna, M.: The AddACO: a bio-inspired modified version of the ant colony optimization algorithm to solve travel salesman problems. Math. Comput. Simul. **218**, 357–382 (2024)
8. Hongjian, L., Jie, S., Lei, Z., et al.: Cost-efficient scheduling algorithms based on beetle antennae search for containerized applications in Kubernetes clouds. J. Supercomput. **79**(9), 10300–10334 (2023)
9. Çavdar, M.C., Korpeoglu, I., Ulusoy, Ö.: A utilization based genetic algorithm for virtual machine placement in cloud systems. Comput. Commun. **214**, 136–148 (2024)
10. Zakarya, M., Gillam, L., Salah, K., Rana, O., Tirunagari, S., Buyya, R.: CoLocateMe: aggregation-based, energy, performance and cost aware VM placement and consolidation in heterogeneous IaaS clouds. IEEE Trans. Serv. Comput. **16**(2), 1023–1038 (2023)
11. Bhaumik, S., et al.: NetStor: network and storage traffic management for ensuring application QoS in a hyperconverged data-center. IEEE Trans. Cloud Comput. **10**(2), 1287–1300 (2022)
12. Li, B., Cui, L., Hao, Z., Li, L., Liu, Y., Li, Y.: eHotSnap: an efficient and hot distributed snapshots system for virtual machine cluster. IEEE Trans. Parallel Distrib. Syst. **34**(8), 2433–2447 (2023)
13. Biçici, E.: A cloud monitor to reduce energy consumption with constrained optimization of server loads. IEEE Access **12**, 25265–25277 (2024)

A Cost-Effective Data Placement Strategy Based on Battle Royale Optimization in Multi-cloud Edge Environments

Sen Zhang, Lili Xiao, Xin Luo, Zhaohui Zhang, and Pengwei Wang(✉)

School of Computer Science and Technology, Donghua University, Shanghai, China
2222753@mail.dhu.edu.cn, {xiaolili,xluo,zhzhang,wangpengwei}@dhu.edu.cn

Abstract. With the advent of big data era, there is a growing demand for users to store data in the cloud. To enhance the availability and privacy of data in the cloud and mitigate risks such as vendor lock-in, distributed multi-cloud storage has attracted widespread attention. However, accessing data from the cloud often encounters issues such as high latency and significant bandwidth costs. Utilizing edge resources can effectively address these problems, but it may potentially reduce data availability and increase storage costs. To this end, we combine multi-cloud and edge resources in this work, taking into account access restrictions on edge resources, and construct a multi-cloud edge storage model. Then, an Opposition-Based Learning Binary Battle Royale Optimizer (OBL-BinBRO) strategy is proposed to find optimal data placement solution, which assists users in determining the cloud and edge service providers for storing and accessing data. Using real-world data for extensive experiments, compared to several representative data placement strategies, the proposed method can effectively reduce the total cost of data placement for users.

Keywords: Multi-cloud · Edge computing · Data placement · Cloud storage · Data availability

1 Introduction

According to predictions by the International Data Corporation (IDC) [1], the global data sphere will grow to 163 ZB by 2025. It is foreseeable that with the rapid growth of the global data sphere, traditional data storage methods can no longer meet the needs of modern users, and future storage demands will further increase. Therefore, cloud storage has emerged in response to these needs.

If users store all their data in a single cloud, they face risks such as low data availability, low data security, and vendor lock-in risks. To mitigate these issues, users have distributed data across multiple cloud providers [2], but this leads to high latency and bandwidth costs. Edge computing has advanced edge cloud storage solutions, offering cheaper bandwidth [3] and lower access latency by

T. Zhu et al. (Eds.): ICA3PP 2024, LNCS 15252, pp. 296–306, 2025.
https://doi.org/10.1007/978-981-96-1528-5_20

storing data closer to users. However, edge storage has lower data availability compared to cloud data centers [4]. Additionally, each cloud provider offers at least two types of storage. These different storage types have varying pricing strategies and levels of availability.

Replication and erasure coding [5] are common data redundancy strategies to enhance availability. Replication duplicates data multiple times, leading to storage wastage and high costs. Erasure coding reduces costs, protects data privacy, and enhances security. It is widely used in cloud storage systems [6]. In edge computing, each edge server's coverage is limited [7], and data can only be transmitted between servers within a limited number of network hops [8]. It is essential to ensure that users can retrieve enough encoding blocks in this network topology to reconstruct data D.

To this end, we propose a multi-cloud edge storage model that leverages the respective advantages of cloud resources and edge resources. An improved data placement strategy based on population-based algorithms was proposed to seek data placement solutions. The main contributions of this work are as follows:

- A multi-cloud edge storage model is presented, including multiple cloud service providers with hot and cold storage tiers and multiple edge service providers. It takes into account a wide array of factors.
- A global discrete optimization model is constructed to address the data placement problem in multi-cloud edge environments. Each data placement solution is represented as a binary string, such that all possible data placement solutions can be mapped to the n-dimensional discrete space $\{0,1\}^n$.
- We adopt an erasure coding data partition strategy and propose an Opposition Based Learning Binary Battle Royale Optimizer (OBL-BinBRO) data placement strategy to help users determine the optimal data placement solution. Using real-world data for extensive experiments, this strategy demonstrates superior performance compared to existing algorithms.

2 System Model and Problem Formulation

The multi-cloud edge storage model consists of geographically distributed Cloud Service Providers (CSPs) and Edge Service Providers (ESPs), as depicted in Fig. 1. Users can access all CSPs and choose different storage types for data blocks. However, each ESP's coverage area is limited, and while data blocks can be transmitted between ESPs via the ESP network topology, the number of network hops for transmission is limited [8]. Thus, users can only choose ESPs within their own region for data placement. In Fig. 1, all ESPs within each region are interconnected within the allowed network hop count.

Definition 1. Service Provider. In our model, each CSP offers two types of storage: Hot and Cool. $SP = \{Z_i, P_{sij}, P_{bij}, P_{oij}, P_{rij}, a_{ij}, F_{ij}\}$ represents a service provider, where i denotes the i-th service provider, and j denotes the storage type ($j = 1$ for Hot, $j = 0$ for Cool). Note that ESPs typically only provide Hot storage, so for ESPs, j is always 1, where:

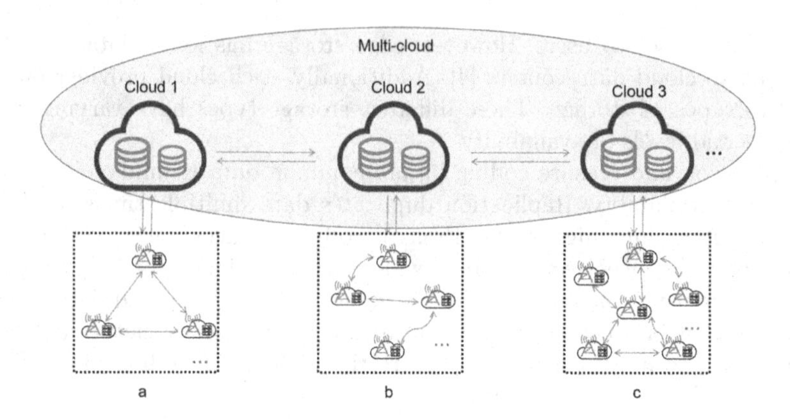

Fig. 1. Multi-cloud edge storage model

- $Z_i = \{x_i, y_i\}$ represents the location of SP_i, x_i represents the longitude of SP_i, y_i represents the latitude of SP_i,
- P_{sij} represents the storage cost per unit size for SP_i storage with type j,
- P_{bij} denotes the bandwidth cost per unit size for SP_i storage with type j,
- P_{oij} is the cost for GET requests for SP_i storage with type j,
- P_{rij} represents the retrieval cost per unit size for SP_i storage with type j,
- a_{ij} indicates the availability with the SP_i store of type j,
- F_{ij} represents the free storage quota for SP_i storage with type j.

The bandwidth cost of ESP is much lower than that of CSP [3]. In addition, we collectively refer to the CSP and ESP of each type of storage as storage nodes.

Definition 2. Data File. For each data file that the user needs to store, it constitutes a data object, represented as $D = \{S, \tau, A_r, L_r\}$. S is the size of the data object the user needs to store, τ is the data access frequency (number of accesses within a time period), A_r is the minimum data availability required by the user, and L_r is the maximum response latency required by the user.

Definition 3. Erasure Coding Parameters. $EC(n, m)$ is an erasure coding strategy where original data is split into m equal-sized blocks and encoded into n blocks, including m original and $(n - m)$ redundant blocks. This allows the system to tolerate up to $(n - m)$ simultaneous failures without data loss. Each storage node has a limitation that it can store at most one block of data.

Definition 4. Data Availability. Assuming each storage node's failure is independent, data availability is the sum of the probabilities of all k storage nodes being simultaneously available [4], where $k \in [m, n]$. We define $C' = \{(SP_{11}, u_{11}), (SP_{10}, u_{10}), \ldots, (SP_{(NC+NE)1}, u_{(NC+NE)1})\}$ is a collection of storage nodes that store the n data blocks. Where NC is the number of CSPs and NE is the number of ESPs in the user's area. We use $u_{ij} \in \{0, 1\}$ to denote whether the data is stored on the j-th storage type of the i-th storage node.

Let $\theta = \binom{|C'|}{k}$ denote the total number of cases where k storage nodes are simultaneously available, and S_v^θ denote the set of storage nodes contained in the v-th case among all cases θ. The formula for data availability is:

$$A = \sum_{k=m}^{n} \sum_{v=1}^{\theta} \left[\prod_{ij_1 \in S_v^\theta} a_{ij_1} \prod_{ij_2 \in C' \setminus S_v^\theta} (1 - a_{ij_2}) \right] \quad (1)$$

Definition 5. Data Response Latency. Let $Distance(d_i)$ be the Euclidean distance between the i-th service provider and the user. Here, λ represents the propagation rate, and b represents the transmission rate. The latency of the i-th service provider is calculated as follows:

$$l_i = \frac{Distance(d_i)}{\lambda} + \frac{S/m}{b} \quad v \in [1, \theta], ij \in S_v^\theta \quad (2)$$

The transmission of data blocks is parallel. So, the data response latency L is the maximum value among the latencies of the m storage nodes.

Definition 6. Storage Cost. The storage cost is the total sum of the storage costs generated by n storage nodes. The calculation method is as follows:

$$P_{storage} = \sum_{ij \in C'} \left(\frac{S}{m} - F_{ij} \right) (P_{sij}) \quad (3)$$

Definition 7. Network and Retrieval Cost. Users only need to access any m data blocks to reconstruct the original data. Therefore, users select the cheapest m storage nodes in terms of bandwidth prices among those that meet the latency requirements for data access. The retrieval cost refers to the cost required for users to recover data blocks. The calculation method is as follows:

$$P_{band} + P_{retr} = \min_{v \in [1,\theta]} \sum_{ij \in S_v^\theta} \frac{S}{m} \tau \left(P_{bij} + P_{rij} \right) \quad (4)$$

Definition 8. Operation Cost. The operation cost is the cost incurred by users when making Get requests to access data files from the cheapest m storage nodes. It can be calculated as follows:

$$P_{oper} = \min_{v \in [1,\theta]} \sum_{ij \in S_v^\theta} \tau \left(P_{oij} \right) \quad (5)$$

The problem to be addressed can be formulated as follows:

$$\text{Minimize } P_{storage} + P_{band} + P_{oper} + P_{retr} \quad (6)$$

Subject to:

$$|C'| = n \quad (7)$$

Algorithm 1. OBL: Opposition-Based Learning

Input: X, N_p, r_1
Output: X
1: **for** $i = 1$ to N_p **do**
2: $\quad \dot{X}_i \leftarrow X_i$
3: $\quad$ **for** bit in $\dot{X}_i$ **do**
4: $\quad\quad r \leftarrow rand(0, 1)$
5: $\quad\quad$ **if** $r > r_1$ **then**
6: $\quad\quad\quad bit \leftarrow 1 - bit$
7: $\quad\quad$ **end if**
8: $\quad$ **end for**
9: $\quad$ Select X_i or $\dot{X}_i$ by formula (10)
10: **end for**
11: **return** X

$$A \geq A_r \tag{8}$$

$$\mathrm{L} \leq \mathrm{L_r} \tag{9}$$

Constraint (7) requires the final data placement solution to have n storage nodes, matching the total data blocks from erasure coding. Additionally, the solution must meet the user's minimum required availability and ensure that the data response latency does not exceed the user's maximum required latency.

3 Low-Cost Data Placement via OBL-BinBRO

The data placement issue is an NP-hard problem [9]. To address this problem, we first construct a global discrete optimization model and then propose an approximate solution algorithm.

In our data placement problem, each data placement scheme is a player. Each data placement solution is represented as a binary string, such that all possible data placement solutions can be mapped to the n-dimensional discrete space $\{0, 1\}^n$. We use X to represent the player population, and N_p to represent the total number of players, where each player is represented by a binary vector: $X_i = (x_{i1}, x_{i2}, \cdots x_{it})$. Where i is the player's serial number. The t is the dimension of the player, representing the number of optional storage nodes in the model. If the k-th storage node is selected, then $x_{ik} = 1$, and $\sum_k x_{ik} = n$. In addition, we use σ to represent the upper limit of the erasure code parameter n, so $0 \leq \sum_k x_{ik} \leq \sigma$.

To evaluate the effectiveness of these data placement schemes, the fitness function is the primary metric. The calculation method is as follows:

$$fitness(x) = \begin{cases} \infty, A < A_r \text{ or } L > L_r \\ P_{storage} + P_{band} + P_{oper} + P_{retr}, \text{ other cases} \end{cases} \tag{10}$$

Algorithm 2. CalSimilarity

Input: X_i, X, N_p
Output: X_j
1: $Sim \leftarrow 0, M_{11} \leftarrow 0, M_{10} \leftarrow 0, M_{01} \leftarrow 0$
2: **for** $j = 1$ to $N_p, j \neq i$ **do**
3: Calculate $M_{11} \leftarrow$ The count of bits where both X_i and X_j are 1
4: Calculate $M_{10} \leftarrow$ The count where the $X_i$bit is 1 while the X_j bit is 0
5: Calculate $M_{01} \leftarrow$ The count where the $X_i$bit is 0 while the X_j bit is 1
6: Calculate $Sim \leftarrow \frac{M_{11}}{M_{11}+M_{10}+M_{01}}$
7: **end for**
8: $X_j \leftarrow$ The player with the maximum Sim value
9: **return** X_j

In the BRO algorithm [10], solutions are often generated randomly. If a solution is near the optimal one, the process is efficient. However, if it's far from the optimal solution, it may consume significant computational resources without finding the optimal solution. Analyzing mirrored positions of candidate solutions can be effective. The core idea of Opposition-Based Learning (OBL) is to analyze both the original and reverse candidate solutions, increasing the likelihood of finding a solution near the global optimum and enhancing convergence speed. However, a complete mirror image of the original solution is unlikely to be optimal. Instead, partial dimensional mirrors are more likely to improve the probability of finding the optimal solution. The algorithm pseudocode for OBL of the whole population X is shown in Algorithm 1. Where r is a random number ranging from 0 to 1, and r_1 is the degree constant of incompleteness. Since the probability of a t-dimensional random binary string having exactly half of its characters different from any optimal solution is maximum, this paper sets r_1 to 0.5, indicating that 50% of the binary encodings need to take their binary reverse numbers.

In the OBL-BinBRO algorithm, each placement scheme is compared with its nearest neighbor in each iteration. The similarity calculation method is shown in Algorithm 2 (Lines 3–6). After comparing, the scheme with the smaller fitness value from the fitness function is the winner, and the other scheme is the loser. Then, our goal is to change the position of the current winning placement scheme to become the ultimate winner, which involves performing a mutation operation on them. In this type of mutation, only two randomly selected bits of the winning schemes will be flipped. We use $mutation1()$ to represent this 2-bit flip mutation (Algorithm 3, lines 14 and 18). As for the losing schemes, they are divided into two categories. One category is when the loser has not reached the maximum loss count, we set the maximum loss count to $Threshold$, and perform a crossover operation between the loser and the current optimal solution. The other category is when the loser has reached the maximum loss count $Threshold$, similar to the mechanism in the Battle Royale game, we respawn them at a random location within a legal area. The loser will randomly generate α number of binary bits

Algorithm 3. Opposition-Based Learning Binary Battle Royale Optimizer

Input: T (itermax), t (current iteration), CSPs, ESPs, A_r, L_r, $Threshold$, N_p, r_1
Output: X_{best} (the best data placement solution), P_T (total cost)

1: Randomly initialize a data placement scheme population X. Initalize all parameters
2: **while** $t \leq T$ **do**
3: Update $X \leftarrow$ **Algorithm 1**. Update X_{best}. Update α by formula (11)
4: **for** $i = 1$ to N_p **do**
5: $X_j \leftarrow$ **Algorithm 2**. $dam \leftarrow j$, $vic \leftarrow i$
6: **if** $fitness(X_i) > fitness(X_j)$ **then**
7: $dam \leftarrow i$, $vic \leftarrow j$
8: **end if**
9: **if** $X_{dam}.damage < Threshold$ **then**
10: $X_{dam} \leftarrow crossover(X_{dam}, X_{best})$, $X_{dam}.damage \leftarrow X_{dam}.damage + 1$
11: **else**
12: $X_{dam} \leftarrow mutation2(X_{dam}, \alpha)$, $X_{dam}.damage \leftarrow 0$
13: **end if**
14: Update $fitness(X_{dam})$, $X_{vic}.damage \leftarrow 0$, $X^*_{vic} \leftarrow mutation1(X_{vic})$
15: **if** $fitness(X^*_{vic}) < fitness(X_{vic})$ **then**
16: $X_{vic} \leftarrow X^*_{vic}$, Update $fitness(X_{vic})$
17: **end if**
18: $X^*_{best} \leftarrow mutation1(X_{best})$
19: **if** $fitness(X^*_{best}) < fitness(X_{best})$ **then**
20: $X_{best} \leftarrow X^*_{best}$, Update $fitness(X_{best})$
21: **end if**
22: **end for**
23: $t \leftarrow t + 1$
24: **end while**
25: **return** X_{best}, P_T

according to the following formula:

$$\alpha = d - (d - 1) * \frac{C_{iter}}{M_{iter}} \tag{11}$$

We use $mutation2()$ to represent this mutation operation (Algorithm 3, line 12), where C_{iter} represents the current iteration number, M_{iter} represents the maximum iteration number, and d represents the dimensionality. Furthermore, after each update of the loser's or winner's solution, a 2-bit mutation test will be performed on the best solution found thus far. If a lower fitness value is obtained after the mutation, the best solution will be changed; otherwise, it will remain unchanged. The complete algorithm process is shown in Algorithm 3.

4 Experimental Results and Analysis

4.1 Experimental Settings

In our experiment, we selected six CSPs and twenty ESPs as the options for users to choose from, with each CSP offering two types of storage. The latest

Table 1. The propagation and transmission rates of CSPs and ESPs

Variable name	Parameter	Value	Unit
CSPs propagation/transmission rate	λ^c	250/400	km/ms
ESPs propagation/transmission rate	λ^e	200/300	km/ms

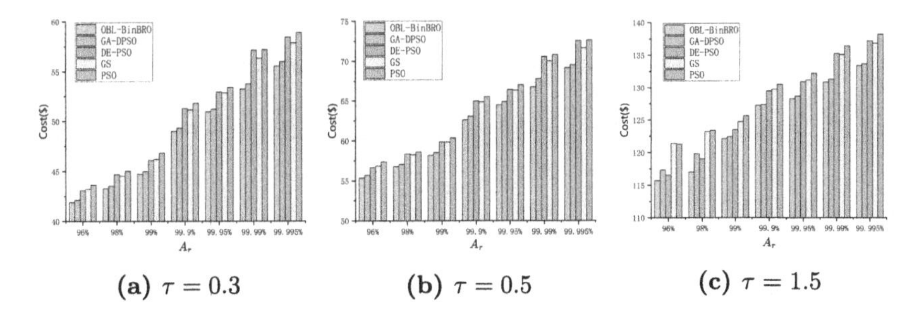

(a) $\tau = 0.3$ **(b)** $\tau = 0.5$ **(c)** $\tau = 1.5$

Fig. 2. Comparison of different strategies with different A_r.

data for CSPs were collected from their respective official websites. The location information for ESPs was obtained from the eua dataset [11]. Due to the lack of appropriate pricing information for ESPs, we referred to the literature [3] and the official websites of the providers to derive a reasonable range of values. We simulated the data availability of the Hot layer of CSPs in the range of [95.0%, 99.0%] [9]; and the data availability of the Cool layer in the range of [90.0%, 95.0%]. As edge is unreliable, we simulated the data availability of ESPs in the range of [70.0%, 75.0%] [4]. We set the value of σ to 12. The parameter settings of the OBL-BinBRO algorithm are: 50 players, 100 maximum iterations, and a threshold of 4. Then, following the literature [12], we set the transmission rate and propagation rate of CSPs and ESPs as shown in Table 1. To evaluate the effectiveness of the proposed data placement strategy, we compared it with the following four representative algorithms. Each experiment result is the average value after running 100 times.

a. Particle Swarm Optimization (PSO).
b. GA-based data placement strategy (GS).
c. A self-adaptive discrete particle swarm optimization algorithm with genetic algorithm operators (GA-DPSO) [13].
d. A differential evolution particle swarm optimization based data placement strategy (DE-PSO) [14].

4.2 Performance Comparison

With S set to 1000GB and no data response latency constraints, Fig. 2 shows the total data placement costs of five algorithms for different A_r values. The OBL-BinBRO algorithm consistently has lower costs than the others. As desired

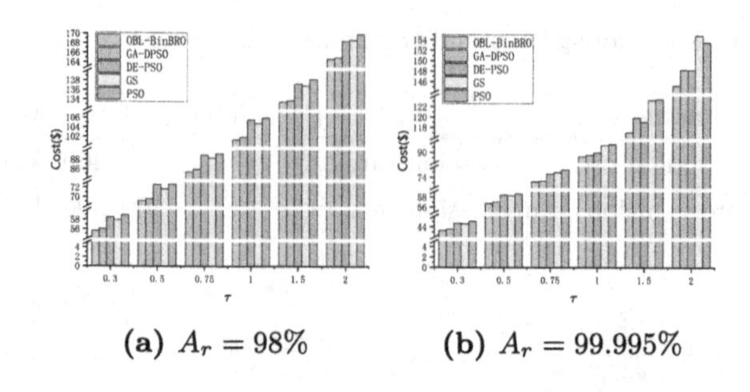

(a) $A_r = 98\%$ (b) $A_r = 99.995\%$

Fig. 3. Comparison of different strategies with different τ.

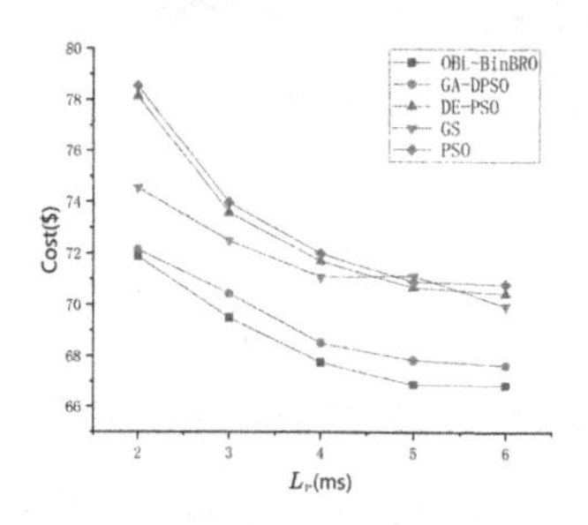

Fig. 4. Comparison of different strategies with different L_r.

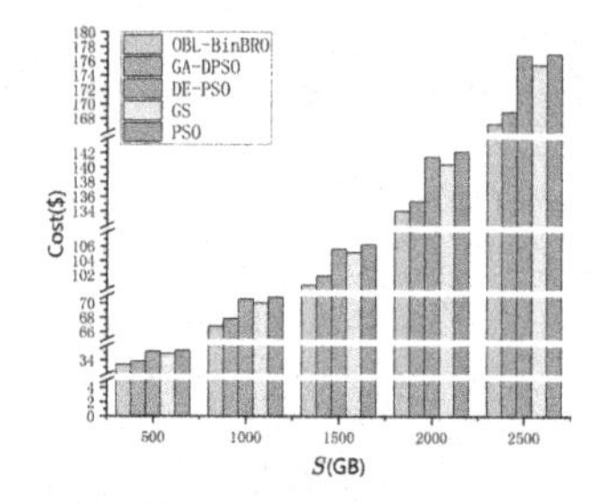

Fig. 5. Comparison of different strategies with different S.

data availability increases, the total costs for all algorithms rise due to the need for more redundant data blocks.

We use $A_r = 98\%$ and $A_r = 99.995\%$ to represent users' low data availability requirements and high data availability requirements, respectively. Here, S is set to 1000GB with no data response latency constraints. In both scenarios, the total data placement costs for five algorithms under different τ values are illustrated in Fig. 3. The OBL-BinBRO algorithm consistently has lower costs than the others. Additionally, as the data access frequency τ increases continuously, the total data placement costs of all five algorithms also increase. This is because frequent data access incurs significant bandwidth costs, which constitute a considerable proportion of the total cost.

Fixed $A_r = 99.99\%$, $\tau = 0.5$, and S defaulted to 1000GB, the total data placement costs for five algorithms under different L_r values are presented in Fig. 4. The graph indicates that as the user relaxes the latency requirements, the total data placement costs for all five algorithms decrease. This is because with lower latency requirements, there are more SPs available for selection, allowing for the choice of lower-priced SPs for data placement. Furthermore, the total data placement costs of the OBL-BinBRO algorithm consistently remain lower than those of the other algorithms.

With A_r fixed at 99.99% and τ fixed at 0.5 without data response latency constraints, the total data placement costs for five algorithms under different S values are presented in Fig. 5. The graph illustrates that as the size of user data objects increases, the total data placement costs for all five algorithms also increase. Among them, the total data placement costs of the OBL-BinBRO algorithm consistently remain lower than those of the other algorithms.

5 Conclusion and Future Work

Multi-cloud data hosting mitigates single cloud storage risks but faces high latency and bandwidth costs. Edge resources reduce bandwidth costs but lack guaranteed data availability. We propose a multi-cloud edge storage model and an OBL-BinBRO data placement strategy. Experiments show our algorithm achieves lower data placement costs. Future improvements will focus on migration costs and predicting dynamic user demand.

Acknowledgement. Pengwei Wang is the corresponding author. This work was partially supported by Ant Group through CCF-Ant Research Fund, the National Natural Science Foundation of China (NSFC) under Grant 61602109.

References

1. IDC, Seagate: Data age 2025: The evolution of data to lifecritical (2017)
2. Wang, P., Chen, Z., Zhou, M., Zhang, Z., Abusorrah, A., Ammari, A.C.: Cost-effective and latency-minimized data placement strategy for spatial crowdsourcing in multi-cloud environment. IEEE Trans. Cloud Comput. **11**(1), 868–878 (2023)
3. Liu, M., Pan, L., Liu, S.: Cost optimization for cloud storage from user perspectives: recent advances, taxonomy, and survey. ACM Comput. Surv. **55**, 1–37 (2023)
4. Kontodimas, K., et al.: Secure distributed storage orchestration on heterogeneous cloud-edge infrastructures. IEEE Trans. Cloud Comput. **11**(4), 3407–3425 (2023). https://doi.org/10.1109/TCC.2023.3287653
5. An, B., Li, Y., Ma, J., Huang, G., Chen, X., Cao, D.: Dcstore: a deduplication-based cloud-of-clouds storage service. In: 2019 IEEE International Conference on Web Services (ICWS), pp. 291–295. IEEE (2019)
6. Konwar, K.M., Prakash, N., Lynch, N., Médard, M.: A layered architecture for erasure-coded consistent distributed storage. In: Proceedings of the ACM Symposium on Principles of Distributed Computing, pp. 63–72 (2017)
7. He, Q., et al.: A game-theoretical approach for user allocation in edge computing environment. IEEE Trans. Parallel Distrib. Syst. **31**(3), 515–529 (2019)
8. Li, Y., Wang, X., Gan, X., Jin, H., Fu, L., Wang, X.: Learning-aided computation offloading for trusted collaborative mobile edge computing. IEEE Trans. Mob. Comput. **19**(12), 2833–2849 (2019)
9. Wang, P., Zhao, C., Liu, W., Chen, Z., Zhang, Z.: Optimizing data placement for cost effective and high available multi-cloud storage. Comput. Inform. **39**(1–2), 51–82 (2020)
10. Akan, T., Agahian, S., Dehkharghani, R.: Binbro: binary battle royale optimizer algorithm. Expert Syst. Appl. **195**, 116599 (2022)

11. Lai, P., et al.: Optimal edge user allocation in edge computing with variable sized vector bin packing. In: Pahl, C., Vukovic, M., Yin, J., Yu, Q. (eds.) ICSOC 2018. LNCS, vol. 11236, pp. 230–245. Springer, Cham (2018). https://doi.org/10.1007/978-3-030-03596-9_15
12. Jia, J., Wang, P.: Low latency deployment of service-based data-intensive applications in cloud-edge environment. In: 2022 IEEE International Conference on Web Services (ICWS), pp. 57–66. IEEE (2022)
13. Lin, B., et al.: A time-driven data placement strategy for a scientific workflow combining edge computing and cloud computing. IEEE Trans. Industr. Inf. **15**(7), 4254–4265 (2019)
14. Zhang, Y., Xu, J., Liu, X., Pan, W., Li, X.: A novel cost-aware data placement strategy for edge-cloud collaborative smart systems. In: 2023 IEEE 16th International Conference on Cloud Computing (CLOUD), pp. 450–456. IEEE (2023)

A Multi-holder Role and Strange Attractor-Based Data Possession Proof in Medical Clouds

Jinyuan Guo[1,2], Lijuan Sun[4], Jingchen Wu[1], Chiawei Chu[5], and Yutong Gao[3,6](✉)

[1] Key Laboratory of Trustworthy Distributed Computing and Service, Ministry of Education, Beijing, China
{guooo_jy,lulu}@bupt.edu.cn

[2] School of Cybersecurity, Beijing University of Posts and Telecommunications, Beijing, China

[3] Key Laboratory of Ethnic Language Intelligent Analysis and Security Governance, Ministry of Education, Minzu University, Beijing, China
ytgao92@muc.edu.cn

[4] Intellectual Property Information Service Center, BHU, Beihang University, Beijing, China
sunlijuan@bupt.edu.cn

[5] City University of Macau, Taipa, Macao, Special Administrative Region of China
cwchu@cityu.edu.mo

[6] Li'an International Education Innovation Pilot Zone, Hainan International College of Minzu University of China, Beijing, Hainan, China

Abstract. Proof of Data Possession is a technique for ensuring the integrity of data stored in cloud storage. However, most audit schemes assume only one role for data owners, which is not suitable for complex Smart Healthcare Network with multiple roles. In such schemes addressing the risk of third-party auditors (TPA) colluding with other entities, blockchain is typically employed. However, this necessitates extensive computation, resulting in unacceptably high energy consumption and prolonged verification times. In this paper, we propose a Proof of Data Possession scheme tailored for medical network environments, which incorporates protocols designed for both healthcare professionals and patients, considering the diverse roles that generate data within medical network applications and the varying performance of the devices they use. During the generation of metadata for verification, the protocol incorporates the varying security risk levels associated with medical data. For subsequent metadata selection, an audit block extraction strategy based on strange attractors is employed, effectively mitigating the risk of collusion by third-party TPA, thereby enhancing the security and efficiency of the scheme. Additionally, we conducted experimental analysis and comparisons, demonstrating higher security and practical applicability.

Keywords: Cloud Data · Proof of Data Possession · Smart Healthcare Network · Strange Attractors

---

J. Y. Guo and L. J. Sun—Contributed equally to this research.

T. Zhu et al. (Eds.): ICA3PP 2024, LNCS 15252, pp. 307–316, 2025.
https://doi.org/10.1007/978-981-96-1528-5_21

1 Introduction

As information technology and healthcare demands grow, smart healthcare networks enhance telemedicine and optimize resources through cloud storage. While cloud storage improves data sharing, it also risks data loss, tampering, and incomplete records. These issues are addressed with Proof of Data Possession (PoDP).

PoDP is a security protocol in cloud storage that verifies if service providers securely store and maintain users' data. It involves the data owner, cloud service provider (CSP) [1, 9], and TPA. The protocol has static and dynamic strategies [7], ensuring reliable audits and secure processes. However, these methods can be inflexible and not lightweight enough for large, sensitive medical data. Additionally, the risk of collusion involving TPA and other entities, like doctors or CSP, can lead to unreliable audits, especially in medical disputes, potentially harming data owners' interests.

The goal of this paper is to design a lightweight and secure data possession proof scheme for medical networks. We propose a method using verification metadata under a data grading system to handle complex medical data and introduce an auditing block extraction strategy based on strange attractors [3] to prevent collusion, offering a more efficient alternative to blockchain. Our contributions are as follows:

- We propose a zero-knowledge-based method for constructing medical verification metadata that considers different risk levels of data
- We introduce an audit block extraction model based on strange attractors, and strong steady-state properties to address the risk of collusion during medical data audits.
- We propose a proof of data possession scheme that addresses the complexities of medical data generation and the differences in device performance among various roles

The rest of this paper is organized as follows. In Sect. 2, we discuss related work. In Sect. 3, we present the data possession proof scheme suitable for medical network scenarios in detail. In Sect. 4, we compare our scheme with others and draw conclusions. Finally, Sect. 5 summarize our work.

2 Related Work

In recent years, numerous experts and scholars have extensively researched Proof of Data Possession. Zhao et al. [4] employed a multi-branch tree for dynamic updates, simplifying the verification structure. Kumar et al. [2] developed a system that enhances data privacy protection using RSA and ASE algorithms for data encryption, offering improved confidentiality. Qi et al. [5] addressed conflicts between adjacent branches in updates and rebalancing of a rank-based Merkle hash balance tree. Guo et al. [6] proposed a batch update algorithm based on the Merkle hash tree, avoiding redundant computations and transmissions. To facilitate multiple users sharing data files in the cloud, Yi et al. [11] used linkable ring signatures to create new metadata supporting anonymous verification and the revocation function for group users. Smita et al. [12] introduced a P2DPDP scheme that balances the costs between TPA and users.

As blockchain technology has rapidly emerged in recent years, it has been applied to Proof of Data Possession. Zhang et al. [10] developed a certificateless public verification scheme using blockchain technology to check whether auditors perform inspections within the stipulated time.

Most prior studies lack integration with practical applications and fail to address data complexity and third-party audit risks, leading to inflexible and insecure strategies in medical contexts. To overcome these limitations, we propose a medical cloud data possession proof scheme using strange attractors and multiple holder roles.

3 Our Protocol

3.1 System Model

In the context of data generation and storage within the medical network, the scheme proposed in this paper involves five entities, as depicted in Fig. 1: healthcare professionals (HP), patients, a Key Generation Center (KGC), a Third-Party Auditor (TPA), and a Cloud Service provider (CSP).

The diagram illustrates the interaction between these five entities. The process described in item 3 includes the construction of verification metadata, while item 5 involves an audit block extraction model based on strange attractors. These components are further elaborated in the subsequent protocol discussion.

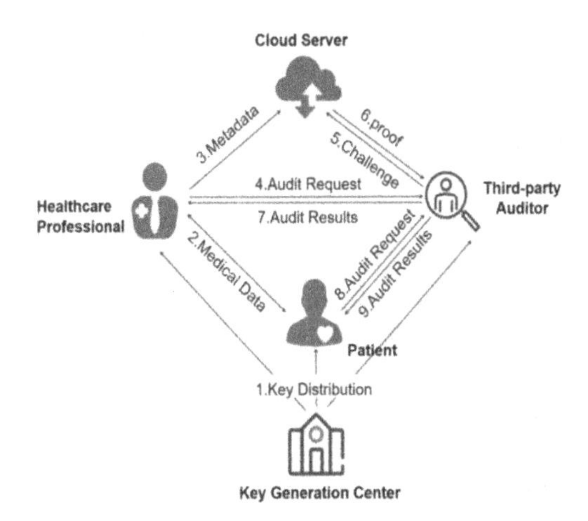

Fig. 1. System model

3.2 Protocol Description for the HP

System Initialization. Input a security parameter 1^k and the KGC system will generate the key associated with the system and the healthcare professional

1. Let the bilinear mapping function $e: G \times G \to G_r$, where G and G_r are multiplicative cyclic groups and the generating element is g, and the large prime order is p Let

H_1, H_2 be the hash functions mapped to the domain of $\{0,1\}^* \rightarrow G, f(\cdot)$ be the one-way hash function mapped to the domain of Z_p: $\{0,1\}^* \rightarrow Z_p$, $h(\cdot)$ is a hash function based on the hash algorithm.
2. KGC selects a random number $\alpha \in Z_p$ as the master private key to compute the public key $v = g^\alpha$, α as the master key of the system and generates the corresponding key $s_i = H_1(UIN_i)^\alpha$ based on the medical member's unique identification number UIN. The medical member receives the key and then carries out the following verification:

$$e(s_i, g) \stackrel{?}{=} e(H_1(UIN_i), v) \tag{1}$$

3. At the same time, KGC randomly generates a pair of signed public-private key pairs (pk, sk), forming the final private key sk = (s_i, sk) and public key pk = (v, pk).

Data Processing and Transmission. The healthcare professional hierarchically categorizes medical documents by sensitivity and generates validation metadata, which is then sent to the CSP and patient, as shown in Fig. 2.

1. Generation of validation metadata. Firstly, the original medical documents F are categorized into sub-table T_i in order based on the sensitivity of the data fields. Thereafter, the healthcare professional's client will randomly generate a number $m \in Z_p$ and chunk the sub-table:

$$d_{ij} = \frac{|T_i|}{m} + sgn(|T_i| mod\ m) \tag{2}$$

where $|\cdot|$ denotes the number of fields in a particular sub-table, sgn($\cdot$) denotes the sign function, and ij denotes the jth block of data in the ith sub-table, so that the total number of blocks n = $\sum_{i,j=1} d_{ij}$ and (i $\in$ [1,6], j $\in$ [1, n]).

2. Select an element u randomly in G, combine the file identification FN, the total number of file data blocks n and the current timestamp t, and then use the additional private key sk to sign this triad to get the file label $ft = FN \| n \| u \| t \| sig_{sk}(FN \| n \| u \| t)$. Sign each data block d_{ij} with private key s_i and element u to get $\sigma_{ij} = \left(H_2(n_{ij}) \cdot u^{(n_{ij} f(n_{ij})) \bmod p}\right)^{s_i} \in G$, which constitutes the homomorphic signature set $\varphi = \{\sigma_{ij} \mid 1 \le i \le 6, 1 \le j \le n\}$.
3. Construct multiple MHT forests with data block hashes $\{h(H_2(d_j)) \mid 1 \le j \le n\}$ as leaf nodes based on multiple T_i, get the set of root nodes R_i of the tree, and compute the total root nodes RT = $H_2(R_1 || \ldots || R_i)$ to the root node signatures to get the $sig_R = (H_2(R_1 \| R_1 || \ldots || R_i))^{s_i}$, and finally send the tuple $\{T_i, ft, \varphi, sig_R\}$ to CSP and the tuple {F, ft, φ, RT, FN} to Patient.

Challenge Message Generation. To ensure data integrity in cloud storage, healthcare professionals can delegate TPA for verification. To prevent collusion by semi-trusted TPAs, a strange attractor-based audit block extraction model is used.

1. CSP carries out data initialization and receives data such as source documents and signature information of healthcare professional, then stores them accordingly, and constructs MHT forest with data block hash value $\{h(H_2(d_j)) \mid 1 \le j \le n\}$ as leaf node when it receives the audit request from TPA, and will save them.

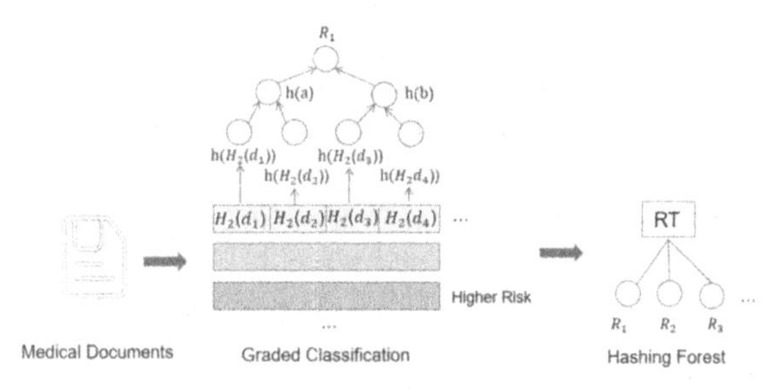

Fig. 2. Building Hashing Forest

2. The healthcare professional initiates an audit request to the TPA and CSP, which includes sending a randomly generated number $w \in N_{+}$, and the role of w will be specified below. The TPA prepares for the subsequent validation by first obtaining the file label *ft* and the MHT structure from the CSP and verifying the label with the additional public key, and then taking out the number of file blocks n and the elements required for the validation u if the validation passes. According to the According to the obtained file structure tree, determine whether the total number of root nodes is equal to n, if not, directly return FALSE to abort the audit.
3. If equal, the completeness of the specific block of data is further verified, at which point the audit block to be extracted will be selected on the basis of the random number generated by the strange attractor. The following are the specific equations for a four-dimensional chaotic system:

$$\begin{cases} \dot{x}_1 = \delta_1 x_2 - \delta_2 x_1 + \delta_3 x_2 x_3 - \delta_4 x_4 \\ \dot{x}_2 = \delta_8 x_1 - x_1 x_4 \\ \dot{x}_3 = x_1 x_2 - \delta_5 \rho x_3 \\ \dot{x}_4 = \delta_6 x_1 + \delta_7 x_2 x_3 \end{cases} \tag{3}$$

where $\dot{x}_i (1 \leq i \leq 4)$ is the four state variables of the system and $\delta_j (1 \leq j \leq 8)$ is the system parameter, and the performance of this chaotic system will be analyzed subsequently. The w received from the healthcare professional will be utilized as the number of iterations of the chaotic system to determine the specific k audit blocks, which constitute an ordered subset of indexes $I = \{s_1, s_2, \cdots, s_k\}$, and finally generate the cover tree T. The TPA selects k random numbers $v_k \in \boldsymbol{Z}_p$ and generates chal $= \{v_k | k > 0\}$.

Chaotic system, due to its sensitivity and unpredictability, generates secure random numbers effectively. Even with the algorithm known, its complexity prevents attackers from reversing or predicting outcomes. By controlling chaotic system parameters, the TPA and CSP can independently generate matching random numbers, mapping them to data blocks and prioritizing sensitive data in the MHT forest. The TPA does not share audit block information with the CSP, preventing collusion. Additionally, as the chaotic system's model is only accessible to the TPA and CSP, healthcare professionals cannot manipulate audit blocks, mitigating joint deception risks.

Proof Information Generation. The CSP receives the challenge details, generates verification evidence, and sends it to the TPA.

1. Based on the parameters w sent by the healthcare professional, the prover CSP iterates the chaotic system and determines the audit block according to the mapping rule, combines the MHT to generate the overlay tree T', and attaches values at and only to the leaf nodes, The leaf node $\{h(H_2(d_i)) \mid s_1 \leq i \leq s_k\}$ corresponding to the data block to be verified is replaced by the corresponding $\{H_2(d_i) \mid s_1 \leq i \leq s_e\}$.
2. Calculate $\mu = \sum_{i=s_1}^{s_k} v_i d_i f(d_i) \in Z_p$ and $\eta = \prod_{i=0}^{N} \sigma_i^{d_i} \in G$, and finally form the evidence $proof = \{\mu, \eta, T', sig_R\}$ to be sent to the verifier TPA.

Proof Verification. The verifier TPA receives the relevant proof from CSP, performs the accounting verification and returns the result to the user.

1. The TPA determines whether the tree structure of the coverage tree T' returned from the CSP side and the coverage tree T computed by itself are the same, if there is any difference, it directly returns FALSE and ends.
2. After extracting $\{H_2(d_i) \mid s_1 \leq i \leq s_c\}$ from the overlay tree T' and storing it in a separate file, calculate $\{h(H_2(d_i)) \mid s_1 \leq i \leq s_c\}$ to replace the corresponding value, and if the leaf nodes are complete, then the root node R can be calculated, and judge whether the value of the root node is correct or not according to the following test formula, and if the validation fails, then it means that the returned $\{H_2(d_i) \mid s_1 \leq i \leq s_c\}$ data or other data block information outside the challenge verification is wrong, directly return FALSE.

$$e(sig_R, g) = e(H_2(R), g^{\alpha}) \tag{4}$$

3. Continue to determine whether the following test equation holds, where $\{H(m_i) \mid s_1 \leq i \leq s_c\}$ has been obtained in the previous. If the same holds, it means that the file data stored in the cloud is correct and passes the integrity audit, and returns TRUE, otherwise outputs FALSE.

$$e(\eta, g) = e\left(\prod_{i=i_1}^{s_k} H_2(d_i)^{v_k} \cdot u^{\mu}, v\right) \tag{5}$$

3.3 Protocol Description for the Patient

System Initialization. It is basically similar to the corresponding part of HP.

1. Consistent with the corresponding section of HP, replace UIN_i with the patient's unique ID number $UMPI_i$:

$$e(s_i, g) \stackrel{?}{=} e(H_1(UMPI_i), v) \tag{6}$$

2. Compute the patient's public key as g^{s_i} and pick an element $u \in G$.

Data Processing and Transmission. The patient will calculate the labels of the corresponding root plants, and the aggregated labels of all MHT forests, based on the medical documents sent by multiple healthcare professional and the root values of the corresponding MHT forests.

1. After receiving the original data and validation metadata sent by P copies of healthcare professional, the patient first confirms the relevant information based on the original medical documents, and subsequently picks an integer $\alpha \in \mathbf{Z}_p$, and computes $\gamma = g^{\alpha}$. Then, the labels of the corresponding root values are computed as follows:

$$\mathrm{Root}_{\mathrm{F_i}} = \mathrm{s_i} \cdot \left(\mathrm{H_2(t \| FN)} \cdot \mathrm{u}^{\mathrm{RT_i}}\right)^{\alpha} (\mathrm{i} \in [1, \mathrm{p}]) \tag{7}$$

In order to alleviate the computation of the patient's device, the tuple $\left(t, \gamma, Root_{FN_i} \cdots Root_{FN_N}\right)$ is then returned to the KGC, which does the next auxiliary computation.

2. KGC receives the patient's hash tree label and obtains the hash forest value from it as follows.

$$V = H_2\left(RT_i \| \cdots \| RT_p\right) \tag{8}$$

From this the aggregated value of hash tree tags of all the files of a patient is calculated and sent to the cloud server. Finally, a randomly generated number $b \in N_+$ is sent to TPA and CSP to identify the audited data.

Challenge. TPA determines the audited data based on the parameters sent by the patient using chaotic system to challenge the CSP.

1. TPA first verifies the validity of the file label, then generates the mapping value according to w using chaotic system, determines the validated data $\mathbb{F}_i \subseteq \left[FN_1, FN_p\right]$, randomly selects the integer $sn_{F_i} \in \mathbf{Z}_p$, and generates $Chal = \left\{sn_{FN_i}\right\}$ and sends it to the CSP.

Proof Information Generation. CSP also determines the audited data based on the parameters sent by the patient using chaotic system, and generates the corresponding proof and returns it to the TPA after receiving the query from the TPA.

1. When the cloud server receives the query, it calculates the aggregation to $\prod_{FN_i \in \mathbb{F}_i} \mathrm{Root}_{FN_i}^{sn_{FN_i}}$ and denote it as $Root_p$.
2. Calculate the linear combination value of the file $\pi = \sum_{FN'_i \in \mathbb{F}_i} sn_{FN'_i} \cdot V_{FN_i}$.
3. Send a data-holding proof $\left\{\pi, Root_p, t\right\}$ to the TPA.

Proof Verification. The TPA receives the proof from the cloud server and verifies it.

1. The TPA verifies that the following test is valid, and if it is valid, then the data in the cloud is intact; if it is not, then the files in the cloud are compromised.

$$\mathrm{e}\left(\mathrm{g}, \mathrm{Root_p}\right) = \mathrm{e}\left(\mathrm{g}^{\mathrm{s_i}}, \mathrm{H_1(t)}^{\sum_{\mathrm{FN_i} \in \mathbb{F}_i} \mathrm{sn_{FN_i}}}\right) \cdot \mathrm{e}\left(\gamma, \left(\prod_{\mathrm{FN_i} \in \mathbb{F}_i} \mathrm{H_2(t \| FN)}^{\mathrm{sn_{FN_i}}}\right) \cdot \mathrm{u}^{\pi}\right) \tag{9}$$

4 Comprehensive Performance Evaluation

In this experiment, we used a Windows system with an Intel i5-8300H CPU @ 2.30 GHz, 16 GB RAM, and 2667 MHz. The main methods include comparative analysis and index testing. We evaluated security and computational overhead by comparing our scheme with MHT-PA, FGUPA, and IPANM [8], the first two being hash-based and IPANM incorporating blockchain. These schemes are supported by major national projects.

4.1 Comparison of Security Aspects

By analyzing our scheme and several representative cloud data integrity auditing schemes, we compare their security features. The detection rate indicates the probability that at least one of the c data blocks extracted from a file with an error rate of θ is detected. The following Table 1 shows that our auditing scheme has higher security.

Table 1. Security Comparison

scheme	Public audit	Data privacy protection	Anti-replacement attack	Anti-Collusion Attack	Detection rate
MHT-PA	√	√	√	×	$1-(1-\theta)^c$
IPANM	√	√	√	√	$1-(1-\theta)^c$
FGUPA	√	√	×	×	$1-(1-\theta)^c$
Our scheme	√	√	√	√	$1-(1-\theta)^c$

4.2 Comparison of Overhead

In the MHT-PA and FGUPA schemes, data blocks are divided into 1 KB segments. First, we compare the communication overhead among the four schemes. The communication overhead for MHT-PA, FGUPA, and our scheme is cO(logn) (considering only healthcare professionals, as this role has the highest complexity, and patient auditing also relies on healthcare professionals to some extent). The communication overhead for IPANM is $3c|q|+(u+3)|G|+u|ID|$, which includes the introduction of blockchain to address the issue of public auditing for a distributed user group.

Regarding the computation time for auditors in the four schemes, the total computation time for IPANM is given by $2T_{Pa}+(c+1)T_{mu}+cT_{Ha}+(c+3)T_{Ex}$, where T_{Pa} represents the time for hash operations, T_{mu} represents the time for multiplication operations, T_{Ha} represents the time for scalar multiplication operations, and T_{Ex} represents the time for modular exponentiation operations. Our scheme, compared to the other two schemes, includes an audit block extraction model based on strange attractors. However, the computation time has not increased significantly. The specific comparisons are presented in Fig. 3, Fig. 4.

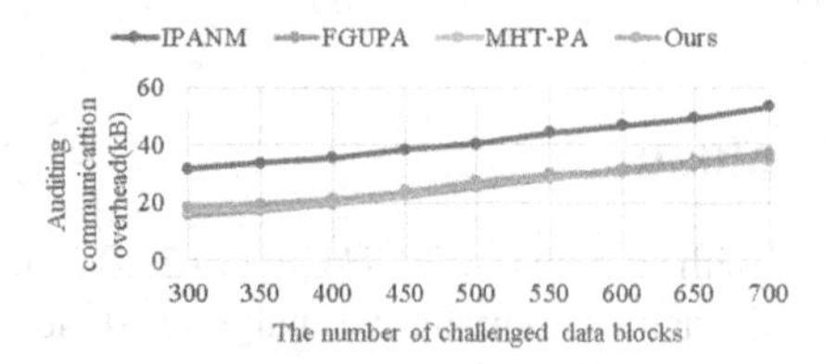

Fig. 3. Auditing Communication Overhead

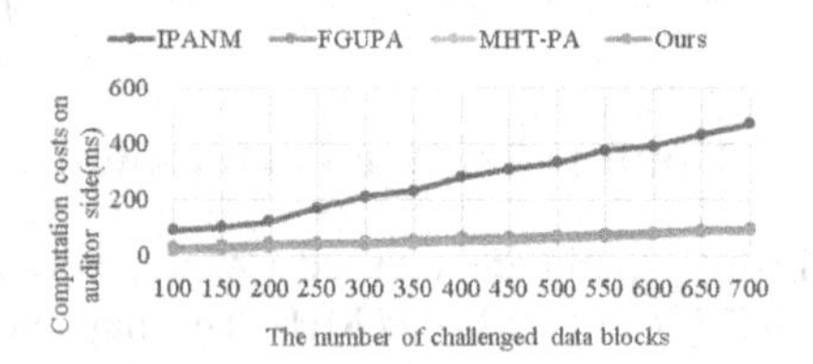

Fig. 4. Auditor Computation Cost

4.3 Performance of the Strange Attractor-Based Model

The chaotic system is generated using the Lorenz attractor, where stable chaotic behavior is identified through the analysis of mathematical models. By applying optimization algorithms and continuously adjusting through experimentation, the initial parameters are determined: initial values set as $x = \{1, 0, 0, 0\}$ and parameters set as $\delta = \{43, 56, 23, 5, 16, 23, 36, 22\}$.

Based on the Jacobian matrix, we can calculate the Lyapunov exponents of the system as $\lambda_1 = 4521.952$, $\lambda_2 = 4521.491$, $\lambda_3 = 4517.536$, $\lambda_4 = 4519.339$. The presence of multiple large positive Lyapunov exponents indicates that the system exhibits chaotic properties and dissipativity. Additionally, we can see the system fractal dimension is $D_L = 2.0706$, confirming its complex fractal nature.

5 Conclusion

In this paper, we propose a Proof of Data Possession scheme tailored for medical network environments, incorporating both healthcare professionals and patient roles. Considering the potential risk of collusion among certain roles during the medical auditing process, we introduce an audit block extraction strategy based on strange attractors, enhancing both the security and efficiency of the scheme. For future work, we plan to explore group-based auditing for healthcare professionals' data to better align the protocol with practical hospital applications.

Acknowledgments. This work is supported by Major Projects of National Natural Science Foundation of China (Grant No. 72293580/72293583), Research on Privacy Data Protection and Iatrogenesis Decision of IHS.

References

1. Kalai Arasan, K., Anandhakumar, P.: Hybrid COOT-reverse cognitive fruit fly optimization-based big data services and virtual machine allocation for cloud storage system. J. Circuits Syst. Comput. **33**(01) (2024)
2. Kumar, A.: A novel privacy preserving HMAC algorithm based on homomorphic encryption and auditing for cloud. In: Fourth International Conference on I-SMAC (I-SMAC), Palladam, India (2020)
3. He, K., Shi, J., Fang, H.: Bifurcation and chaos analysis of a fractional-order delay financial risk system using dynamic system approach and persistent homology. Math. Comput. Simul. **223**, 253–274 (2024)
4. Zhao, Y.L., Li, S.Q.: Dynamic flexible multiple-replica provable data possession in cloud. In: 17th International Computer Conference on Wavelet Active Media Technology and Information Processing (ICCWAMTIP), Chengdu, China (2020)
5. Zhang, K., et al: Revocable certificateless provable data possession with identity privacy in cloud storage. Comput. Stand. Interfaces **90** (2024)
6. Guo, W., et al.: Outsourced dynamic provable data possession with batch update for secure cloud storage. Future Gener. Comput. Syst. **95**, 309–322 (2019)

7. Pawar, A.B., Ghumbre, S.U., Jogdand, R.M.: Privacy preserving model-based authentication and data security in cloud computing. Int. J. Pervasive Comput. Commun. **19**(2), 173–190 (2023). https://doi.org/10.1108/IJPCC-11-2020-0193
8. Huang, et al.: IPANM: incentive public auditing scheme for non-manager groups in clouds. IEEE Trans. Dependable Secure Comput. **19**(2), 936–952 (2022)
9. Nie, Z.X., Li, J., Duan, F.H., et al.: A collaborative ledger storing model for lightweight blockchains based on Chord Ring. J. Supercomput. **80**(4) (2024)
10. Zhang, et al.: Blockchain-based public integrity verification for cloud storage against procrastinating auditors. IEEE Trans. Cloud Comput., 1 (2019)
11. Qi, Y., Luo, et al.: Blockchain-based privacy-preserving group data auditing with secure user revocation. Comput. Syst. Sci. Eng. **45**(1), 183–199 (2023)
12. Chaudhari, et al.: Privacy preserving dynamic provable data possession with batch update for secure cloud storage. Int. J. Adv. Comput. Sci. Appl. **12**(6) (2021)

Federated Learning and Parallel Prompt Scheduling Strategies for Large Language Models

Guangtong Lv[1,2], Bruce Gu[1,2](✉), Xiaocong Jia[1,2], Longxiang Gao[1,2](✉), Youyang Qu[1,2], and Lei Cui[1,2]

[1] Key Laboratory of Computing Power Network and Information Security, Ministry of Education, Shandong Computer Science Center (National Supercomputer Center in Jinan), Qilu University of Technology (Shandong Academy of Sciences), Jinan, China
bruce.qu@yu.edu.au , longxiang.qao@deakin.edu.au

[2] Shandong Provincial Key Laboratory of Computer Power Internet and Service Computing, Shandong Fundamental Research Center for Computer Science, Jinan, China

Abstract. Large language models have garnered significant attention and are widely utilized across different fields due to their impressive performance. However, centralized training of these models can pose privacy risks like data leakage, hindering their application and advancement. As a decentralized learning paradigm, Federated Learning(FL) can effectively address data security in large language models and has specific application scenarios. In addition, the fine-tuning process of large language models also incurs significant communication expenses and leads to the loss of computational resources. The utilization of prompt tuning in federated learning is anticipated to decrease training costs, accelerate the training process, and enhance model performance. In this paper, we propose 'FedLLM-PPS' to explore a novel approach for large language models within federated learning. This method address data security concerns associated with large language models while enhancing performance and efficiency. Experiments based on several datasets for various Natural Language Processing(NLP) tasks demonstrate superior performance and efficiency compared to baseline methods.

Keywords: Large Language Models · Federated Learning · Prompt-Tuning

1 Introduction

With the continuous development of artificial intelligence, large language models [1] (e.g., ChatGPT) have entered the public's field of vision. Their ability to converse with human beings is impressive, and the emergence of this technology has also attracted the attention of various industries to these large language models. Large Language Models have a powerful ability to handle complex tasks and

T. Zhu et al. (Eds.): ICA3PP 2024, LNCS 15252, pp. 317–326, 2025.
https://doi.org/10.1007/978-981-96-1528-5_22

have been more widely used in various industries such as education [2], finance [3], and law [4]. Large language models can be used for natural language processing tasks, including but not limited to machine translation, human-machine dialogue, and text summarization. However, the centralized training model will inevitably pose particular risks to privacy and security [5].

Addressing the privacy concerns associated with centralized training of large language models is crucial. Recently, a growing societal and individual focus on data security has been growing [6]. Federated learning [7], a novel machine learning paradigm, differs from traditional machine learning methods that require data to be centralized to train models. In federated learning, individual participants do not need to transfer data to a central server; model parameters are exchanged for training and updating [8]. Federated learning is anticipated to offer a framework for addressing data security concerns in the context of large language models.

Moreover, training a large language model requires many computational resources. However, it is challenging to replicate the necessary conditions in natural production environments. Efficient parameter fine-tuning involves adjusting a small number of model parameters or adding new ones while keeping most large language model parameters fixed. There are numerous methods for efficiently fine-tuning model parameters. The adapter method [9] was first applied in the field of NLP. Each transformer block is enhanced by incorporating two adapter modules. These modules are positioned after the multi-head attention mapping and after the two-layer feed-forward neural network, respectively. The Prefix Tuning [10] introduces a continuous and task-specific vector that is added to all Transformer layers before the key κ and the value υ.

In this paper, we propose a parameter-efficient fine-tuning method for enhancing large language models using federated learning. In the framework of federated learning, we propose a local parallel prompt strategy by optimally selecting the prompt that is suitable for the current task. The method achieves high accuracy. Our main contributions are as follows:

- We propose FedLLM-PPS, a novel approach for fine-tuning federated large language models. By aggregating and tuning a small number of prompts in addition to those already present in the training model, we can significantly reduce communication costs.
- The parallel prompt strategy can help the model acquire and utilize various prompt information more effectively, thereby enhancing the model's performance.
- We conducted experiments on NLP tasks to measure the performance of FedLLM-PPS. The experiments show that FedLLM-PPS reduces communication costs without any degradation in accuracy.

2 Related Work

Recently, there has been a growing interest in issues related to fine-tuning large language models in a federated learning framework. In the following, we briefly discuss relevant research on this issue and the most relevant approaches to our work.

2.1 Federated Learning

Federated learning [11] is a distributed machine learning framework with the primary objective of safeguarding privacy. In this framework, clients can construct and train models without revealing the local raw data, ensuring that the data is not directly shared during communication. The global model training is performed by transmitting the encrypted model parameters from the local model training to the central server. The trained global model parameters are sent to each client to participate in the next round of training, and the iteration is repeated until the model converges.

In order to ensure the security of the federated learning architecture [12], extensive research has been conducted. [13] enhanced the homomorphic encryption method by incorporating Paillier and other encryption algorithms to safeguard the training model and data. Perturbation methods are more efficient techniques that involve adding arbitrary integers as noise to affect the model's parameters. Differential privacy [14] is commonly implemented through perturbation by adding an appropriate amount of noise. However, the randomness inherent in this method can affect the accuracy and usability of the model.

2.2 Fine-Tuning in LLM

Large language models [15] exhibit more vital language comprehension and generation capabilities compared to smaller-scale models, primarily due to their vast parameter sizes. Mitigating the substantial communication and computational costs incurred by updating the parameters during the training process is of paramount importance. Prompt-Tuning [16], a method for fine-tuning pre-trained language models, modifies the model by adjusting the input cues rather than modifying the model parameters. Formally, the mathematical formula can be expressed as:

$$X^l = [p_1^l, \ldots, p_{N_p}^l, x_1^l, \ldots, x_{N_x}^l] \tag{1}$$

where X^l represents the sequence of input tokens for layer l, including the input prompt tokens p and the original input tokens x. N_p indicates the number of prompt input tokens, and N_x indicates the number of original input tokens.

Aprompt [17] employs another strategy. In addition to inserting input hints at the beginning of the input sequence of each Transformer layer, additional learnable hints are added in front of the corresponding query, key, and value matrices in the Self-Attention block to learn the attentive attention pattern.

2.3 Federated Learning in LLM

Recently, several studies have been conducted on federated learning and the modeling of large language models. FATE-LLM [18] explores the framework of federated large language models and provides some research on how large language models can be fine-tuned in the framework of federated learning. OpenFedLLM [19] has conducted extensive research on federated learning setups and pointed out two fundamental processes: joint instruction tuning and joint value alignment. FedPrompt [20] enabled FL based on prompt tuning. However, it involves communicating the entire set of parameters in the prompts, leading to significant communication costs.

In previous work, little discussion has been given to optimizing large language models within a federated learning framework. Unlike previous work, we have enhanced the approach to fine-tuning large language models within a federated learning framework by developing parallel prompt strategies. We fine-tune the model by designing multiple parallel prompts locally to improve the its performance while reducing communication costs. In addition, we introduce relevant measures for parameter management on the client side to further enhance the model performance.

3 Methodology

In this section, we first discuss the problem of efficiently fine-tuning large language models in a Federated Learning framework. Then we state the system model structure of FedLLM-PPS. Finally, we elaborate on how to implement the parallel prompt strategy and its optimisation process under the Federated Learning framework.

3.1 Problem Formulation

The problem addressed in this paper is how to efficiently tune a large language model in FL based on Prompt-tuning. Given a large language model $\mathcal{M}$ with L layers, let p^l represent the set of parameters for generating prompts, where l is the number of layers in $\mathcal{M}$. Denote all the hints as P. During the tuning process, all the parameters in the model $\mathcal{M}$ are frozen, and only the parameters in P are updated to improve the performance of the model $\mathcal{M}$.

For the federated learning scenario, we need to consider the following considerations. We consider a FL scenario with a parameter server and K devices in the setting. We assume that the participants involved in the fine-tuning of model $\mathcal{M}$ are distributed across multiple devices, each having data $D_i = (X_i, Y_i)_{N_i}$ on each device i. Here, N denotes the total number of samples on all devices, and n_i denotes the total number of samples on device i. X represents all samples, x_i represents samples on device i, and Y represents labels. The problem addressed in this paper can be formulated as follows: how to efficiently generate P in order to minimize the global loss.

$$\arg\min_{P} \mathcal{L}(\mathcal{M}, P) = \sum_{i=1, p_i \in P}^{M} \frac{n_i}{N} \mathcal{L}_i(\mathcal{M}, p_i) \tag{2}$$

where $\mathcal{L}_i(\mathcal{M}, p_i)$ is the empirical loss of client K:

$$\mathcal{L}_i(\mathcal{M}, p_i) = \mathbf{E}_{(x_i, y_i) \in D_i} \ell_i(\mathcal{M}, p_i, x_i, y_i) \tag{3}$$

It refers to the local loss, which is computed at device i with $\ell_i(\mathcal{M}, p_i, x_i, y_i)$ for the large language model $\mathcal{M}$ and the prompt parameter P on $\{x_i, y_i\}$

For the NLP task, each sample $x \in X$ represents the input, and $y \in Y$ represents the corresponding label. Each sample x consists of multiple tokens, i.e., $X = (x_1, x_2, \ldots, x_n)$ The prompt P is also composed of multiple tokens, $P = (p_1, p_2, \ldots, p_n)$, referring to the corresponding cue parameters that can be trained. We use $\mathcal{T}(\cdot)$ to denote the cue template that defines how to connect the cue information with the data. For example, $s_x^p = \mathcal{T}(\cdot)$ represents the sample after the cue is combined with the data, which also includes a $[mask]$ token. Inputting this into the bigram model yields the output label Y. A mapping relation is needed since we are only interested in some of the words in the specified part. For example, if the word predicted by $[mask]$ is 'fantastic', it is considered Positive. This can be achieved through $\mathcal{V}$ (tagged word mapper), i.e., $Y = \mathcal{V}(Great)$, where 'Great' represents the output of the model and Y denotes the predicted label.

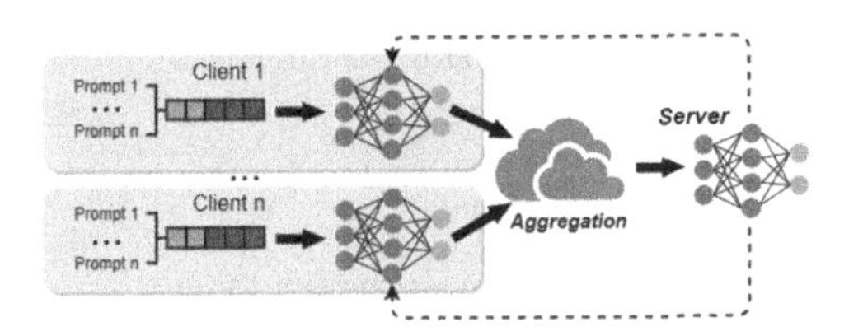

Fig. 1. A schematic diagram of FedLLM-PPS.

3.2 System Model

As shown in Fig. 1, based on the overall architecture of federated learning, we utilize servers and multiple devices for the entire tuning process. We assume that a large language model is deployed on each local device, and each local client is set up with multiple parallel prompts $(p_1, p_2 \ldots, p_n)$ to process the task. Each client uses local data for model training. Clients generate independent outputs for each sample using each prompt separately during training. For the input samples, each client calculates a weighted average of the outputs of multiple prompts to derive the local prediction results. The loss is calculated using the fused results, and the gradient is calculated based on the loss. The local model parameters are updated based on the obtained gradient and then uploaded to the central server. The central server uses an aggregation algorithm to calculate the weighted average of the model parameters uploaded by all clients in order to obtain the global model. The central server distributes the updated global model parameters to all clients, who receive and perform local model updates.

3.3 Optimization

At the initial stage, the server and clients randomly select pre-set prompts and weight combinations, which will serve as inputs for the initial evaluation of the local language models. For instance, by inputting multiple prompts $(p_1, p_2, \ldots, p_n)$, we obtain independent outputs $(O_1, O_2, \ldots, O_n)$. Each prompt output is assigned a weight $(w_1, w_2, \ldots, w_n)$. Each local client evaluates the performance of the local language model on their local validation set using these initial weight combinations. We use a weighted average to fuse the outputs:

$$O_{final} = w_1 O_1 + w_2 O_2 + \ldots + w_n O_n \tag{4}$$

To further evaluate model performance and minimize the gap between the output results and the true values, we calculate the cross-entropy loss as a performance metric:

$$\mathcal{L}(\widehat{y}, y) = -\sum_i y_i log(\widehat{y}_i) \tag{5}$$

where $\widehat{y} = O_{final}$. In order to update the weights, the gradient of the loss function concerning these weights must be calculated. Using the chain rule, the gradient will be computed in two steps: first, the gradient of the loss for the fusion result, and then the gradient of the fusion result to the weights. For the cross-entropy loss, the gradient of the loss to the fusion result is:

$$\frac{\partial \mathcal{L}}{\partial O_{final}} = \widehat{y} - y \tag{6}$$

The gradient of the fusion result O for each weight w is :

$$\frac{\partial O_{final}}{\partial w_i} = O_i \tag{7}$$

Combining the last two parts of the gradient gives the gradient of the loss against weight:

$$\frac{\partial \mathcal{L}}{\partial w_i} = \frac{\partial \mathcal{L}}{\partial O_{final}} \cdot \frac{\partial O_{final}}{\partial w_i} = (\widehat{y} - y) \cdot O_i \tag{8}$$

The weights are updated based on the gradient. Assuming a learning rate of η, the weight update formula is:

$$w_i \leftarrow w_i - \eta \cdot \frac{\partial \mathcal{L}}{\partial w_i} \tag{9}$$

To safeguard data security and prevent malicious attacks on the parameter server, we construct a Gaussian process model to serve as a surrogate model for our server based on Bayesian Optimization [21]. The server aggregates evaluations from clients and updates model parameters by training a Gaussian process surrogate model with a Matern kernel. Next, a selection of the next set of weight combinations for evaluation is made using an acquisition function:

$$\mathcal{F}(w) = \mu(w) + \beta \cdot \nu(w) \tag{10}$$

where β is a tuning parameter that balances the mean ($\mu(\cdot)$) and variance ($\nu(\cdot)$) in the Gaussian process model.

The client receives the new set of weights, evaluates its performance on the local validation set, and sends the evaluation results back to the server. The server updates the Gaussian process model with the new evaluation results and iterates the acquisition function optimization and client evaluation processes until reaching a predefined maximum number of iterations or until there is no significant improvement in model performance. The server aggregates all client model parameters updates the global language model parameters, and sends the updated model parameters back to the clients.

4 Experiments

In this section, we present the experimental results over two baselines and two NLP tasks to demonstrate the sophistication of FedLLM-PPS

4.1 Experimental Setup

Datasets: We utilize the SST-2 and IMDB datasets for the sentence categorization task. SST-2 is akin to the IMDB dataset and includes human annotations of sentences and associated sentiments from movie reviews. The categories are divided into positive and negative sentiments. For the sentence inference task, we utilize the QNLI dataset, which comprises question-paragraph pairs that help ascertain the implicit nature of questions and sentences.

Modeling and Training Details: We select the most representative and widely used PLMs, such as the base versions of BERT, for our experiments. In the main experiments, we use multiple cue templates. Based on previous work, we set the learning rate to 0.2. We consider a FL environment with 10 devices and a parameter server and we implement the FedAvg system for the FL setup.

Baselines and Metrics: We use two baseline methods, including, P-tuning v2 and FedPrompt. We measure the communication cost of the model based on the number of communication parameters, and we evaluate the model's performance on the task based on accuracy(ACC).

4.2 Results

In FedLLM-PPS, the parameters involved in training are the same as the communication expenses. Full-Parameter Fine-Tuning requires 110M parameters and FedLLM-PPS requires about 239K parameters, significantly reducing communication costs and accelerating of local model training and aggregation in federated learning scenarios. The reduced parameter size effectively alleviates the limitation on computational and storage resources, which allows some resource-constrained edge devices to participate in the federated learning process and makes federated learning better applied to real-world scenarios.

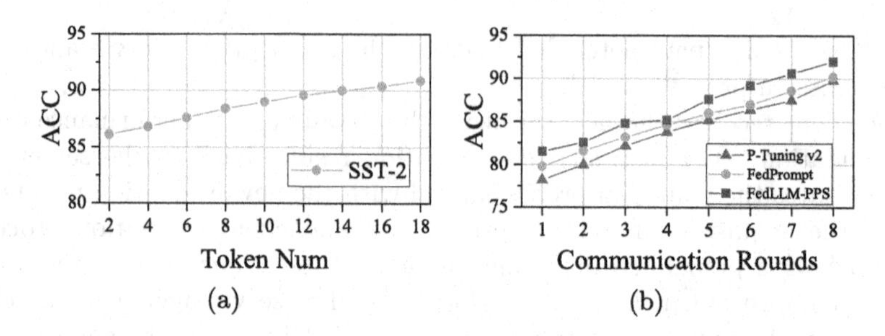

Fig. 2. ACC (%) results with different token numbers (a). ACC (%) results with our proposed approach, compared with two baselines on SST-2 task (b)

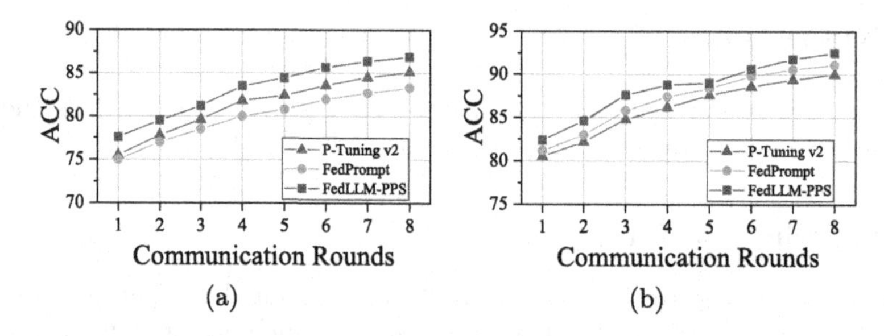

Fig. 3. ACC (%) results with our proposed approach, compared with two baselines on QNLI task (a) and IMDB task (b)

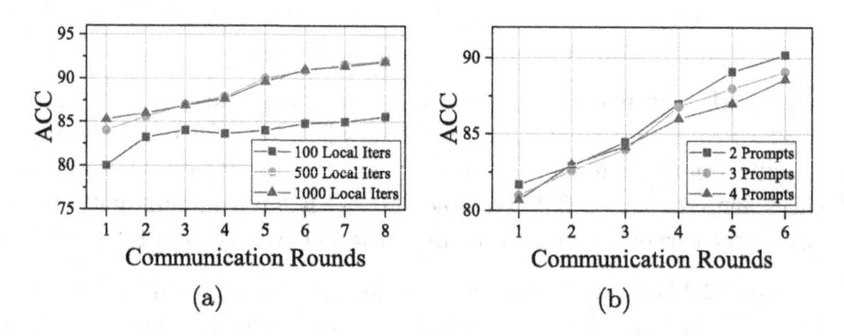

Fig. 4. ACC (%) results with different local iterations (a) and different prompts (b) on SST-2 task

We tested the results of different numbers of token settings based on SST-2 task. As shown in Fig. 2(a), more tokens will perform better. The performance is more pronounced when the number of tokens reaches the interval of 10 to 16. In order to trade off model performance and communication expenses, we control the number of tokens within this interval in the following experiments. We tested the results for the number of parallel prompt settings. As shown in Fig. 4(b), the input of two prompts performs best for the model performance, so in the later experiments, we set the prompts similarly. The main results of the performance of FedLLM-PPS are shown in Fig. 2(b), Fig. 3(a) and Fig. 3(b). Compared to the two baseline models, FedLLM-PPS has an ACC of 91.03% and 91.57% on the two sentence categorization tasks and 86.92% on the complex sentence inference task.

We investigated the effect of the number of local iterations per round. In a federated learning scenario, there is a trade-off between the number of client iterations. Too many iterations can lead to local overfitting and too few iterations can lead to poor model performance. We conducted experiments with 100, 500 and 1000 local iterations. As shown in Fig. 4(a), 100 iterations model performance is poor, 500 and 1000 iterations may lead to local overfitting and model performance improvement is not significant enough.

5 Conclusion

In this work, we propose "FedLLM-PPS" to investigate the parallel prompt adjustment strategy of large language models in federated learning scenarios. By employing this strategy, we can avoid the need for extensive parameter fine-tuning. Parallel prompts are beneficial for discovering and validating the best prompt design, thereby enhancing the model's performance. Our experiments have proven that FedLLM-PPS offers significant advantages in terms of accuracy and other aspects. Although FedLLM-PPS has achieved some results, its parallel prompt strategy may need better performance as prompt size increases. Practical methods are still needed to balance between prompt size and performance. Next, we will continue conducting relevant research.

Acknowledgement. This paper is sponsored by Research Fund for International Young Scientists, project number: 62350410478.

References

1. Devlin, J., Chang, M.-W., Lee, K., Toutanova, K.: Bert: pre-training of deep bidirectional transformers for language understanding. arXiv preprint arXiv:1810.04805 (2018)
2. Wang, S., et al.: Large language models for education: a survey and outlook. arXiv preprint arXiv:2403.18105 (2024)
3. Lee, J., Stevens, N., Han, S.C., Song, M.: A survey of large language models in finance (finllms) (2024)
4. Lai, J., Gan, W., Wu, J., Qi, Z., Philip, S.Y.: Large language models in law: a survey (2023)
5. Gao, L., Luan, T.H., Gu, B., Qu, Y., Xiang, Y.: Privacy issues in edge computing. In: Privacy-Preserving in Edge Computing. WN, pp. 15–34. Springer, Singapore (2021). https://doi.org/10.1007/978-981-16-2199-4_2
6. Gao, L., Luan, T.H., Gu, B., Qu, Y., Xiang, Y.: Privacy-Preserving in Edge Computing. Springer, Cham (2021)
7. Rafi, T.H., Noor, F.A., Hussain, T., Chae, D.-K.: Fairness and privacy preserving in federated learning: a survey. Inf. Fusion **105**, 102198 (2024)

8. Bruce, G., Gao, L., Wang, X., Youyang, Q., Jin, J., Shui, Yu.: Privacy on the edge: customizable privacy-preserving context sharing in hierarchical edge computing. IEEE Trans. Network Sci. Eng. **7**(4), 2298–2309 (2019)
9. Houlsby, N., et al.: Parameter-efficient transfer learning for NLP. In: International Conference on Machine Learning, pp. 2790–2799. PMLR (2019)
10. Li, X.L., Liang, P.: Prefix-tuning: optimizing continuous prompts for generation. arXiv preprint arXiv:2101.00190 (2021)
11. McMahan, B., Moore, E., Ramage, D., Hampson, S., y Arcas, B.A.: Communication-efficient learning of deep networks from decentralized data. In: Artificial Intelligence and Statistics, pp. 1273–1282. PMLR (2017)
12. Liu, Y., Youyang, Q., Chenhao, X., Hao, Z., Bruce, G.: Blockchain-enabled asynchronous federated learning in edge computing. Sensors **21**(10), 3335 (2021)
13. Kim, A., Song, Y., Kim, M., Lee, K., Cheon, J.H.: Logistic regression model training based on the approximate homomorphic encryption. BMC Med. Genomics **11**, 23–31 (2018)
14. Ho, S., Youyang, Q., Bruce, G., Gao, L., Li, J., Xiang, Y.: DP-GAN: differentially private consecutive data publishing using generative adversarial nets. J. Netw. Comput. Appl. **185**, 103066 (2021)
15. Zhao, W.X., et al.: A survey of large language models. arXiv preprint arXiv:2303.18223 (2023)
16. Lester, B., Al-Rfou, R., Constant, N.: The power of scale for parameter-efficient prompt tuning. arXiv preprint arXiv:2104.08691 (2021)
17. Wang, Q., et al.: Aprompt: attention prompt tuning for efficient adaptation of pre-trained language models. In: Proceedings of the 2023 Conference on Empirical Methods in Natural Language Processing, pp. 9147–9160 (2023)
18. Fan, T., et al.: Fate-llm: a industrial grade federated learning framework for large language models (2023)
19. Ye, R., et al.: Openfedllm: training large language models on decentralized private data via federated learning. arXiv preprint arXiv:2402.06954 (2024)
20. Zhao, H., Du, W., Li, F., Li, P., Liu, G.: Fedprompt: communication-efficient and privacy-preserving prompt tuning in federated learning. In: ICASSP 2023-2023 IEEE International Conference on Acoustics, Speech and Signal Processing (ICASSP), pp. 1–5. IEEE (2023)
21. Kusner, M.J., Paige, B., Hernández-Lobato, J.M.: Grammar variational autoencoder (2017)

Privacy-Preserving Federated Learning Framework in Response Gaming Systems

Qiong Li[1,2,3], Yizhao Zhu[1,2](✉), and Kaio Leong[4]

[1] City University of Macau, Macau 999078, China
zhuyizhao@gmail.com
[2] Hunan Engineering Technology Research Center of Digitalization of CNC Machining Process for Precision Parts, Changsha 410208, China
[3] Hunan Industry Polytechnic, Changsha 410208, China
[4] Macao Polytechnic University, Macau 999078, China

Abstract. Responsible gaming (RG) promotes safe gambling practices and is essential for the sustainable development of the gaming industry. Current RG systems typically utilize setting gaming limits and maintaining blocking lists. Static and rigid gaming limits can adversely affect both customer experience and operator business development due to their inflexibility. While machine learning could enable tailor-made limits, it requires exchanging sensitive data among multiple parties, such as casino operators, financial institutions, and third parties, raising data privacy and confidentiality concerns due to the sensitive nature of personal and financial information. This work proposes a pioneering privacy-preserving federated learning framework for response gaming systems to address these challenges. During model training, our approach leverages federated learning to mitigate sensitive data exchange issues. Additionally, we incorporate Labeled Private Set Intersection (LPSI) to enhance privacy protection during gradient exchanges in federated learning. Extensive experiments demonstrate that our approach is both effective and privacy-preserving.

Keywords: Response Gaming · Federated Learning · Privacy-Preserving · Private Set Intersection

1 Introduction

Responsible gaming (RG) promotes safe gambling practices through various strategies aimed at protecting players from the potential harms of excessive gambling [3]. Key RG practices include setting gambling limits and implementing self-exclusion options [3,5,9]. Current responsible gaming systems often rely on inflexible approaches, implementing uniform, non-adaptive gaming limits or

This work was supported by the Hunan Industry Polytechnic Project for 2024 under Grant GYKYWT202403.

T. Zhu et al. (Eds.): ICA3PP 2024, LNCS 15252, pp. 327–336, 2025.
https://doi.org/10.1007/978-981-96-1528-5_23

requiring manual adjustments for individual players. This rigidity hampers real-time limit adjustments and personalized gaming experiences. Such static limits impact customer satisfaction and constrain operators' business growth potential. While machine learning offers promising solutions for creating tailored gaming limits, it introduces new challenges. The process necessitates exchanging sensitive information among various stakeholders, including casino operators, financial institutions, and third-party entities. This data sharing raises significant concerns regarding privacy and confidentiality, particularly given the sensitive nature of personal and financial data involved in the process.

Federated learning has been widely adopted to leverage data from various sectors while preserving data privacy [11]. However, practical implementations face challenges such as model inversion attacks and membership inference attacks [19]. Studies have shown that gradient information can compromise user privacy [7,8,10,17,23,25], it is challenging to determine the intentions of the clients involved in training, and ensuring the reliability of the central server is also tricky, more than simply updating the model to safeguard user privacy is required. Private Set Intersection (PSI) is a cryptographic protocol designed to allow two or more parties to compute the intersection of their private datasets without revealing any information about the items not in the intersection [27]. Labeled Private Set Intersection (LPSI) extends the basic Private Set Intersection (PSI) protocol. It adds a layer of functionality by transferring associated labels or values along with the intersecting elements [4,7]. Our previous work proposed efficient PSI protocols with improved features [27]. In our previous work [27], we proposed several practical and malicious PSI protocols with improved efficiency. The primary motivation for our work is to address the limitations of existing Responsible Gaming (RG) systems by integrating Labeled Private Set Intersection (LPSI) into a Federated Learning framework. We introduce a novel privacy-preserving federated learning approach incorporating LPSI to mitigate privacy concerns during gradient exchanges. Our main contributions are as follows:

1. We proposed an innovative privacy-preserving federated learning framework for responsible gaming systems. This framework addresses the challenges of sensitive data exchange and potential attacks on federated learning, enhancing data privacy and confidentiality in RG systems.
2. We extended the Private Set Intersection (PSI) protocols from our previous work [27] to support Labeled Private Set Intersection. This extension maintains the practical and malicious-resistant properties of the original PSI protocols.
3. We conducted comprehensive experiments demonstrating our approach's effectiveness and privacy-preserving capabilities.

Our proposed approach advances the field of responsible gaming by providing a secure, efficient, and privacy-focused solution that overcomes the limitations of current RG systems.

2 Related Work

Developed by Google in 2016 [18], federated learning has gained widespread adoption in various sectors, addressing data fragmentation and privacy issues. However, it faces several challenges, particularly in security and efficiency. Researchers have proposed blockchain-based solutions to enhance federated learning's security and reliability. Majeed *et al.* [15] introduced FLchain, while Sharma *et al.* [22] utilized multi-layer and multi-chain structures. Arachchige *et al.* [2] developed PriModChain, integrating differential privacy and smart contracts. Lu *et al.* [14] proposed a hybrid blockchain architecture. Various approaches have been proposed to address malicious attacks and improve model reliability. Preveneers *et al.* [20] introduced a blockchain-based audit scheme, while Zhu *et al.* [26] created a secure collaborative training mechanism. Zhao *et al.* [24] combined federated learning with local differential privacy, and Li *et al.* [12] proposed BOppCL to address data privacy and trust issues. Deng *et al.* [6] introduced FAIR, incorporating data quality parameters, and Li et al. [13] developed the PR-OppCL framework, a reputation-based opportunistic federated learning approach to mitigate issues such as multiple iterations, extended training periods, and inefficiency.

To the best of our knowledge, existing approaches have not yet provided a comprehensive solution that simultaneously addresses privacy concerns, enhances model performance and improves training efficiency. Our paper investigates these challenges and introduces a novel labeled private set intersection mechanism. This mechanism is designed to safeguard privacy during gradient-sharing processes, particularly in the context of financial, gaming, and industrial enterprises.

3 System Model

3.1 System Workflow

Figure 1 presents our privacy-preserving federated learning framework for responsible gaming systems. The system operates as follows:

① **Task Publication**: The task publisher disseminates a comprehensive task list such as bet limits, loss limits, time limits, deposit limits prediction, model training, and fine-turning on existing models. This list specifies the anticipated gradients from local training data and the desired training accuracy for each task. The central server utilizes this detailed task list to establish communication and coordinate with each participating node in the federated learning network.

② **Local Gradient Upload**: Involved Parties upload gradient data as per the task list. PSI is employed to address privacy concerns during gradient sharing, preventing malicious actors from identifying the precise gradient sources.

③ **Model Training**: The central server refines the model using the received gradients and computes the global gradient.

④ **Global Gradient Download**: Involved Parties download global gradients based on their local requirements, update model parameters, and proceed to the next training round.

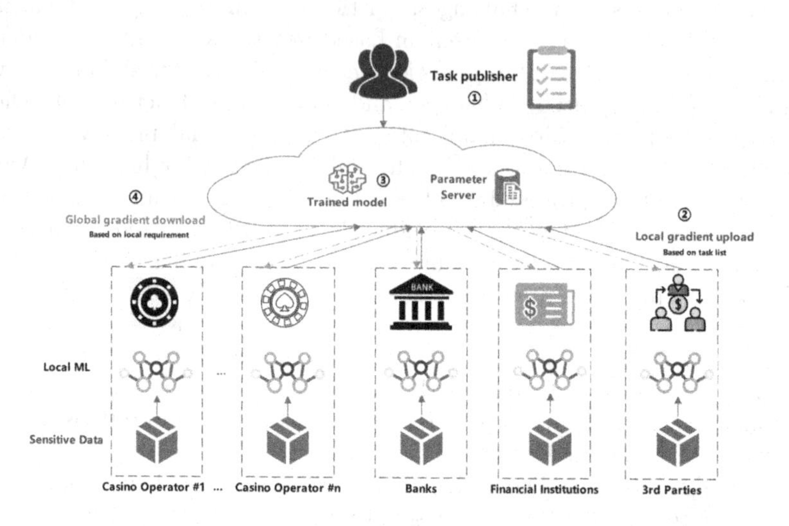

Fig. 1. System Model of Our Privacy-Preserving Federated Learning Framework

3.2 Federated Learning Algorithms

A typical federated learning scenario trains a global model ω across numerous client devices, with clients processing local data and uploading only gradient information. The central server's objective function $F(\omega)$ is:

$$min_{\omega}F(\omega), F(\omega) = \sum_{k=1}^{m} \frac{n_k}{n} F_k(\omega). \tag{1}$$

Where m is the total device count, n the aggregate data amount, n_k the data quantity from the k_{th} client, and $F_k(\omega)$ the local objective function:

$$F_k(\omega) = \frac{1}{n_k} \sum_{i \in d_k} f_i(\omega). \tag{2}$$

Here, d_k is the local dataset of client k, and $f_i(\omega) = \alpha(x_i, y_i, \omega)$ is the loss function for instances (x_i, y_i) in d_k.

Federated learning typically uses large-batch stochastic gradient descent (SGD) for optimization. Local client model weights are updated as follows:

$$\omega_{t,k} = \omega_{t-1,k} - \eta \nabla F_k(\omega). \tag{3}$$

The regulatory server aggregates the model in the t_{th} round as:

$$\omega_t = \sum_{k=1}^{K} \frac{n_k}{n} \omega_{t,k}. \tag{4}$$

3.3 Labeled Private Set Intersection Mechanism

Building on our previous work [27], we extend our efficient PSI protocol to support labeled private set intersection, as depicted in Fig. 2.

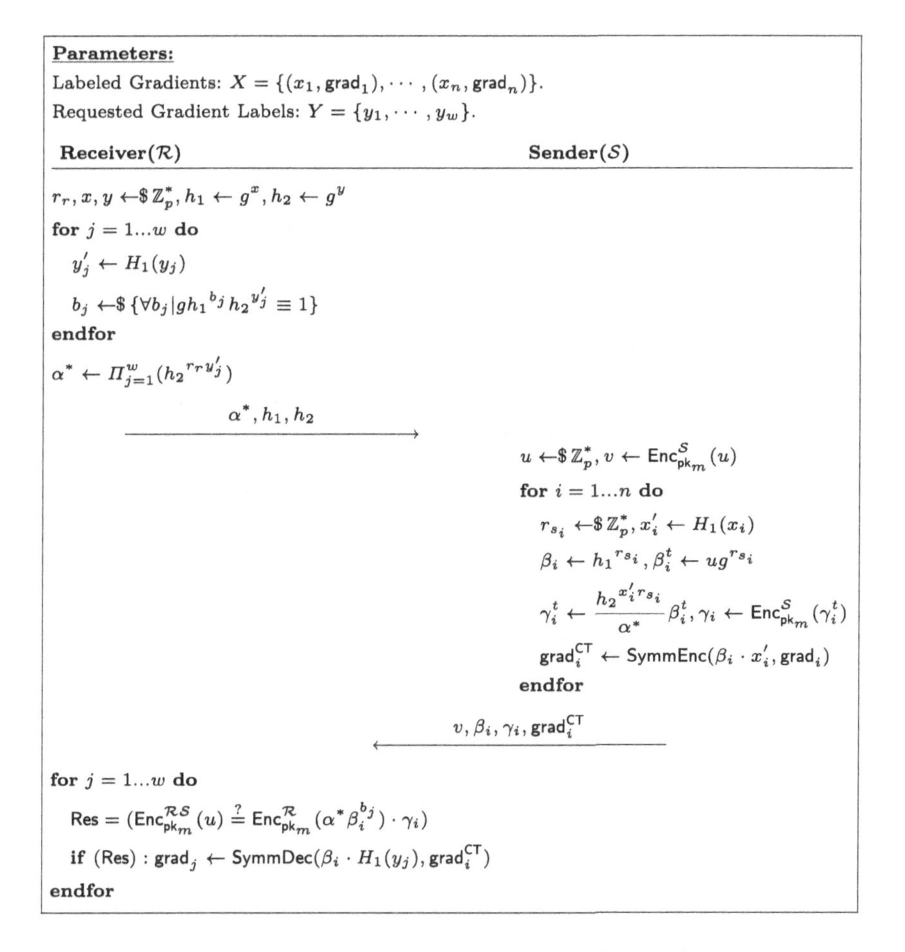

Fig. 2. Our Size-Hiding Labeled PSI Protocol

Our Labeled PSI protocol (Fig. 2) extends [27], using symmetric encryption functions satisfying IND – CPA security [21] and random key robustness [1]. For detailed notation, parameters, and security analysis, refer to [27], which applies to this protocol.

3.4 Labeled Private Set Intersection Integration

This subsection outlines integrating labeled PSI into our privacy-preserving federated learning framework.

1. **Local Gradient Upload**: The Casino Operator, Bank, Financial Institutions, and third parties act as the sender, while the Centralized Server is the receiver. The privacy protection measures focus on local receiver gradient labels of the receiver from the sender and disclose only requested task-related gradients to the receiver. (Details in Fig. 4)
2. **Global Gradient Download**: The Casino Operator, Bank, Financial Institutions, and third parties act as the receiver, while the Centralized Server is the sender. The protocol protects receiver gradient labels from senders and discloses only requested task-related gradients to the receiver. The privacy protection measures focus on global gradient labels of the receiver from the sender, which are kept confidential from the centralized server based on local machine learning (ML) training requirements and only disclose the requested task-related gradients to the receiver. (Details in Fig. 3)

Input: Labeled Local Gradients: $X = \{(x_1, \mathsf{grad}_1), \cdots, (x_n, \mathsf{grad}_n)\}$, Requested Gradient Labels: $Y = \{y_1, \cdots, y_w\}$ from Casino Operators, Banks, Financial Institutions, and third parties (Users) and Centralized Server (Server).
Output: No output $\perp$ for Casino Operators, Banks, Financial Institutions, and third parties, $\mathcal{O} = \{\mathsf{grad}_j\}|X \cap Y$ for Centralized Server.

1: Server obfuscates and processes its private set Y, denotes as Y'.
2: Server → Users: Server sends Y' and related auxiliary information $\mathsf{Aux}_{\mathsf{Server}}$ to Users.
3: Users obfuscates and processes its private set X with Y' and $\mathsf{Aux}_{\mathsf{Server}}$, and then denotes it as X'.
4: Users → Server: Users sends X' and related auxiliary information $\mathsf{Aux}_{\mathsf{Server}}$ to Server.
5: Server computes $\mathcal{O}$ with X', $\mathsf{Aux}_{\mathsf{Users}}$ and secret keys of Server.

Fig. 3. Global Gradient Download with Our LPSI

3.5 Evaluation and Analysis

In federated learning, clients train local models to minimize local errors. Ideally, local gradients should align with the optimal global update direction. Let θ_i represent the angle between local and global update directions for client i. In targeted attacks, malicious clients aim to align their gradients with poisoned targets, defined by angle δ_j. Clustering-based detection methods assume $\theta_i \neq \delta_j$ to

Input: Global Gradients: $X = \{(x_1, \mathsf{grad}_1), \cdots, (x_n, \mathsf{grad}_n)\}$, Requested Gradient Labels: $Y = \{y_1, \cdots, y_w\}$ from Centralized Server (Server) and Casino Operators, Banks, Financial Institutions, and third parties (Users).
Output: No output $\perp$ for Centralized Server (Server), $\mathcal{O} = \{\mathsf{grad}_i\}|X \cap Y$ for Casino Operators, Banks, Financial Institutions, and third parties (Users).
1: Users obfuscates and processes its private set Y, denotes as Y'.
2: Users $\rightarrow$ Server: Users sends Y' and related auxiliary information $\mathsf{Aux}_{\mathsf{Users}}$ to Sever.
3: Server obfuscates and processes its private set X with Y' and $\mathsf{Aux}_{\mathsf{Users}}$, and then denotes it as X'.
4: Server $\rightarrow$ Users: Server sends X' and related auxiliary information $\mathsf{Aux}_{\mathsf{Server}}$ to Users.
5: Server computes $\mathcal{O}$ with X', $\mathsf{Aux}_{\mathsf{Users}}$ and secret keys of Server.

Fig. 4. Local Gradient Upload with Our LPSI

distinguish honest from malicious clients. The angle γ between malicious clients' gradients is expected to be smaller than between malicious and honest clients or between honest clients. Pairwise cosine similarity assesses client alignment. An alignment-based penalty mechanism within a reputation system mitigates malicious impacts by reducing suspicious clients' learning rates and selection probability. This server-side detection mechanism ensures high model accuracy and attack failure rates, even with numerous malicious users, without introducing client-side changes or overhead.

4 Simulation and Performance Analysis

4.1 Simulation Settings

We implemented our solution in Python using LeNet on MNIST and CIFAR-10 datasets, running experiments on an Intel i5-8250U CPU with 40GB RAM and NVIDIA GTX 1660 GPU. We compared our framework's performance against FedAvg [16] and FAIR, evaluating training rounds, local training time, and overall training time.

4.2 Simulation Results Analysis

The data presented in Table 1 demonstrates that the proposed technique achieves a reduction in training iterations exceeding 10% compared to FedAvg and FAIR methods.

Using MNIST and CIFAR-10 datasets, we compare our method with FedAvg and FAIR. Figures 5(a) and 5(b) show that our framework requires fewer training iterations to achieve 90%

Table 1. Training Rounds Needed to Reach The Desired Accuracy

Scheme	MNIST(90%)/turns	CIFAR-10(60%)/turns
FedAvg	23	95
FAIR	17	88
Ours	14	83

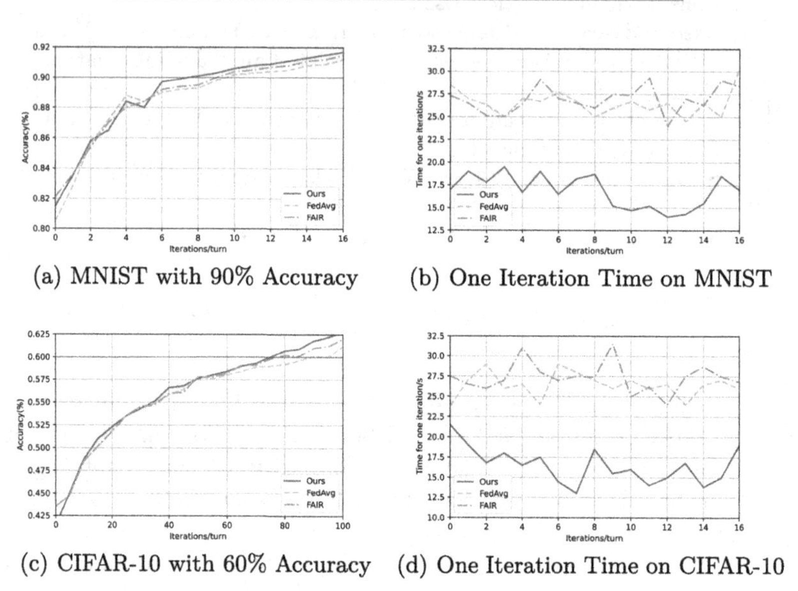

(a) MNIST with 90% Accuracy (b) One Iteration Time on MNIST

(c) CIFAR-10 with 60% Accuracy (d) One Iteration Time on CIFAR-10

Fig. 5. Training Iterations and One Iteration Time

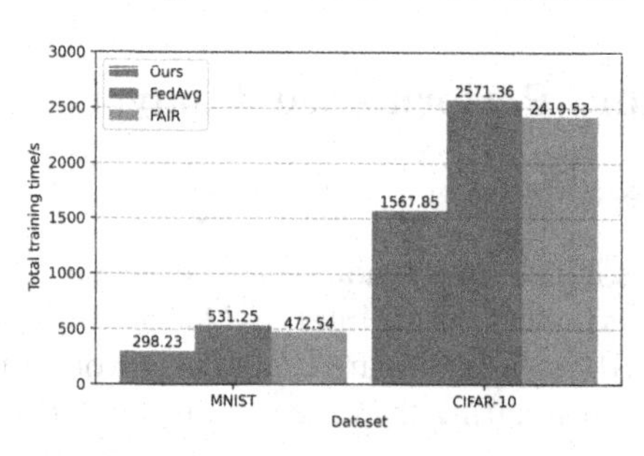

Fig. 6. Total Training Time

Figure 6 shows our approach's overall training duration. Our method achieves the shortest total training time by minimizing iterations, local training time, and inter-client waiting periods. Using CIFAR-10, at 60%.

5 Conclusion

In conclusion, we proposed a privacy-preserving federated learning framework for responsible gaming systems. By combining federated learning with Labeled Private Set Intersection, our approach enables personalized gaming limits while protecting sensitive data. Experimental results confirm the framework's effectiveness and privacy-preserving capabilities, contributing to safer gambling practices and sustainable industry development. This work balances personalized gaming experiences with data confidentiality, addressing key challenges in responsible gaming implementation. Future research will explore optimizing gradient transmission, accelerating model convergence, and implementing continuous learning techniques.

Acknowledgements. This work was supported by the Hunan Industry Polytechnic Project for 2024 under Grant GYKYWT202403.

References

1. Abdalla, M., Bellare, M., Neven, G.: Robust encryption. In: Micciancio, D. (ed.) TCC 2010. LNCS, vol. 5978, pp. 480–497. Springer, Heidelberg (2010). https://doi.org/10.1007/978-3-642-11799-2_28
2. Arachchige, P.C.M., Bertok, P., Khalil, I., Liu, D., Camtepe, S., Atiquzzaman, M.: A trustworthy privacy preserving framework for machine learning in industrial IoT systems. IEEE Trans. Industr. Inf. **16**(9), 6092–6102 (2020)
3. Blaszczynski, A., et al.: Responsible gambling: general principles and minimal requirements. J. Gambl. Stud. **27**, 565–573 (2011)
4. Cong, K., et al.: Labeled psi from homomorphic encryption with reduced computation and communication. In: Proceedings of the 2021 ACM SIGSAC Conference on Computer and Communications Security, pp. 1135–1150 (2021)
5. Currie, S.R., Hodgins, D.C., Wang, J., El-Guebaly, N., Wynne, H.: In pursuit of empirically based responsible gambling limits. Int. Gambl. Stud. **8**(2), 207–227 (2008)
6. Deng, Y., et al.: Fair: quality-aware federated learning with precise user incentive and model aggregation. In: IEEE INFOCOM 2021-IEEE Conference on Computer Communications, pp. 1–10. IEEE (2021)
7. Fredrikson, M., Lantz, E., Jha, S., Lin, S., Page, D., Ristenpart, T.: Privacy in pharmacogenetics: an {End-to-End} case study of personalized warfarin dosing. In: 23rd USENIX Security Symposium (USENIX Security 2014), pp. 17–32 (2014)
8. Geiping, J., Bauermeister, H., Dröge, H., Moeller, M.: Inverting gradients-how easy is it to break privacy in federated learning? Adv. Neural. Inf. Process. Syst. **33**, 16937–16947 (2020)
9. Hing, N., Cherney, L., Gainsbury, S.M., Lubman, D.I., Wood, R.T., Blaszczynski, A.: Maintaining and losing control during internet gambling: a qualitative study of gamblers' experiences. New Media Soc. **17**(7), 1075–1095 (2015)
10. Hitaj, B., Ateniese, G., Perez-Cruz, F.: Deep models under the GAN: information leakage from collaborative deep learning. In: Proceedings of the 2017 ACM SIGSAC Conference on Computer and Communications Security, pp. 603–618 (2017)

11. Li, L., Fan, Y., Tse, M., Lin, K.Y.: A review of applications in federated learning. Comput. Ind. Eng. **149**, 106854 (2020)
12. Li, Q., Wang, W., Zhu, Y., Ying, Z.: BOppCL: blockchain-enabled opportunistic federated learning applied in intelligent transportation systems. Electronics **13**(1), 136 (2023)
13. Li, Q., Yi, X., Nie, J., Zhu, Y.: PR-OppCL: privacy-preserving reputation-based opportunistic federated learning in intelligent transportation system. IEEE Trans. Veh. Technol. (2024)
14. Lu, Y., Huang, X., Zhang, K., Maharjan, S., Zhang, Y.: Blockchain empowered asynchronous federated learning for secure data sharing in internet of vehicles. IEEE Trans. Veh. Technol. **69**(4), 4298–4311 (2020)
15. Majeed, U., Hong, C.S.: Flchain: federated learning via MEC-enabled blockchain network. In: 2019 20th Asia-Pacific Network Operations and Management Symposium (APNOMS), pp. 1–4. IEEE (2019)
16. McMahan, B., Moore, E., Ramage, D., Hampson, S., y Arcas, B.A.: Communication-efficient learning of deep networks from decentralized data. In: Artificial Intelligence and Statistics, pp. 1273–1282. PMLR (2017)
17. Melis, L., Song, C., De Cristofaro, E., Shmatikov, V.: Exploiting unintended feature leakage in collaborative learning. In: 2019 IEEE Symposium on Security and Privacy (SP), pp. 691–706. IEEE (2019)
18. Mohri, M., Sivek, G., Suresh, A.T.: Agnostic federated learning. In: International Conference on Machine Learning, pp. 4615–4625. PMLR (2019)
19. Mothukuri, V., Parizi, R.M., Pouriyeh, S., Huang, Y., Dehghantanha, A., Srivastava, G.: A survey on security and privacy of federated learning. Futur. Gener. Comput. Syst. **115**, 619–640 (2021)
20. Preuveneers, D., Rimmer, V., Tsingenopoulos, I., Spooren, J., Joosen, W., Ilie-Zudor, E.: Chained anomaly detection models for federated learning: an intrusion detection case study. Appl. Sci. **8**(12), 2663 (2018)
21. Rogaway, P.: Nonce-based symmetric encryption. In: International Workshop on Fast Software Encryption, pp. 348–358. Springer (2004)
22. Sharma, P.K., Park, J.H., Cho, K.: Blockchain and federated learning-based distributed computing defence framework for sustainable society. Sustain. Urban Areas **59**, 102220 (2020)
23. Song, M., et al.: Analyzing user-level privacy attack against federated learning. IEEE J. Sel. Areas Commun. **38**(10), 2430–2444 (2020)
24. Zhao, Y., et al.: Local differential privacy-based federated learning for internet of things. IEEE Internet Things J. **8**(11), 8836–8853 (2020)
25. Zhu, L., Liu, Z., Han, S.: Deep leakage from gradients. In: Advances in Neural Information Processing Systems, vol. 32 (2019)
26. Zhu, X., Li, H., Yu, Y.: Blockchain-based privacy preserving deep learning. In: Guo, F., Huang, X., Yung, M. (eds.) Inscrypt 2018. LNCS, vol. 11449, pp. 370–383. Springer, Cham (2019). https://doi.org/10.1007/978-3-030-14234-6_20
27. Zhu, Y., Chen, L., Mu, Y.: Practical and malicious private set intersection with improved efficiency. Theor. Comput. Sci. 114443 (2024)

Author Index

T. Zhu et al. (Eds.): ICA3PP 2024, LNCS 15252, pp. 337–338, 2025.
https://doi.org/10.1007/978-981-96-1528-5

Printed in the United States
by Baker & Taylor Publisher Services